FOREWORD

The deregulation of electricity markets is leading to increased pressure on nuclear energy to stay competitive. The economic performance of nuclear energy can be improved by extending plant life and combining this with an increase of ^{235}U enrichment and a higher discharge burn-up of fuel. An additional benefit of this strategy is a decrease in the amount of discharged spent fuel. High burn-up does, however, place stricter demands on fuel performance, although the nuclear fuel burnt in today's reactors is extremely reliable and has a very low failure rate.

Nuclear fuel is licensed in reactor-operating countries for discharge burn-up below the level forecast for future fuel cycles. Before such licensing is approved, the performance of fuel in the burn-up range beyond that used today needs to be investigated and the predictive models benchmarked with experiments. Five countries have licensed the loading of mixed-oxide (MOX) fuel in some of their reactors. Specificity and similarity of this fuel with UO_2 when considering higher burn-up is also being addressed and is becoming of interest to countries considering reactor-based, excess weapons plutonium disposal.

A number of high burn-up experiments have been carried out, and one area considered to be of great importance is the degradation of fuel thermal conductivity as burn-up increases. The need for an international seminar reviewing recent progress and aiming to improve the models used in codes predicting thermal performance was expressed by several national and international institutions. The Commissariat à l'énergie atomique (CEA) agreed to organise the seminar from 3-6 March 1998 at the Château de Cadarache. The seminar was co-organised by the OECD Nuclear Energy Agency (NEA), in co-operation with the International Atomic Energy Agency (IAEA), and hosted by the Département d'études des combustibles (DEC). The general chairman of the seminar was Clément Lemaignan; the co-chairman was Carlo Vitanza of the Halden Reactor Project; and the secretary was Pierre Chantoin, assisted by Régine Bousquet. The organising committee, which reviewed the presentations proposed by participants and shaped the technical programme, was made up of the following people:

- Daniel Baron, EDF;
- Pierre Beslu, CEA;
- Patrick Blanpain, Framatome;
- Pierre Chantoin, CEA;
- Yannick Guérin, CEA;
- Clément Lemaignan, CEA;
- Patrick Menut, IAEA;
- Enrico Sartori, OECD/NEA;
- Maria Trotabas, Cogema.

Sixty-six experts from nineteen countries and four international organisations attended the seminar. The current proceedings provide a summary of the discussions and conclusions of the seminar, together with the text of the presentations made.

The text is published on the responsibility of the Secretary-General of the OECD. The views expressed do not necessarily correspond to those of the national authorities concerned.

Acknowledgements

We would like to express our thanks to the organising committee, the technical programme chairs and all those who contributed to the success of the seminar by presenting their work and competent views. Our gratitude goes to CEA for hosting the seminar and to CEA, Cogema, EDF and Framatome for their generous hospitality. Special thanks go to Régine Bousquet for taking care of the burdensome local arrangements and contacts as well as to Amanda McWhorter for her dedication in editing these proceedings.

TABLE OF CONTENTS

EXECUTIVE SUMMARY

M. Pierre Beslu opened the seminar and welcomed participants. He stressed the great importance of understanding fuel performance in view of using higher enrichment fuel, achieving higher burn-up and for introducing innovative fuel cycles. He announced that the CEA envisages to hold a series of three such seminars, the present one addressing thermal fuel performance, a second fission gas release and a third concerned with pellet clad interaction.

Thermal performance occupies the most important aspect of the fuel performance modelling. Not only is it extremely important from a safety point of view, but also many of the material properties of interest and behaviour, such as transport properties like fuel creep and fission gas release are thermally activated processes. Thus, in order to model these processes correctly, it is critical to calculate temperatures and their distribution as accurately as possible.

In recent years, considerable progress has been made in the understanding of fuel rod thermal performance. This is primarily due to the development of suitable in-core instrumentation and its application to the measurement of fuel centreline temperature during irradiation. There is now a substantial database on in-pile thermal performance covering a wide range of design parameters and material characteristics at both the beginning of life and at high burn-up. In addition post irradiation examination (PIE) techniques have been developed to measure thermal diffusivity and heat capacity on previously irradiated UO_2 fuel, thus allowing a measurement of fuel thermal conductivity at high burn-up.

As an introduction to the seminar, J.A. Turnbull reviewed the thermal behaviour of nuclear fuel and addressed the factors affecting gap conductance and fuel pellet conductivity. This review is the first paper in these proceedings.

The first two sessions were devoted to *Fuel Thermal Conductivity Data* and *Thermal Conductivity Modelling*. This was followed by a first panel discussion, chaired by Hubert Bairiot and Louis-Christian Bernard addressing *Solving and Emerging Thermal Conductivity Questions*. The third session covered *Fuel Clad-Gap Evolution Modelling* followed by a second panel discussing *Gap Evolution and Heat Transfer Questions*, chaired by Gary Gates and Marc Lippens. The two final sessions covered *Experimental Databases* and *Advances in Code Development on Thermal Aspects*.

A final panel, chaired by J.A. Turnbull and Daniel Baron, reviewed the conclusions of the two panels on thermal conductivity and gap conductance and summarised – after discussion among all participants – the key conclusions of the seminar. These conclusions immediately follow this Executive Summary.

Clément Lemaignan closed the seminar thanking participants for their valuable and competent contributions.

CONCLUDING PANEL SUMMARY
J.A. Turnbull, Daniel Baron

Fuel thermal conductivity

Fuel thermal conductivity correlations

The initial thermal conductivity correlations are well established for almost all fuel oxides loaded in commercial reactors. The influence of parameters such as temperature, stochiometry, plutonium content and gadolinium content are relatively well known. Some models are able to account for all these parameters simultaneously.

MOX fuels

The non-homogeneity of MOX fuel can be accounted for, using mathematical homogenisation techniques. However, even for high Pu content the fuel thermal conductivity is close to the uranium oxide fuel conductivity providing O/M close to 2.000. Participants seem to agree on a degradation of 4 to 5% per 10% Pu. However, it was shown that a deviation from stochiometry has a stronger effect.

The burn-up effect

The burn-up effect on fuel thermal conductivity has been assessed up to 80 MWd/KgU, thanks to in-pile centreline temperature measurements and out-of-pile thermal diffusivity measurements. There is some agreement between these two methods.

Cp variation with burn-up

Analysing shut-down temperature records in the Halden reactor, a slight increase of the Cp with burn-up is likely to happen. Nevertheless the discussion outlined that the Cp variation with burn-up could be neglected. The experimental data expected in the near future will certainly allow checking this assumption.

Further needs

A lack of data is identified on the burn-up degradation at temperatures higher than 1800 K. A decrease of the conductivity improvement due to the electronic heat transport is likely to happen with burn-up. This effect was already observed when increasing the gadolinium content.

Heat transfer in the rim

The rim thermal conductivity evolution is not yet clearly known. The thermal degradation due to the onset of numerous micrometer porosities could be balanced by the cleaning of the matrix subsequent to this porosity build-up. In order to investigate the net effect of the rim structure, samples irradiated to high burn-up and temperatures lower than 800 K are required. Other questions are still open about the rim fission gas release and the rim volume variation during the porosity build-up.

Gap conductance

Pellet fragmentation and relocation

The stochastic nature of this phenomenon prohibits analytical modelling of gap conductance. Reliance must be placed on benchmarking models against in-pile data. There is no general consensus as to a definitive formulation for gap conductance. Fortunately a large database exists and providing this is employed, there are limited difficulties in formulating an adequate gap conductance model which correctly reflects the effects of gap dimension, fill gas composition and pressure.

Surface roughness

In principle this should be an important contributor to gap conductance by its effect of inhibiting heat flow. In practice, at both the beginning of line and at high burn-up, there appears to be a little effect on heat transfer between the fuel pellet and the cladding.

Formation of inner clad oxide layer

At high burn-up, when the fuel and clad have been in intimate contact for some time, an inner oxide layer is formed, 6-10 microns thick. It has been shown that this has a complex structure, which can include ZrO_2, uranium and caesium, depending on the heat rating and burn-up. It is supposed that its contribution to gap heat transfer is small and if anything, beneficial, as it would tend to eliminate effects due to surface roughness and misaligned pellet fragments.

Closed gap conductance

There is evidence to suggest that under these conditions, heat transfer is good and substantially independent of the fill gas composition and pressure, since it corresponds to a solid bound between fuel and cladding. There is some debate as to whether or not the conductance depends on interfacial pressure.

SESSION I

Fuel Thermal Conductivity Data

Chairs: S.K. Yagnik, D. Baron

A REVIEW OF THE THERMAL BEHAVIOUR OF NUCLEAR FUEL

J.A. Turnbull

Abstract

In order to provide a reliable assessment of fuel performance it is critical to calculate temperatures and their distribution as accurately as possible. For this reason, the nuclear industry has devoted much time and effort to provide data and evaluations of fuel temperatures during steady state and transient conditions.

The main thermal resistance in the fuel rod is that of the fuel to clad gap and the thermal conductivity of the pellet. Some of the factors affecting gap conductance and fuel pellet conductivity are given below:

Gap conductance	fuel-to-clad gap separation surface roughness of fuel and cladding fill gas composition (helium, argon, xenon) fill gas pressure
Fuel conductivity	density cracking composition, additives, MOX, Gd, etc. irradiation

With the drive for higher discharge burn-up, experiments to measure fuel centreline temperatures demonstrated that the UO_2 thermal conductivity decreased with burn-up. This is attributed to the build-up of irradiation damage and fission products within the lattice. In this way, the effect of irradiation is very similar to the addition of impurities and "alloying" components to the fuel e.g. $(U,Gd)O_2$ and MOX.

Whereas in-pile measurements of centreline temperature can only measure the integrated effect of burn-up on the conductivity in the temperature range across the pellet, recent measurements using laser flash on small scale specimens have shown some of the complexities of the degradation process including recovery at high temperatures. It is clear that the conductivity is a function of the state of point defects, interstitials and vacancies, gas and fission product "atoms" and their agglomeration into secondary defects, dislocation loops, precipitates and bubbles. In other words, models of thermal conductivity and fission gas release have become inextricably mixed. Modelling of all these features continues to present a challenge from both a mathematical as well as physical point of view.

Introduction

Thermal performance occupies the most important aspect of fuel performance assessment. Not only is it extremely important from a safety point of view, particularly in the avoidance of fuel melting and loss of geometry, but many of the material properties of interest e.g. transport properties like fuel creep and fission gas release are exponentially dependent on temperature. Thus, in order to provide a reliable assessment of fuel performance it is critical to calculate temperatures and their distribution as accurately as possible.

In discussing the factors which are important in the thermal performance of fuel rods, it can be assumed that the thermal hydraulic heat transfer on the outer surface of the cladding are known and obey standard equations. The main thermal resistance in the fuel rod is that of the fuel to clad gap and the thermal conductivity of the pellet. Any oxide formed on the outside of the clad will also contribute, and this effect becomes important at high burn-up.

The purpose of this paper is to review our current understanding of thermal performance and to illustrate the various separate effects with data where available. Unfortunately it is very difficult to separate the contributions of the gap and fuel conductance to the overall behaviour, although some degree of separation can be made through the time dependence of the temperature evolution. The factors that influence these two phenomena are given below:

Gap conductance	fuel-to-clad gap separation surface roughness of fuel and cladding fill gas composition (helium, argon, xenon) fill gas pressure
Fuel conductivity	density cracking composition, additives, MOX, Gd etc. irradiation

Gap conductance

The gap conductance, h_{gap} responsible for the temperature drop between the fuel surface and the inner bore of the cladding is made up as the sum of the three parallel conduction routes: conduction through the gas gap, h_{gas}, conduction through areas of contact, h_{cont} and radiative heat transfer, h_{rad}:

$$h_{gap} = h_{gas} + h_{cont} + h_{rad}$$

In terms of the gap width d, the roughness of the surfaces r and the "temperature jump distance" g, conduction through a gas of conductivity k_{gas} can be expressed by:

$$h_{gas} = k_{gas} / (d + r + 2g)$$

The temperature jump distance is an engineering solution to the imperfect energy transfer between gas molecules and a solid surface and is considered as the extra separation of the surfaces necessary for a linear temperature gradient across the gap.

$$2g = Const. \cdot F(Mol.Wt) \cdot \frac{\sqrt{T(K)}}{\Pr ess}$$

From this expression, it can be seen that this "extra gap width" is proportional to the square root of temperature and inversely proportional to pressure; it is also greater for light atoms e.g. helium than for heavy atoms like xenon. Figure 1 shows data from two centreline thermocouples situated in nominally identical rods as the helium fill gas pressure was increased. It can be seen that the temperatures reduced with increasing pressure in accord with the expression above. Also, the temperature reduction increased with increasing power, i.e. as the gap temperature increased.

Until pellet-clad contact, the most important term with respect to gap conductance is h_{gas} where it can be seen that its value depends critically on gas conductivity and the gap dimension. Despite the uncertainties in gap dimension due to pellet fragmentation, experiments performed with rods fitted with centreline thermocouples and fabricated with different gap sizes have verified the form of this equation. The effect of different gap size can be seen from centreline temperature data shown in Figure 2 whilst the effect of different fill gas compositions can be seen in Figure 3. It is to be noted that when filled with helium gas, the relationship between centreline temperature and power is essentially linear.

The predominant gas atoms generated by fission are krypton and xenon, both of which have a much lower thermal conductivity than helium. Thus when fission gases are released to the fuel-clad interspace, the addition of Kr and Xe to the original fill gas causes a reduction of gap conductance resulting in higher fuel temperatures. This has been confirmed by experiment as shown in Figure 3. As is evident in the figure, the temperature/power relationship for xenon filled rods is highly non-linear.

The influence of gap size on fuel temperature is summarised in Figure 4 where temperatures at 20 kW/m and 30 kW/m for both helium and xenon filled rods are shown. A linear relationship between fuel temperature and gap size is apparent for both gas compositions.

For intermediate xenon/helium compositions, representative of differing degrees of fission gas release, a linear relation between temperature and xenon content is observed for all other conditions remaining constant.

It should be noted that extrapolation of the relationships shown in Figure 4 to very small gap sizes indicates that the difference between xenon and helium filled rods vanishes. This result implies that heat conductance through a tightly closed gap must occur predominantly through solid-solid contact spots and less through roughness pockets where the fill gas plays a role. The gap conductance must therefore allow for a progressive flattening of asperities of fuel and cladding with increasing contact pressure; likewise, any waviness should be assumed to have disappeared.

In order to model gap conductance it has proved necessary to introduce fitting parameters in order to account for the stochastic nature of pellet fragmentation and eccentricity. This is done by artificially reducing the fuel-clad gap by an empirical constant or function, or by combining contributions from gas conduction and contact conduction with an appropriate function F derived from experimental data:

$$h = F \cdot h_{gas} + (1 - F) \cdot h_{contact}$$

17

UO$_2$ thermal conductivity

Heat transport through solids takes place by both electronic conduction and by the propagation of elastic waves through the crystal lattice; the total conductivity k is the sum of k$_{ph}$ and k$_{el}$, the phonon and electronic conductivity. In UO$_2$ the electronic mobility is too slow to provide a significant contribution at temperatures lower than around 1600°C and consequently the most important mode of heat transfer with regard to fuel rod performance is that of conduction through the lattice. In this regime, the thermal conductivity can be derived by considering the elastic waves as particles or "phonons" with their transport properties evaluated by kinetic theory in the same way as those of an ideal gas. Resistance to heat transfer occurs through phonon-phonon collisions, the so-called "Umklapp Process", and scattering of phonons at local discontinuities in the lattice order. The general formula describing phonon conductivity in terms of temperature T is given as:

$$k_{ph} = \frac{1}{(A + B.T)}$$

where B describes the phonon-phonon collisions and is characteristic of the host material and A is dependent on impurity or alloying additions.

Because UO$_2$ fuel pellets are manufactured by sintering, they are inevitably less than the theoretical density due to the presence of residual porosity. Consequently, the thermal conductivity is reduced by the presence of these pores. There are several expressions relating the conductivity to the theoretical conductivity k(100%) and the fractional porosity P such as:

$$k = k(100\%) \cdot (1 - P)^{2.5}$$

or that by Kampf and Karsten [1]:

$$k(P) = k(100\%) \cdot (1 - P^{2/3})$$

The effect of this porosity is shown in Figure 5 for 100%, 95% and 90% dense material.

The addition of impurities or additions to the UO$_2$ matrix will also reduce the conductivity by increasing the degree of phonon scattering. This can be accommodated via the "A" term in the expression:

$$k_{ph} = \frac{1}{A_0 + \sum A_i \cdot Conc_i + B \cdot T}$$

The effect is clearly demonstrated in data from the GRIMOX experiments [2], and Figure 6 shows a comparison of measured centreline temperatures for UO$_2$ and MOX rods. A comparison between UO$_2$ and (Gd,U)O$_2$ is shown in Figure 7 for rods irradiated in the Halden HBWR. In both cases, the additions cause a measurable decrease in thermal conductivity and hence an increase in fuel temperatures.

It comes as no surprise therefore that during irradiation, the thermal conductivity of UO_2 is reduced by irradiation damage to the lattice and the conversion by fission of uranium atoms to twice the number of lighter and dissimilar fission product atoms. An example demonstrating the increase in centreline temperature at constant power as a function of burn-up is shown in Figure 8. This is for a small gap helium-filled rod where the temperatures were at all times too low to promote any fission gas release. Although the gap was closing progressively due to fuel swelling, the degradation in conductivity due to the build up of fission products causes a gradual increase in measured temperature. In this case, the effect of burn-up can be expressed as:

$$k_{ph} = \frac{1}{A_0 + A_1 \cdot BU + B \cdot T}$$

A further example of the effect of burn-up was demonstrated in the third and final Risø Project [3]. Here, several fuel rods which had been previously irradiated were re-fabricated with pressure transducers and new centreline thermocouples and ramped to powers approaching 40 kW/m. Figure 9(a) shows the data from a PWR test, AN3, previously irradiated to 36 MWd/kgUO$_2$, where the power was raised in short steps and the resulting increase in temperature and fission gas release recorded. Figure 9(b) shows the same data along with predictions of temperature *without* account taken for the effect of burn-up on conductivity. Here it can be seen that even allowing for released fission gas, the measured temperatures are between 200 and 400 °C higher than the predictions.

A nominally identical fuel rod designated AN4 was refabricated and back filled with 1 bar xenon instead of the 15 bar helium for AN3. Profilometry before and after the tests demonstrated that the gaps were closed, at least at the maximum powers. Comparison of the measured temperatures, Figure 10 shows the effect of the change in fill gas in a closed gap condition. It is interesting to note that the difference is very small, ~80°C, showing that most of the heat was conducted through areas of contact with little effect of gas conduction through residual gas gaps. This is an important result since it shows that at high burn-up, when the fuel-to-clad gap is closed, there is little effect of released fission gas on fuel temperatures, i.e. there is no appreciable "thermal feedback" as observed in cases when the gap remained open. A contributor to this is the formation of a "bonding layer" formed after the fuel and cladding have made intimate contact. This has been studied by Une and co-workers, and Figure 11 illustrating the conditions for bonding layer formation is taken from Reference [4].

The degradation of conductivity with burn-up affects both transient and steady-state thermal behaviour, and Figure 8 shows how, even at constant power, fuel temperatures gradually rise as a function of burn-up. Degradation also affects fuel rod transient behaviour. Experiments performed in the Halden HBWR by logging temperature response during reactor scrams have shown how the fuel rod time constant increases with burn-up (Figure 12) as to be expected with a reducing pellet thermal conductivity.

The experiments described above provide a measurement of the effect of burn-up integrated over the range of temperature between the periphery and the centre of the pellet, typically 300 to 1000°C. Recently there have been a number of studies reported where the thermal conductivity of irradiated fuel has been measured by a laser flash technique. This is performed on small 1 mm thick samples of fuel cut from irradiated pellets and examined in an out-of-pile experiment where the temperature of the sample and hence the measurement is known very accurately. By subjecting the sample to repeated up and down temperature ramps, it is now obvious that the degree of degradation is influenced by temperature, and that a certain amount of recovery occurs at high temperatures.

Figure 13 shows data from two samples taken from different locations in a fuel pellet which had been irradiated to ~38 MWd/kgUO$_2$ [5]. Initially the conductivity of the samples differed, reflecting different local in-pile conditions, but as the temperature was raised from 400 to 1400°C and then returned to 400°C there was a clear recovery with the final data from both samples lying along the same curve.

Both irradiation damage and fission product atoms are expected to contribute towards the degradation. However, the slow evolution of degradation with burn-up suggests that the build-up of fission products has the dominant effect. In which case, the recovery is understandable in terms of the agglomeration of previously isolated atoms into secondary defects like fission gas bubbles and solid fission product precipitates. In which case, the out-of-pile data cannot be transferred directly to in-pile situations. There must be a model which describes the balance between diffusion and irradiation induced re-solution and hence, for local conditions, the fraction of fission products that exist in isolation and in secondary defects. Thus the model for thermal conductivity becomes intimately associated with models of fission product release and all their attendant complications.

The high burn-up rim structure

Examination of fuel irradiated to high burn-up has shown the presence of a restructured zone close to the pellet periphery. This zone or "rim structure" is characterised by a recrystallisation with the appearance of grains or sub-grains of around 0.25-0.5 microns and the formation of large pores ~0.5 microns diameter, presumably filled with fission gases. Studies with fuel of different initial grain sizes suggests that the restructuring occurs earlier in small grain size fuel. In 10-12 micron grain size fuel restructuring appeared to start at a local burn-up of 70 MWd/kgU, whilst for a grain size of 30 microns, restructuring was delayed to in excess of 100 MWd/kgU [4]. Typically, the thickness of this restructured layer is 100-200 microns although additional isolated areas are sometimes seen remote from the pellet surface, but always in the vicinity of a grain boundary.

The thickness of this restructured region is of the same order as that for enhanced plutonium production from thermal neutron capture by ^{238}U, though no cause and effect relationship is conclusive. The current theory is that the generation of plutonium leads to an enhanced fission density and defect production rate. Dislocations and dislocation loops formed by the rapid accumulation of point defects will eventually, with increasing burn-up, become entangled and organised into sub-grains with high angle boundaries; Figure 14 is taken from Une and co-workers [3] and illustrates the conditions required for restructuring. The driving force for the ordering process and sub-grain formation is the reduction of lattice strain introduced by the irradiation damage. The movement of dislocations are also thought to sweep fission gas atoms and intragranular bubbles to form the large pores observed by optical microscopy and Scanning Electron Microscopy (SEM).

There is every reason to suspect that this restructuring will have an impact on the thermal performance of fuel rods in addition to any other phenomena concerning the gap or fuel conductance. The nature of enhanced plutonium production and fission means that within this rim region the burn-up is substantially in excess of the pellet average burn-up [6]. The extra fission products associated with this burn-up will considerably degrade the local thermal conductivity. However, the restructuring as seen by SEM examination could lead to a certain amount of recovery if the sub-grain structure reduced the lattice damage caused by irradiation damage, and the pores contained fission gas atoms which otherwise would reside in the lattice. However, the net effect of the rim structure must await experiments to determine the local conductivity.

Discussion and conclusions

Gap conductance

It is concluded that although modelling of gap conductance is well understood, it is acknowledged that the stochastic nature of pellet and fragment relocation means that the final predictions must be tuned to in-reactor data. Fortunately, there is a wealth of such data and providing these are used, there are no outstanding problems to adequate modelling.

It appears that under closed gap conditions, which occurs at high burn-up, there is little residual effect of surface roughness and, unlike open gap situations, temperatures are substantially independent of fill gas pressure and composition. A contributor to this observation is the formation of a fuel-clad bonding layer between surfaces in intimate contact.

Fuel conductivity

The formulation of thermal conductivity in terms of phonon interaction with lattice works well for UO_2 including the effects of additives and degradation due to burn-up. Further data are required at high burn-ups for standard fuel, $(Gd,U)O_2$ and MOX. In particular, the effect of the restructured rim region on thermal performance requires investigation.

Further studies are required using the laser flash technique to evaluate the recovery processes and to extend the modelling to a mechanistic description of degradation. In support of these studies, heat capacity measurements are necessary to interpret the laser flash measurements as well as providing additional information on the kinetics of point defect mobility and thermal conductivity recovery.

REFERENCES

[1]   H. Kampf and G. Karsten, *Nucl. Appl. Technol.*, 9, 288, 1970.

[2]   L. Caillot, M. Charles, C. Lemaignan, A. Chotard and F. Montagnon, "Analytical Measurement of the Thermal Behaviour of MOX Fuel", ANS Topical Meeting on LWR Fuel Performance, West Palm Beach, Florida, USA, 17-21 April 1994.

[3]   The Third Risø Fission Gas Release Project, Final Report, Risø Internal Document, March 1991.

[4]   K. Une, K. Nogita, S. Kashiba, T. Toyonaga and M. Amaya. "Effect of Irradiation-Induced Microstructural Evolution on High Burn-up Fuel Behaviour", ANS Topical Meeting on LWR Fuel Performance, Portland, Oregon, USA, 2-6 March 1997.

[5]   R. Gomme, "Thermal Properties Measurements on Irradiated Fuel: An Overview of Capabilities and Developments at AEA Technology Windscale", this meeting.

[6]   K. Lassmann, C.T. Walker and J. van de Laar, "Thermal Analysis of Ultra High Burn-up Irradiations Employing the TRANSURANUS Code", this meeting.

Figure 1. The decrease in temperature of two thermocouples as a function of helium pressure at different power levels

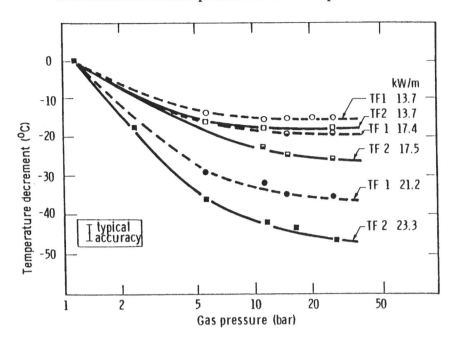

Figure 2. Fuel centreline temperatures as a function of power for two helium-filled rods of different gap size

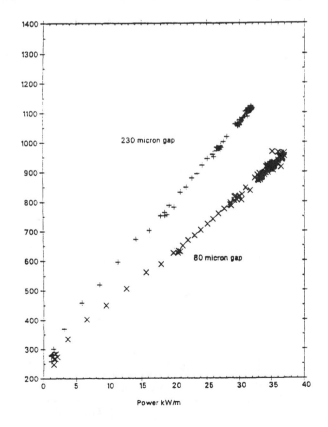

**Figure 3. Start-up data for 8 thermocouples in 4 rods showing the trend
of measured temperatures as a function of power with different fill gases in the gap**

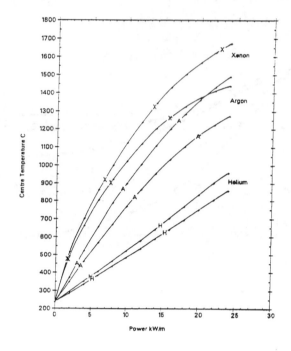

**Figure 4. The trend of measured fuel temperatures as a function
of gap size at two different power levels when filled with helium or xenon**

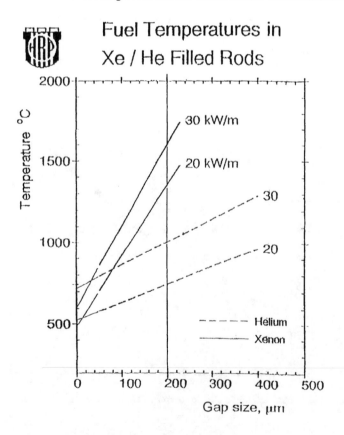

Figure 5. The effect of density on the thermal conductivity of UO$_2$

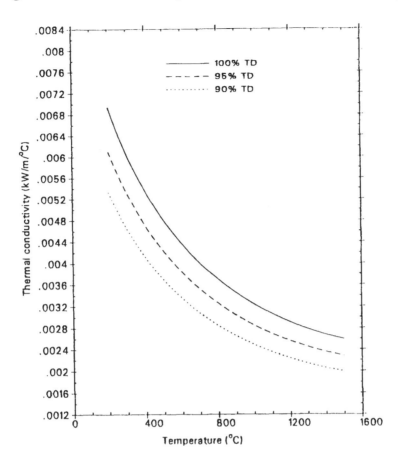

Figure 6. A comparison of centreline temperature for MOX and UO$_2$ fuel as a function of linear heat in the GRIMOX experiment

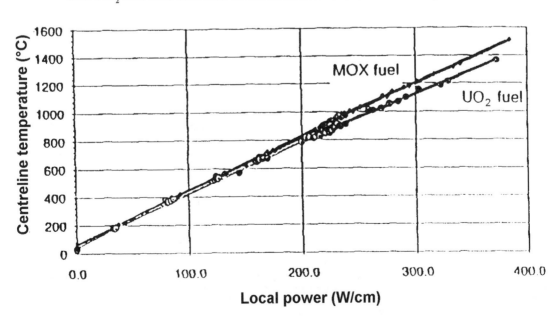

Figure 7. A comparison of temperatures in standard and gadolinia-doped UO₂ fuel rods

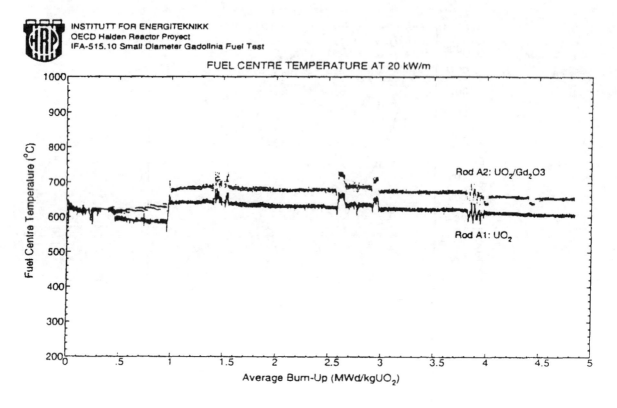

INSTITUTT FOR ENERGITEKNIKK
OECD Halden Reactor Proyect
IFA-515.10 Small Diameter Gadolinia Fuel Test

FUEL CENTRE TEMPERATURE AT 20 kW/m

Figure 8. Measured fuel temperatures at 20 and 30 kW/m for a small helium-filled rod as a function of burn-up

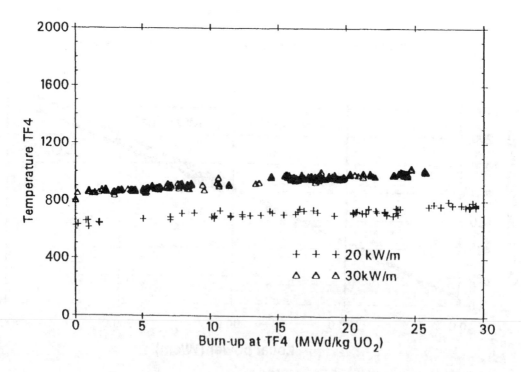

Figure 9(a). Data from test AN3 of the Third Risø Fission Gas Release Project; the figure shows power, FGR and temperature during the test performed at 36 MWd/kgUO$_2$

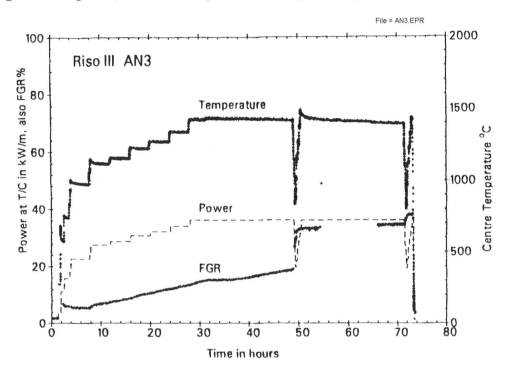

Figure 9(b). Measured temperature during the AN3 test compared to predictions made without allowance for conductivity degradation

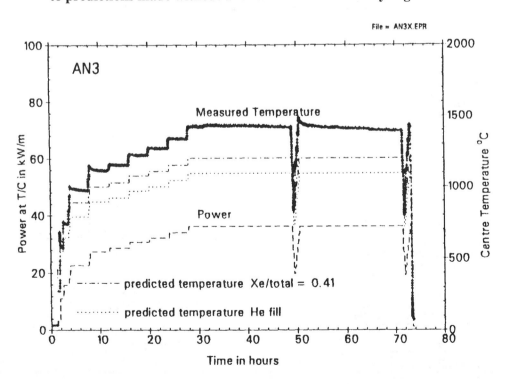

Figure 10. Comparison of measured temperatures
for the tests AN3 (helium-filled) and AN4 (xenon-filled)

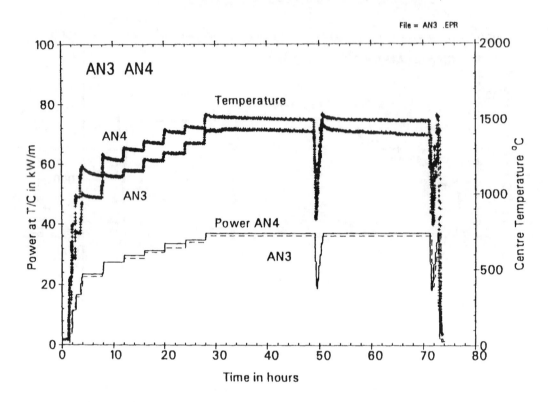

Figure 11. Conditions and mechanism for the formation
of the fuel-clad bonding layer according to Une *et al.* [4]

Formation mechanism of the bonding layer   *Formation conditions of bonding layer*

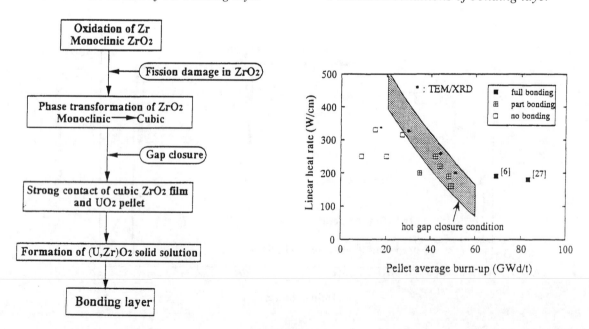

Figure 12. Fuel temperature versus time after a reactor scram at different burn-ups and derived time constants as a function of burn-up

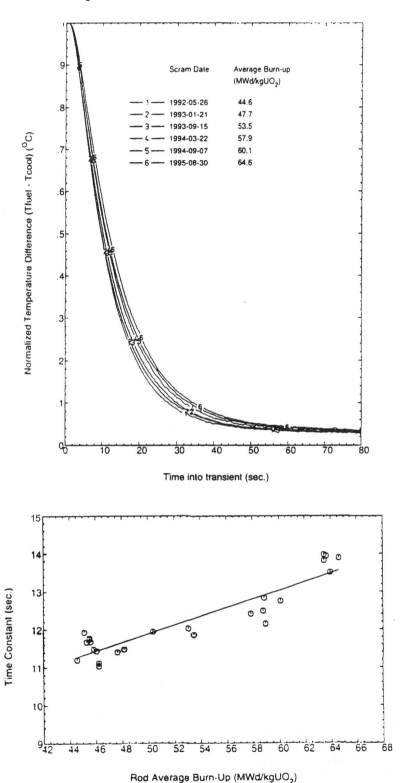

Figure 13. Thermal conductivity of two samples taken from the same pellet and measured at ~38 MWd/kgUO$_2$ by laser flash; note how the conductivities start at different values but recover to identical values

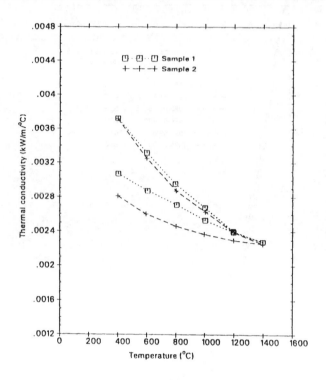

Figure 14. Conditions and mechanism for fuel restructuring at the pellet rim according to Une *et al.* [4]

Formation mechanism of the rim structure

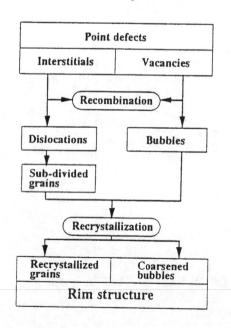

Burn-up dependence of rim structure width obtained from XE profiles of EPMA

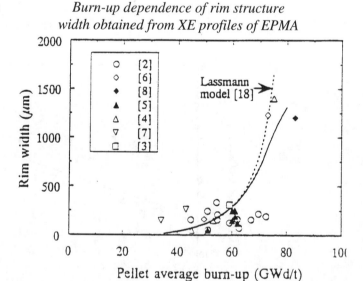

THERMAL PROPERTIES MEASUREMENTS ON IRRADIATED FUEL: AN OVERVIEW OF CAPABILITIES AND DEVELOPMENTS AT AEA TECHNOLOGY, WINDSCALE

Robin Gomme
AEA Technology plc
Fuel Performance Group
Windscale, Cumbria, UK

Abstract

This presentation provides an overview of the capabilities at AEA Technology Windscale for the measurement of thermal diffusivity and specific heat, by reference in particular to results obtained from measurements on fuel from Halden assembly IFA 558. Developments in measurement techniques, apparatus and sample preparation methods are also highlighted.

Introduction

Laser flash thermal diffusivity measurements have been performed at AEA Technology Windscale for a number of years. More recently, this capability has been supplemented by the acquisition of a differential scanning calorimeter, specially adapted for use in a shielded facility. This paper summarises the Windscale capability in the thermal properties and related areas, highlighting recent developments and illustrating sample data.

Capabilities

Thermal property measurements on irradiated and unirradiated fuels are now performed routinely at AEA Technology Windscale. A wide range of fuel types of burn ups to 80 GWd/tU, including UO_2, MOX, Gd-doped, Nb-doped and other "special" fuels have been studied for a wide range of UK and international commercial customers, including BNFL, UK Health and Safety Executive, Rolls Royce and Associates, Studsvik Nuclear, as well as for multi-national collaborative programmes such as the NFIR Project and High Burn-Up Chemistry Programme. This, of course, has the implication that the vast majority of the data obtained is proprietary to these customers, and so cannot be explicitly referred to here. In this respect we acknowledge the permission given by the UK Health and Safety Executive, and the UK Nuclear Industry Management Committee, to present data obtained from studies of fuel from the Halden assembly IFA 558.

Over the past twelve months, the Windscale facilities have been relocated and upgraded. Following considerable investment, a dedicated shielded facility has now been incorporated into the main Windscale cave-line in Building 13 (this consists of a series of thirteen caves, linked by a shielded railway system, allowing transfer of samples/components between caves). As part of this refurbishment exercise, the opportunity was taken to install a new thermal diffusivity rig. The previously existing specific heat rig has been transferred to the new facility, but has also undergone a programme of improvements.

The ultimate aim in measuring these thermal properties is, of course, the determination of thermal conductivity values for use in safety assessments. Thermal conductivity (k) is related to diffusivity (α) and specific heat (Cp) by the relation $k = \alpha.Cp.\rho$, where ρ is the density. Installation of a mercury pycnometer into the thermal properties cell is scheduled to take place before the end of 1998, whereupon the Windscale facility will be uniquely placed to measure all three parameters required for the determination of thermal conductivity.

Also of interest as part of the measurement of thermal properties are the changes which may occur, particularly in irradiated samples, in these parameters as a result of the heating/annealing cycle to which the samples are subjected during the measurement process. These property changes have been attributed to microstructural changes arising from the annealing of irradiation damage and redistribution/precipitation of fission product atoms. AEA Technology Windscale possess the capabilities – EPMA, SEM, XRD – to study all these aspects, with typical work programmes involving the characterisation pre- and post-measurement of the samples of interest.

Thermal diffusivity

Apparatus

The previously used custom-built rig has now been replaced by a Netzsch LFA-427. This offers considerable improvements over the original rig. In particular the vertical arrangement of the laser/sample/detector, which are now all contained in-cell, allows for the detector to be much closer to the sample, and thus gives greater sensitivity at lower temperatures. This, coupled with the fact that the furnace can operate routinely to 1700°C, or higher with an argon atmosphere, greatly increases the temperature range available for measurement. Further improvement is offered by the new system having a programmable temperature control. One other important difference between the old and new rigs is that the sample is now held in a horizontal rather than vertical orientation. This has facilitated measurements on highly delicate (very high burn-up) samples.

Heat loss corrections

Following the laser flash, heat losses will affect the back surface temperature response, and particularly for materials of a low thermal conductivity (e.g. ceramics such as UO_2) this can result in an incorrect determination of the thermal diffusivity if not appropriately corrected. Irradiated ceramic fuel is extensively cracked, due to the large thermal stresses generated in-reactor, and the amount of cracking increases progressively with fuel burn-up. Consequently, often only small sample fragments can be obtained (typically 3-4 mm×1 mm) Such small samples do not possess the ideal (~10:1) aspect ratio to allow the "standard" analysis techniques to be used with any degree of confidence, since there are additional routes for heat loss via radial conduction to the outer rim of the sample, and to the sample holder.

Of the "standard" techniques available, the Cowan method, which assumes that the heat loss is by radiation alone, will be superior to the standard logarithmic technique at high temperatures where radiation heat loss dominates. However, at low temperatures, where radial heat conduction contributes significantly to the total heat loss, the logarithmic method is more suitable, because during short periods, the effects of radial heat losses are less important. A unified analysis route [1] has therefore been developed at Windscale to combine the beneficial features of the two "standard" methods, so that it can be applied to both low and high temperature data. This "modified standard logarithmic method" essentially comprises the logarithmic method, modified to include a heat loss correction term, derived from fits to idealised data over the temperature range of interest.

Typical data

Figure 1 shows some typical thermal diffusivity data, obtained from a sample taken from Rod 1 of Halden Assembly IFA 558, irradiated to ~40 GWd/tU. This shows features which are now known to be typical of irradiated fuel, namely: a degradation compared with values obtained from unirradiated fuel, and a hysteresis or recovery effect, where the extent of recovery as measured at any one temperature typically increases as the maximum temperature encountered in the heating cycle increases, until some "saturation" point is reached (although the recovery is never such as to return the diffusivity to the unirradiated value).

The degradation of conductivity (diffusivity) is attributable to both irradiation damage and the presence of fission products, with the latter thought to be the major contributor (see e.g. [2]). The recovery on annealing has been interpreted qualitatively in terms of the re-distribution of isolated fission product atoms, second phase precipitates and gas bubbles. The majority of fission products are typically in the form of isolated atoms distributed uniformly throughout the lattice, causing strong scattering of phonons and, and hence reduction in conductivity. During the annealing cycle of the thermal diffusivity measurement, a temperature is reached at which the isolated atoms are able to diffuse to form precipitates or bubbles – the degree of agglomeration increasing with temperature – decreasing the number of scattering centres, leading to a recovery of diffusivity. Therefore, during the annealing cycle, there will exist a range of structures, as the ratio of isolated atoms to bubbles varies with temperature.

Quantification of the relationship between microstructural change and diffusivity recovery presently forms the basis of an extensive programme of work being performed under the NFIR Project.

Sample preparation

As described above, the typical fuel fragment size available from a sample of irradiated fuel is of cross-section 3×3 mm. For specific cases it is often of value to know the exact positional origin of the sample, so that the observed thermal properties can be related, for example to operating temperatures.

A technique has been developed at Windscale, whereby the origin of a sample may be recorded by slitting a fuel segment axially, followed by a series of mounting and polishing operations to identify appropriately sized fragments.

The availability of fragments is of course ultimately governed by the cracking pattern in the fuel, and for high burn-up/ramped fuel, it is possible that no fragments of appropriate dimension can be obtained.

To overcome this, a technique is being developed where a transverse disc sample is prepared, complete with its adherent cladding. By an appropriate choice of sample holder, the area on the back face which is monitored, can itself be targeted.

Figure 2 compares data obtained from such a sample, with that obtained from the more "idealised" geometry of a disc sample of comparable burn-up. The agreement of the two data sets is very good for this initial test. Work is continuing to further refine/validate this technique.

Specific heat

Apparatus/techniques

The basic equipment comprises a standard Netzsch DSC-404 high temperature Differential Scanning Calorimeter as described by Kaiserberger [3], although in order to allow measurements on irradiated samples, the operation of the calorimeter has been adapted for remote use inside a shielded cell [4,5].

Specific heat determination is achieved by performing consecutive runs on baseline, reference and samples. Only a comparison of the deviations of the DSC responses from the baseline for the reference standard and the sample is necessary. Sapphire is one of the few calibrants available for specific heat determination via high temperature DSC, and its use is now well established in providing a calibration of the signal displacement by the sample. A typical measurement sequence employed at Windscale is as follows:

Run	Sample Crucible	Reference Crucible
1. Baseline	Empty	Empty
2. Repeat baseline	Empty	Empty
3. Sapphire-1	Sapphire std.	Empty
4. Repeat Sapphire-1	Sapphire std.	Empty
5. Sample	Sample	Empty
6. Repeat sample	Repeat sample	Empty

Observations have shown, however, the existence of baseline shifts between successive runs, and thus considerable efforts have been devoted to developing baseline correction techniques [6].

Baseline corrections

It is stated in the literature that for the DSC method to produce reliable and reproducible results, every effort must be made to duplicate experimental conditions in each of the successive measurements. However, in practice, it is impossible to completely meet this requirement with the result that there will always be some degree of baseline shift or scatter between measurement runs.

Specific heat determination in scanning mode is most commonly achieved by applying a controlled heating ramp to the sample over the temperature range of interest. It is also advisable to record data on cooling, to determine whether any irreversible changes have occurred to the sample or apparatus during measurement runs. Early work at Windscale (on unirradiated material) showed that in some instances there could be a significant deviation between the heating (up-ramp) and cooling (down-ramp) specific heat results. As no irreversible sample changes were expected (or observed) the cause was attributed to changes in experimental conditions between runs. Two key experimental observations led to the development of the correction methods presently used: i) the deviation of the up-ramp specific heat results from the accepted standard values was of opposite sign to the deviation of the down-ramp results; ii) more often than not, the data could be adequately corrected by simply taking an average of the up-ramp and down-ramp results, weighted by the magnitude of the ramp rate (i.e. the faster the ramp rate the smaller the deviation).

In developing the correction methods it is assumed that: i) the baseline shifts seen are due to slight changes in thermal resistance in the system, so that the sign of the correction term is the same, irrespective of whether it is applied to up-ramp or down-ramp data; ii) the baseline shifts observed are small enough so that they do not have a significant effect on the instrument sensitivity.

In the method developed at Windscale, the aim was to define a scanning temperature programme with a series of fast (e.g. 20°C/min) up and down ramps, separated by a small number of isothermal holds at equally spaced temperatures. (This has been much facilitated by the upgrading of the controlling software from "12 segment" to "96 segment".) By performing a regression analysis of the isothermal DSC voltages as a function of temperature, it is possible to estimate any underlying thermal resistive baseline shifts that have occurred between the various runs. A feature of the method is that the effect of any applied correction on derived specific heat values is of a different sign for up- and down-ramp data, and provided that the sample is not altered as a result of the heating ramps, then corrected up- and down-ramp results can be checked against one another for consistency. In this way, a comparison of corrected up- and down-ramp data can provide a very sensitive test of how well the analysis method is performing.

Typical data

Figures 3 and 4 illustrate the temperature/time histories which are now routinely used for irradiated samples at Windscale. Changes in the DSC voltage responses clearly show the presence of thermal events during the temperature cycle.

Figures 5 and 6 compare the data obtained from "Run 1" and "Run 4" (i.e. effectively annealed) on a sample from IFA 558. For "Run 1", the shape of the traces for the up-ramps suggests there are exothermic processes occurring, which appear to be annealed out following a down-ramp and return to the previous temperature – i.e. an effect directly comparable to that observed during the thermal diffusivity measurements. By the time of the fourth cycle, these effects have been almost completely annealed out, and a much "smoother" trace is observed.

Figure 6 also shows a regression fit to the "Run 4" data, and compares this with the MATPRO recommendation for *unirradiated* UO_2. Agreement is observed to within ±5%. The data also show evidence of endothermic processes above 1300°C, which have been found to relax out on a time scale of ~1 hour. The origin of these effects has still to be determined, but are likely to be associated with restructuring effects in the fuel.

As part of the programme of continuing improvements to the facilities, gas flow lines have been incorporated to the rig, the intention then being to monitor for evidence of fission gas release accompanying the apparent microstructural changes. These modifications to the rig are anticipated to be implemented within the next few months.

Thermal conductivity

As discussed previously, the ultimate aim of these measurements is the determination of thermal conductivity values. The annealing effects referred to result in a continuing change in microstructure during the temperature cycle, until the "fully annealed" state is reached. In deriving a conductivity value, it is therefore most appropriate to combine the fully annealed values of diffusivity and specific heat.

Figure 7 compares the thermal conductivity derived from: the diffusivity values from the "down" part of the cycle; specific heat values taken from the regression fit to "Run 4"; and an assumed density of 95% of theoretical, with values derived from the Halden correlation [2,7]. The values derived from the measurements are typically seen to be 8% higher.

This observation should not, of course, be taken to necessarily imply that the Halden correlation is pessimistic for the in-reactor case, where other processes not present in the out-of-pile tests are occurring.

Concluding remarks

- The capability to measure thermal diffusivity and specific heat of irradiated fuel at AEA Technology Windscale has been reviewed.

- This capability has recently been upgraded by the acquisition of a new thermal diffusivity rig and improved software for specific heat measurement.

- A key element of both techniques has been the development of "in-house" codes to assess heat loss effects during diffusivity measurement and baseline correction factors for specific heat.

- Sampling techniques presently being developed should facilitate measurements on high burn-up/ramped fuel.

- Thermal diffusivity data from IFA 558 fuel has shown features typical of irradiated fuel: significant degradation compared with unirradiated values, and recovery following heating to elevated temperatures.

- Specific heat data from fuel from the same source appears to show analogous annealing effects. These data show values comparable to the MATPRO recommendation for unirradiated fuel. It would, however, appear to be prudent to obtain data to higher burn-up and for the presence of additives/dopants.

- Thermal conductivity values derived from the measured data are typically 8% higher than given by the Halden in-pile correlation.

Acknowledgements

The author wishes to acknowledge the contributions of Dr. T.L. Shaw and Mr. J.C. Carrol to this paper.

REFERENCES

[1]   Shaw, T.L. and Ellis, W.E., "Heat Loss Corrections Applied to the Measurement of Thermal Diffusivity of Small Samples by the Laser Flash Technique," High Temperatures – High Pressures, 30, 127 (1998).

[2]   Turnbull, J.A., "A Comparison Between In-Pile and Laser Flash Determinations of UO_2 Thermal Conductivity", HWR-484.

[3]   Kaiserberger, E., Janoschek, J. and Wassmer, E., *Thermochimica Acta*, 148, 499 (1989).

[4]   Shaw, T.L., Carrol, J.C. and Gomme, R.A., "Specific Heat Measurements on Irradiated UO_2 Fuel from the Halden IFA 558 Experiment", Paper presented to the Enlarged Halden Programme Group Meeting, Loen, May 1996, HRP-347/16.

[5]   Shaw, T.L., Carrol, J.C. and Gomme, R.A., "Thermal Conductivity Determinations for Irradiated Urania Fuel", High Temperatures - High Pressures, 30, 133 (1998).

[6]   Shaw, T.L. and Carrol, J.C., "Application of Baseline Correction Techniques to the "Ratio Method" of DSC Specific Heat Determination", Paper presented to the 13th Symposium on Thermophysical Properties, Boulder, Colorado, USA, 22-27 June 1997.

[7]   Wiesenack, W., Vankeerberghen, M. and Thankappen, R., "UO_2 Conductivity Degradation Based on In-Pile Temperature Data", HWR-469.

Figure 1. Comparison of thermal diffusivity data from IFA 558 fuel with unirradiated data

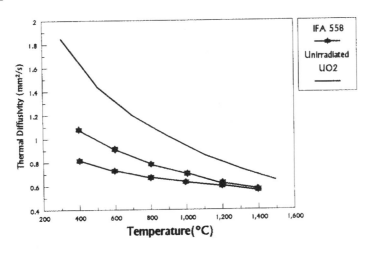

Figure 2. Comparison of data from targeted disc with data from idealised geometry

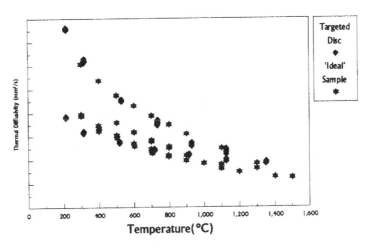

Figure 3. IFA 558: DSC voltage response for first cycle (Runs 1 & 3)

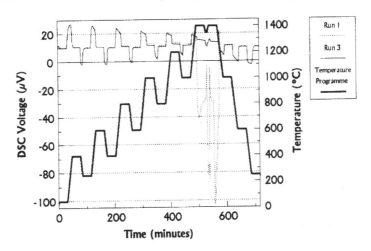

Figure 4. IFA 558: DSC voltage response for second cycle (Runs 2 & 4)

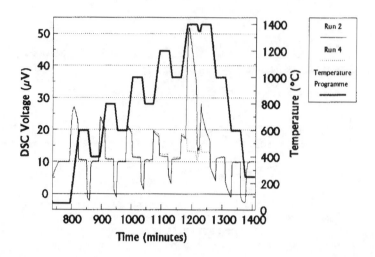

Figure 5. IFA 558 Run 1 Cp results using isothermal baseline correction method

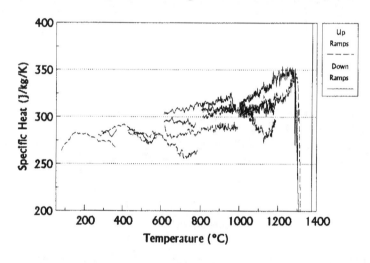

Figure 6. IFA 558 Rune 4 Cp results using isothermal baseline correction method

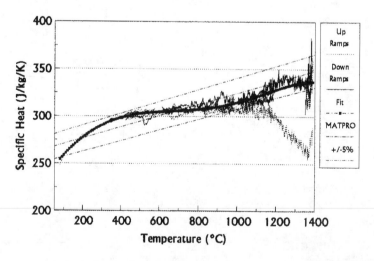

Figure 7. Comparison of thermal conductivity values derived from measurements with values from Halden correlation

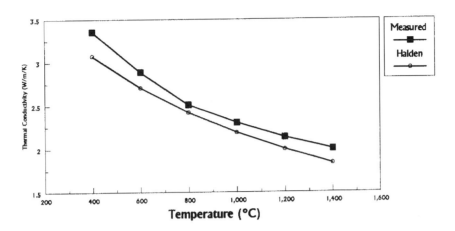

THERMAL DIFFUSIVITY OF HIGH BURN-UP UO$_2$ PELLET IRRADIATED AT HBWR

Jinichi Nakamura

Japan Atomic Energy Research Institute, Japan

Abstract

Thermal diffusivity of high burn-up UO$_2$ (63 MWd/kgU) irradiated at HBWR was measured from 290 to 1794 K by laser flash method. The thermal diffusivity of high burn-up UO$_2$ was lower than half that of unirradiated UO$_2$ at room temperature and the difference between them decreased as the measurement temperature increased. The measurements were repeated three or four times on the same sample, with increasing the maximum measurement temperature. Then, thermal diffusivity gradually increased at low temperature region. It was estimated that this increase of thermal diffusivity was mainly caused by the recovery of radiation damage. The thermal diffusivity data of the samples were separated into two groups. The difference of the thermal diffusivity of these groups was mostly explained by the effect of density difference. The present results on the samples measured after annealing at temperature between 700 and 1300 K were a little smaller than those of SIMFUEL, which chemically simulated the effects of burn-up by adding solid FPs. The relative degradation of thermal conductivity with burn-up estimated from the present data agreed well with that derived from fuel centre temperature measurement by expansion thermometer at HBWR.

Introduction

It is estimated that the thermal conductivity of UO_2 degrades with burn-up increase due to FP accumulation, radiation damage, density change and so on. The degradation of thermal conductivity of pellets results in the increase of fuel centre temperature and strongly affects fuel behaviour such as fission gas release and PCMI. Therefore, the evaluation of thermal conductivity degradation of the fuel pellet is very important, especially at high burn-up.

The effect of solid FP accumulation on thermal conductivity was evaluated by the thermal diffusivity measurement of simulated high burn-up UO_2, added with solid FPs [1,2]. The in-pile measurement of thermal conductivity of irradiated fuel was attempted in the 60s, but the fuel burn-up was relatively low [3]. Recently, the Halden Project has performed some in-pile measurement on fuel centre temperature during irradiation to high burn-up level, and the results showed a gradual increase of fuel centre temperature with a burn-up increase at the same power level [4,5].

Thermal conductivity can also be derived from the thermal diffusivity data combined with heat capacity and density data. Several studies concerning thermal diffusivity data have recently been published [6-11], and models of thermal conductivity of irradiated UO_2 [12,13] were proposed based on this diffusivity data. This paper summarises JAERI's measurement results of thermal diffusivity on high burn-up fuel irradiated at HBWR up to 63 MWd/kgU and the comparison will be made with other diffusivity data and in-pile measurement data.

Experiment

Samples

The UO_2 disk samples were prepared from a fuel pin(Rod 201 of IFA-401) irradiated at HBWR up to 63 MWd/kgU. The pre-irradiation characterisation of the rod is shown in Table 1. The initial density of the pellet was 94.5% TD. The fuel rod was sliced with diamond wheel slicing machine to obtain disk-shaped specimens of about 1 mm thickness. There were many cracks in the pellet and it was difficult to get a large size sample. The shapes of the disk specimens were irregular and four specimens of about 3×3 mm size were obtained. These specimens were washed in acetone to remove the oil used for cooling during cutting. Visual inspection was carried out to check free of cracks. The weight and thickness of the specimen were measured by balance and dial gauge.

Measurements and analysis method

The thermal diffusivity of irradiated UO_2 was measured over the temperature range from 290 to 1794 K by the laser flash method. The specimen was installed in the apparatus developed by JAERI [14,15]; Figure 1 shows a block diagram of the apparatus. It is composed of a specimen holder, heating furnace, laser oscillator, vacuum pumps, infrared detector for sample temperature measurement, radiation shielding, transfer cask and data acquisition systems. The specimen is prepared and set in the specimen holder in a lead cell and transferred to the apparatus in the transfer cask. The sample holder is automatically set into the heating furnace with remote control systems. Ruby laser(6J/pulse) is used as a heat source. The rear surface temperature of the sample is measured by In-Sb infrared detector. The thermal diffusivity was determined by "logarithmic method" [16] instead of the conventional "half time method". The logarithmic method has the advantage of reducing the experimental error even under non-ideal boundary conditions compared with the half time method [16].

Experimental results

Thermal diffusivity of four samples was measured. The weight, thickness and the measured temperature range of each sample are shown in Table 2. The thickness and weight of the samples are about 1 mm and 100 mg, respectively. The measured temperature range of the samples is from room temperature up to a maximum of 1794 K. As an example of the experimental results, the measured thermal diffusivity of sample No. 3 is depicted in Figure 2 together with the data of SIMFUEL [1], which simulates high burn-up UO_2 chemically by adding soluble FPs. Four runs were repeated while increasing the maximum temperature of measurement. The thermal diffusivity of high burn-up UO_2 was lower than half that of unirradiated UO_2 at room temperature and the difference between them decreased as temperature increased. The thermal diffusivity increased gradually as the experienced maximum temperature increased from about 870 K up to about 1300 K. The data measured in the fourth run at room temperature is only a little larger than that in the third run at room temperature. These data suggest that the recovery of thermal diffusivity between 1300 K and 1793 K is relatively small. The temperature dependency of thermal diffusivity of the third run of sample No. 3 after having experienced about 1300 K is very similar to that of 8% at SIMFUEL. The thermal diffusivity of sample No. 2 is shown in Figure 3. The tendency of recovery of thermal diffusivity is very similar to that of sample No. 3. However, the absolute value of thermal diffusivity is much smaller than that of sample No. 3, and it is about three-fourths that of sample No. 3 at room temperature.

The thermal diffusivity data of four measured samples were classified into two groups, although all four samples showed a quite similar tendency of the recovery of the thermal diffusivity between 700 K and 1300 K. Figure 4 shows the thermal diffusivity data of four different samples after heating up to about 1300 K, which correspond to the data after the recovery. The data of sample No. 1 are a little higher than that sample No. 3. On the other hand, the data of sample No. 4 are almost the same as that for sample No. 2. These samples were obtained from a 10 cm-long segment in the same rod, therefore the difference of the thermal diffusivity may be attributed to the radial structure difference in each sample because of small power history change in axial direction.

Discussions

Recovery of the thermal diffusivity of irradiated UO_2

Many factors affect thermal diffusivity and conductivity, for example, soluble FP, precipitated metallic FP, fission gas bubble, O/U ratio, density/porosity and radiation damage. In these factors, only precipitated metallic FP increases the thermal diffusivity and the other factors decrease it. The O/U ratio has strong effects on thermal diffusivity. However, the O/U ratio of UO_2 irradiated under LWR condition is reported to remain close to stoichiometry even at relatively high burn-up [17,18]. This suggests that the effect of the O/U ratio could be small in the present measurement. The present results showed the increase of the thermal diffusivity when the measurement was repeated, especially in the temperature range between 700 K and 1300 K. The possible reasons for such a recovery of thermal diffusivity are recovery of irradiation damage (point defects or defect cluster) and/or the change of fission gas bubbles in the matrix or on the grain boundary.

Nogita et al. [19] measured the recovery of irradiated UO$_2$ by means of X-ray diffraction and TEM observation. They observed the recovery of lattice parameter at 723-1123 K, recovery of broadening of X-ray reflection peak above 1423 K and the coalescence of small bubble above 1873 K. They estimated the possible point defect recovery for the lattice parameter at this stage is recovery of U-vacancy and the recovery of the defect cluster is responsible for that of the broadening of the X-ray reflection peak. Matzke and Turos [20] have observed the recovery of the UO$_2$ implanted with Xe ions between 600 and 1900 K (depending on the doses of Xe ions), and they attributed the recovery to U-vacancy and the defect cluster.

The increase of thermal diffusivity in the present test was observed between 700 K and 1300 K, which almost corresponds to the temperature range of the recovery of lattice parameter reported by Nogita et al. [19]. This suggests that the increase of the thermal diffusivity between 700 K and 1300 K is caused by the recovery of radiation damages (perhaps, uranium vacancy and may include the effects of defect cluster and the change of fission gas bubble). Based on the observation of Nogita et al., it is suggested that the recovery of the defect cluster and the coalescence of small gas bubbles occurs at a higher temperature than that of U-vacancy. However, differences of the irradiation history of the samples used in the present study and that of Nogita et al. make it difficult to compare the recovery behaviour directly.

Effect of microstructure and density

The thermal diffusivity data of four samples were separated into two groups. Samples No. 3 and No. 1 showed relatively high thermal diffusivity, and samples No. 2 and No. 4 showed low thermal diffusivity. To clarify the cause of thermal diffusivity scatter among the samples, the micro structure of the samples was examined. Extensive columnar grain growth and some metallic FP were observed in sample No. 3, which showed higher thermal diffusivity. On the other hand, columnar grain growth was not observed in sample No. 2. The cross-section of Rod 201, from which the samples are derived, is shown in Figure 5. Extensive columnar grain growth was observed in the area $r/r_0 \leqq 0.6$. The power history of Rod 201 is depicted in Figure 6. The rod power was very high (510 W/cm) at the beginning of irradiation, and it decreased almost linearly with burn-up. The rod power at the end of the irradiation was below 170 W/cm. From this power history, the columnar grain growth in Rod 201 was estimated to have occurred during the high power operation period at the beginning of irradiation. After the columnar grain growth, the density of the columnar gain region is estimated to increase to nearly 100% TD, because the pore in the columnar grain region migrated to centre and formed the centre hole. From these data, it is estimated that the pellet of Rod 201 consists of two regions with different density, i.e. columnar grain region and non-columnar grain region. The initial density of Rod 201 was 94.5% TD and the final density measured by PIE was 91.4% TD. These data indicate that swelling has occurred during irradiation after the columnar grain growth.

To estimate the local density, porosity measurement and density measurement of a small sample were performed. The scatter of the density and porosity was relatively large, but it was observed that the density at the columnar grain region was generally higher than that of the non-columnar grain region. Based on these observations and the density of 91.4% TD for a whole pellet determined by PIE, the density of the columnar and non-columnar regions was estimated to be 92-96% TD and 83-89% TD, respectively.

To see the effect of the density on thermal diffusivity, the thermal diffusivity corrected to 95% TD of samples No. 2 and No. 3 was estimated by the procedures discussed elsewhere [10]. The relation of the thermal conductivity and porosity is given by the Loeb equation:

$$k_m = k_{th}(1 - \eta P)$$ (1)

k_m :thermal conductivity of sample
k_{th} :thermal conductivity of sample with theoretical density
η :an experimental parameter, [MATPRO: $\eta = 2.6 - 5 \times 10^{-4}(T-273.15)$]
P :porosity of sample

Then, the thermal diffusivity normalised to 95% TD, α_{95} is given by:

$$\alpha_{95} = \alpha_m \left[(1 - 0.05\eta)(1 - P)\right] / \left[(1 - \eta P)(1 - 0.05)\right]$$ (2)

The thermal diffusivity normalised to 95% TD of samples No. 3 and No. 2 is depicted in Figure 7 on the assumption that the densities of samples No. 3 and No. 2 are 94% TD and 86% TD, which are the average values of the estimated densities at columnar grain region and at non-columnar grain region, respectively. The thermal diffusivity of the two samples normalised to 95% TD agreed well each other, and the differences between the samples can be explained mainly by the difference in the sample density.

Thermal conductivity of irradiated UO_2

The thermal conductivity is derived from thermal diffusivity, density and heat capacity. The heat capacity of irradiated UO_2 is not yet fully clear, and the effects of soluble FPs or single-component additives were investigated by several authors [21-23]. However, the measured results show different tendencies at high temperature region. Recently, Verrall and Lucuta [23] reported the heat capacity measurement of SIMFUEL. Their results did not show any anomalous increase relative to UO_2. Therefore, we adopted the heat capacity value of unirradiated UO_2 evaluated by Kerrisk and Clifton [24] for the calculation of thermal conductivity of irradiated UO_2. Figure 8 shows the thermal conductivity of sample No. 3 calculated from thermal diffusivity data using the estimated density (94% TD) and heat capacity of unirradiated UO_2. The general tendency of the temperature dependence of thermal conductivity is similar to that of thermal diffusivity. The thermal diffusivity data show a small peak around 400 K. This peak is caused by the large increase of heat capacity in this temperature region.

To precisely evaluate the thermal conductivity of irradiated UO_2, the heat capacity data of irradiated UO_2 is important. Shaw et al. [9] reported the heat capacity measurement on irradiated UO_2, and some exothermic reaction was observed in high temperature regions. Accumulation of heat capacity data on irradiated UO_2 is necessary.

Comparison between thermal diffusivity data and fuel centre temperature data

The thermal diffusivity data [6-11], which were recently published, showed quite similar recovery behaviour in high temperature regions, and general temperature dependence of these data showed relatively good agreement. Lucuta *et al.* [12] and Amaya and Hirai [13] proposed models of thermal conductivity of irradiated UO_2 based on these diffusivity data and early in-pile data.

Halden project measured the thermal conductivity change using the expansion thermometer [4,5]. The burn-up dependence of the fuel centre temperature normalised to 25 kW/m showed a gradual increase of fuel centre temperature with burn-up increase, that means the degradation of thermal conductivity of fuel. Halden project evaluated the centre temperature change with burn-up, and proposed the following formula for thermal conductivity degradation of UO_2 with burn-up [5]:

$$k = \frac{1}{A + B \times Bu + (C - D \times Bu) \times T} \qquad (3)$$

Figure 9 shows the relative thermal conductivity degradation at 700 K, 1000 K and 1500 K derived from the above equation. The thermal conductivity at 1000 K decreased about 34% at 55 MWd/kgUO$_2$ (63 MWd/kgU). At high temperature region, the burn-up dependency of thermal conductivity becomes smaller compared with that at low temperature. The thermal conductivity was also evaluated from the present thermal diffusivity data and specific heat capacity of unirradiated UO_2 [24]. The thermal conductivity degradation evaluated from the present results corrected for 95% TD is also plotted in Figure 9. The present results derived from sample No. 3 data (first measured data at the temperature) agreed well with the Halden Eq. (3) to evaluate the thermal conductivity degradation based on fuel centre temperature measured by extension thermometer. However, the measurement conditions of in-pile and out-pile tests are quite different. In the case of in-pile measurement, there exist fission damage, fission gas generation, formation of gas bubbles, annealing of damage, temperature gradient and so on, and the fuel is under non-equilibrium conditions. Therefore, we need more detailed data and understanding of the fuel behaviour under irradiation to compare the in-pile and out-pile data.

Conclusion

- The thermal diffusivity of high burn-up UO_2 was lower than half that of unirradiated UO_2 at room temperature and the difference between them decreased as the measurement temperature increased.

- The thermal diffusivity increased gradually with temperature increase between 700 K and 1300 K. These increases were estimated to be mainly caused by the recovery of radiation damage.

- The thermal diffusivity data of four samples were separated into two groups. The difference of thermal diffusivity of these groups was mostly explained by the effect of density difference.

- It was estimated that the reduction of thermal diffusivity of irradiated UO_2 in this study was caused mainly by soluble FP, density reduction, radiation damage (point defect and defect cluster) and may also be affected by fission gas bubbles.

- The relative reduction of thermal conductivity with burn-up at the temperature region between 700 K and 1500 K derived from thermal diffusivity data in the present test agreed well with that derived from the fuel centre temperature data by the Halden Project.

Acknowledgements

The author would like to thank Mr. I. Owada and Mr. S. Miyata for their experimental work, and also the Halden Project for offering the high burn-up fuel.

REFERENCES

[1]   P.G. Lucuta *et al.*, *J. Nucl. Mater.*,188 (1992), 198.

[2]   S. Ishimoto *et al.*, *J. Nucl. Sci. Technol.*, 31 (1994), 796.

[3]   "Thermal Conductivity of Uranium Dioxide", Tech. Rep. Ser. No. 59, IAEA(1966)

[4]   E. Kolstad and C. Vitanza, *J. Nucl. Mater.*, 188 (1992), 104.

[5]   W. Wiesenack, Proc. 1997 International Topical Meeting on LWR Fuel Performance, 2-6 March 1997, Portland, Oregon, USA (1997), 507.

[6]   J. Nakamura *et al.*, Proc. 1995 Fall Meeting at Energy Soc. Jpn., p. 590 (in Japanese).

[7]   J. Nakamura *et al.*, IAEA TCM on Advances in Pellet Technology for Improved Performance at High Burn-up, 26 Oct.-1 Nov. 1996, Tokyo, Japan (1996).

[8]   M. Hirai *et al.*, IAEA TCM on Advances in Pellet Technology for Improved Performance at High Burn-up, 26 Oct.-1 Nov. 1996, Tokyo, Japan (1996).

[9]   T.L. Shaw *et al.*, IAEA TCM on Advances in Pellet Technology for Improved Performance at High Burn-up, 26 Oct.-1 Nov. 1996, Tokyo, Japan,(1996)

[10]   J. Nakamura *et al.*, Proc. 1997 International Topical Meeting on LWR Fuel Performance, 2-6 March 1997, Portland, Oregon, USA (1997), 499.

[11]   K. Ohira and N. Itagaki, 1997 International Topical Meeting on LWR Fuel Performance, 2-6 March 1997, Portland, Oregon, USA (1997), 541.

[12]   P.G. Lucuta *et al.*, *J. Nucl. Mater.*, 232 (1996), 166.

[13]   M. Amaya and M. Hirai, *J. Nucl. Mater.*, 247 (1997), 76.

[14]   I. Owada *et al.*, JAERI-M 93-244 (1994).

[15] T. Yamahara *et al.*, IAEA Technical Committee Meeting on Recent Development on PIE Techniques for Water Reactor Fuel, Cadarache, France, 1994.

[16] Y. Takahashi *et al.*, *Netsu Sokutei*, 15 (1988), 103.

[17] K. Une *et al.*, *J. Nucl. Sci. Technol.*, 28 (1991), 409.

[18] Hj. Matzke, *J. Nucl. Mater.*, 208 (1994), 18.

[19] K. Nogita and Une, *J. Nucl. Sci. Technol.*, 30 (1993), 900.

[20] Hj. Matzke and A. Turos, *J. Nucl. Mater.*, 188 (1992), 285.

[21] T. Matsui *et al.*, *J. Nucl. Mater.*, 188 (1992), 205.

[22] Y. Takahashi and M. Asou, *J. Nucl. Mater.*, 201 (1993), 108.

[23] R.A. Verrall and P.G. Lucuta, *J. Nucl. Mater.*, 228 (1996), 251.

[24] J.F. Kerrisk and D.G. Clifton, *Nucl. Technol.*, 16 (1972), 531.

Table 1. Pre-irradiation characterisation of fuel Rod 201

Fuel pellet		Cladding	
composition	UO_2	material	Zry-2
enrichment	7.0%	outer diameter	14.29 mm
density	94.5% TD	inner diameter	12.67 mm
diameter	12.35 mm		
length	15.0 mm	Fuel rod	
dish depth	0.6 mm	gap	0.35 mm
shoulder width	1.0 mm	filler gas	He
sintering temp.	1600°C	fill pressure	0.1 MPa
sintering time	5 h	free volume	20.2 cc
grain size	25 μ m	stack length	226 mm

Table 2. Sample size and measured temperature range

Sample no.	Thickness(mm)	Weight(mg)	Measured temperature range(K)
No. 1	1.119	113.4	292-1288
No. 2	0.955	100.9	291-1496
No. 3	0.925	103.5	290-1794
No. 4	0.947	83.1	290-1590

Figure 1. Thermal diffusivity measurement apparatus

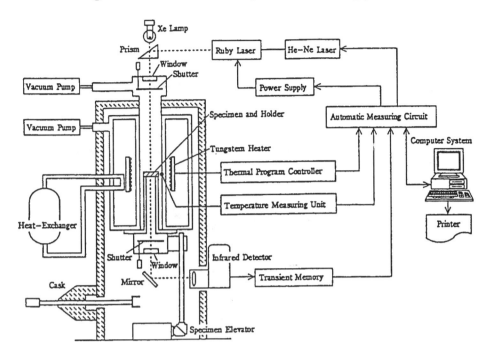

Figure 2. Thermal diffusivity of sample No. 3 (63 MWd/kgU)

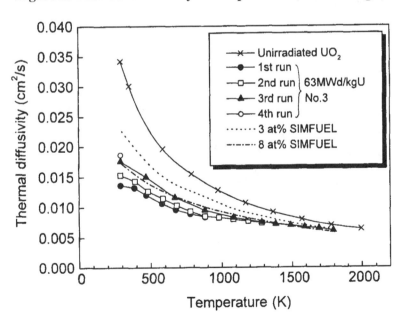

Figure 3. Thermal diffusivity of sample No. 2 (63 MWd/kgU)

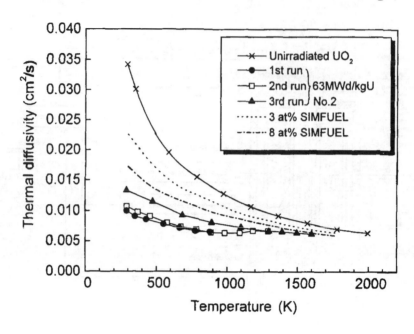

Figure 4. Thermal diffusivity of different samples after heating up to 173 K

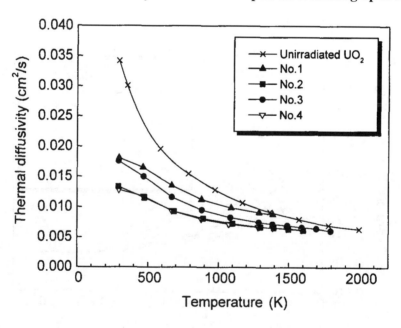

Figure 5. Cross-section of Rod 201

(a) Before etching

(b) After etching

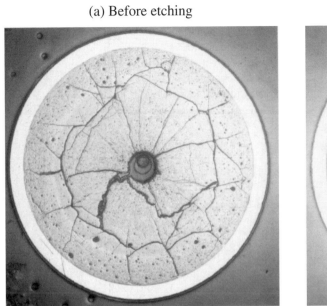
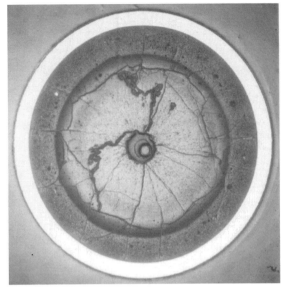

Figure 6. Power history of Rod 201

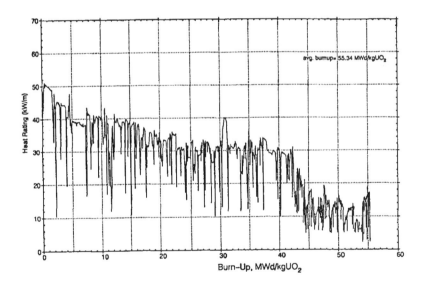

Figure 7. Thermal diffusivity of samples No. 3 and No. 2 corrected to 95% TD

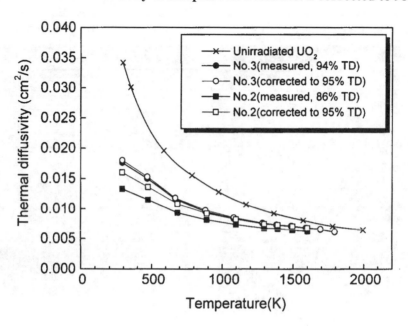

Figure 8. Thermal conductivity of sample No. 3 calculated from thermal diffusivity data using the heat capacity of unirradiated UO$_2$ in literature

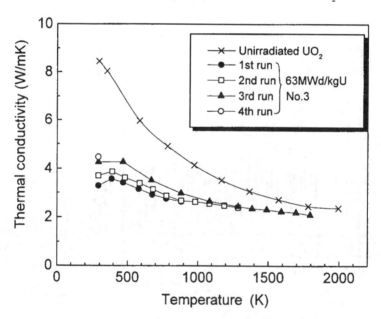

Figure 9. Comparison of relative thermal conductivity degradation at different temperatures derived from thermal diffusivity data and fuel centre temperature data [5]

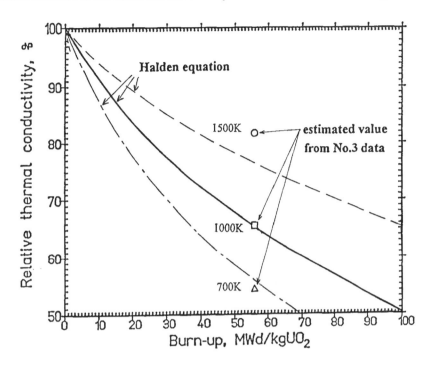

REVIEW OF FUEL THERMAL CONDUCTIVITY DATA AND MODELS

C.E. Beyer, D.D. Lanning
Pacific Northwest National Laboratory

Abstract

A review is presented of in-reactor and ex-reactor data that support degradation of fuel thermal conductivity due to burn-up. The limitations of the in-reactor fuel centreline temperature data are that they are only a measure of the integrated effects of degradation and do not provide a measure of degradation as a function of temperature. Examination of the in-reactor data demonstrates that there is a large degree of uncertainty in degradation at low to moderate burn-up levels due to large uncertainties in gap conductance. However, these uncertainties are reduced at high fuel burn-ups (> 50 GWd/MTU). Examination of the ex-reactor thermal diffusivity data from high burn-up fuel samples demonstrates that there is considerable variability of the data between samples at similar burn-up levels. It is proposed that different irradiation temperatures of these samples is one of the causes for the variability of the diffusivity data between different samples at the same burn-up level. Three recent open literature thermal conductivity models with burn-up degradation are compared to both the in-reactor and ex-reactor data at moderate-to-high burn-up levels.

Introduction

A review is presented of the theory of the physical mechanisms that are thought to be responsible for fuel (UO_2) thermal conductivity degradation due to both irradiation and temperature. Based on this theory it appears that out-of-reactor thermal diffusivity measurements only at the sample irradiation temperature are applicable to in-reactor fuel thermal conductivity. The limitations of current out-of-reactor diffusivity and in-reactor centreline temperature data are discussed with suggestions for the focus of future experimental efforts to quantify the effects of burn-up degradation.

Three recent thermal conductivity models will be presented and compared to both out-of-reactor thermal conductivity (based on thermal diffusivity measurements of high burn-up fuel samples) data and in-reactor fuel centreline temperature data from experimental rods irradiated to moderate-to-high burn-up levels. The limitations of each model are also discussed.

Background on theory of thermal conductivity

In order to understand the limitations of applying out-of-reactor thermal diffusivity data to in-reactor fuel thermal conductivity, it is helpful to first review qualitatively the theory of which parameters affect fuel thermal conductivity. Fuel thermal conductivity in light water reactors (LWRs) during normal operation is dominated by a phonon heat transport mechanism that is typically described at moderate-to-high temperatures by the simple relationship:

$$\lambda_{phonon} = \frac{1}{A + BT}$$

where A and B are material constants and T = Temperature.

The above temperature dependence is referred to as the Umklapp process where thermal conductivity is inversely proportional to temperature. Electron heat transport mechanism that is typically described by a temperature to power, or by an exponential temperature dependence, only becomes important at temperatures greater than 1700 K and will not be discussed further.

Phonon heat transport is sometimes thought of as a particle representing a quantum of thermal vibration is a solid crystal that moves down a thermal gradient in the crystal. The phonon is given wave characteristics such as vibrational frequency and wave length. At low absolute temperatures (no longer in the Umklapp regime) and for a perfect crystal lattice (no defects) the conductivity of a solid can be very high. The introduction of lattice perturbations by either thermal or point defects interferes with the man free path of the wave. Increased temperatures increase the random vibration of the crystal lattice, which interferes with the phonon wave frequency thus increasing the resistance to the phonon heat flow. Phonon resistance is also increased by a) stress in the lattice by either point defects, i.e. host atoms (U or O) out of their normal lattice positions or foreign atoms (fission products) stuffed into the lattice; b) line defects, i.e. dislocations; or c) agglomerates, i.e. precipitates such as bubbles or solid fission products. Generally point defects have the greatest scattering ability for phonons at moderate-to-high temperatures. Line defects are less effective because they impose less strain on the lattice than point defects. Larger agglomerates such as porosity, bubbles, solid precipitates and grain boundaries have the least scattering ability. The phonon scattering can also be looked at in terms of the thermodynamic energy of the lattice; i.e. higher lattice energy created by the point defects results in higher lattice resistivity.

The impact of both irradiation and temperature on phonon heat transport in fuel is complex because they introduce several competing mechanisms that can either increase or decrease the thermal conductivity. Irradiation creates point defects by 1) displacement of non-fissionable host atoms out of their normal crystal lattice positions by neutrons or other particles (measured as displacements per atom, dpa) and will be referred to as Frenkel defects, and 2) the destruction of fissionable uranium atoms during nuclear fissioning and the simultaneous creation of new fission product atoms that distort the crystal lattice. The fissioning process creates greater damage than Frenkel defects. The Frenkel defects are also more easily annealed out because the displacement distances are smaller. Given the radius of damage created by one fission and its track length it can be calculated that every atom within the UO_2 matrix will be moved from its original position every few hours in an LWR. Therefore, this irradiation damage (including fission products in the matrix) should significantly increase the resistance to heat transfer. The fissioning process also destroys small bubbles and solid precipitates, and removes gas atoms from larger bubbles, i.e. this reintroduces fission products back into the matrix as point defects, when they are in the path of a fission track.

Increasing the fuel temperature in an irradiation environment will have a beneficial effect to (increase) thermal conductivity because it will allow host atoms to diffuse back to their original lattice positions, i.e. anneal out host atoms defects, and will allow fission products to precipitate into bubbles and solid fission products, i.e. annealing out of fission products, that offer less resistance than the fission product point defect. It should also be noted that the fission process reintroduces precipitated fission products back into the matrix. Therefore, under irradiation at a given temperature there will be a dynamic equilibrium between the creation of point defects and the annealing of these defects. This pseudo equilibrium number of point defects will be dependent on burn-up and irradiation temperature with the equilibrium changing when either of these variables change. From this it can be concluded that fuel thermal conductivity will be dependent on burn-up and irradiation temperature.

For example, heating irradiated UO_2 during out-of-reactor thermal diffusivity measurements will change the number of defects created by the dynamic equilibrium between creation and annealing during the last few days of the irradiation. This is because it will anneal out the defects without ongoing irradiation to create additional defects. This points to the problem of trying to apply the results of out-of-reactor measurements of thermal diffusivity from high burn-up UO_2 samples to in-reactor thermal conductivity. The annealing effect is probably limited for temperatures below the most recent irradiation temperature, but annealing becomes more important when the irradiation temperature is exceeded because the higher temperature anneals out point defects without the dynamic presence of irradiation. The number of point defects will be lower and thermal conductivity higher above the irradiation temperature than if the sample were under irradiation. This leads to the conclusion that out-of-reactor thermal diffusivity of high burn-up samples is only applicable to in-reactor thermal conductivity when measured at the recent irradiation temperature of the sample. The thermal diffusivity data measured at temperatures below and above the irradiation temperature are not applicable to in-reactor thermal conductivity at this burn-up. In order to obtain valid thermal diffusivity measurements as a function of burn-up and temperature, several UO_2 samples should be irradiated at different isothermal temperatures (between 700 to 1100 K) and burn-ups. Thermal diffusivity measured from each sample will provide one data point for the relationship between thermal conductivity and irradiation temperature and burn-up.

From the above discussion is hypothesised that some of the variability seen in out-of-reactor thermal diffusivity data is due to the differences in the irradiation temperatures of the samples. It has been proposed [1,2] that some of the variability is due to differences in bubble shapes, sizes and densities; but another likely mechanism is the differences in irradiation temperature of the fuel samples and the number of point defects in these samples. It has been previously proposed by Turnbull at the 1996 Halden meeting in Loen, Norway, that the annealing out of fission as atoms into bubbles is the primary mechanism for the increase in thermal diffusivity with temperature from out-of-reactor measurements of high burn-up samples. This seems to be a very likely scenario based on the fact that xenon and krypton atoms in the UO_2 matrix, i.e. in solution, create large lattice strains and the discussions on the theory of thermal conductivity degradation above. However, this still needs to be confirmed based on thermal diffusivity measurements on low burn-up, moderate burn-up and high burn-up samples irradiated at different isothermal (known) temperatures and by examination of the fuel microstructure.

Thermal conductivity (phonon) models

Three UO_2 thermal conductivity correlations as functions of temperature, porosity and burn-up from the open literature are presented in this section. These models are all derived empirically from various ex-reactor and in-reactor data. Following sections will evaluate these models in comparison to ex-reactor (thermal diffusivity) and in-reactor (centreline temperature) data. The models considered are those proposed by Lucuta-Matzke-Hastings [1], by Halden Reactor Project [3] and by Nuclear Fuel Industries Ltd. (NFI) [4]. Only the phonon terms of each model are presented here because, as noted earlier, the electronic contribution becomes significant at temperatures greater than 1700 K. The model predictions as a function of temperature at burn-ups of 1, 30 and 60 GWd/MTU are shown in Figures 1, 2 and 3, respectively, for comparison.

**Figure 1. Comparison of thermal conductivity predictions from
NFI, Lucuta and Halden models at a burn-up of 1 GWd/MTU**

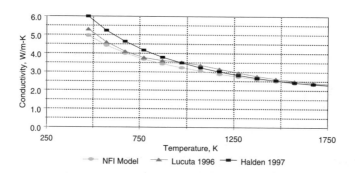

**Figure 2. Comparison of thermal conductivity predictions from
NFI, Lucuta and Halden models at a burn-up of 30 GWd/MTU**

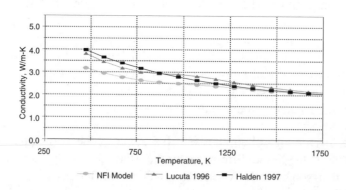

Figure 3. Comparison of thermal conductivity predictions from NFI, Lucuta and Halden models at a burn-up of 60 GWd/MTU

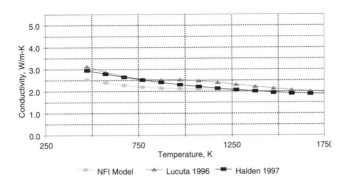

The Lucuta *et al.* model [1] is based on thermal diffusivity measurements of simulated burn-up fuel (SIMFUEL) data [5] to account for the effects of soluble rare earth fission products, early in-reactor data from Daniel and Cohen [6] to account for the effects of radiation defects at low burn-up and temperature, and theoretical considerations for the effects of solid fission products. However, this model does not account for the effects of soluble fission gas atoms which, as shown in the discussion of theory, may have a significant effect on thermal conductivity. The phonon model has several complicated equations for each term and, therefore, the general form of the expression is given as:

$$\lambda_{phonon} = K_{1d} K_{1p} K_{2p} K_{3x} K_{4r} \lambda_0$$

where: K_{1d} = effect of fission products in crystal matrix (solution);

K_{1p} = effect of precipitated solid (metal) fission products (increases λ);

K_{2p} = effect of porosity based on Maxwell-Euken factor;

K_{3x} = effect of stoichiometry (assumed = 1 for LWR fuel);

K_{4r} = effect of radiation damage (function of temperature only), and;

λ_0 = unirradiated thermal conductivity of 100% dense UO_2 from Harding and Martin [7].

The Halden model [3] is based on in-reactor measurements of fuel centreline temperature in experimental high burn-up fuel rods in the Halden reactor. The majority of these data come from fuel irradiated at low temperatures, i.e. less than 1073 K. The phonon model is expressed as:

$$\lambda_{phonon} = \frac{1}{A + B \bullet burn + C(burn) \bullet T}$$

where: C = D(1-H $\cong$ burn);

A, B, D and H are constants;

burn = burn-up, MWd/kgUO$_2$, and;

T = temperature, °C.

The NFI model [4] is based on the out-of-reactor measurement of fuel diffusivity data from one high burn-up (61 GWd/MTU) specimen and also benchmarked against RISØ III fuel centreline temperature data [8] using the NFI fuel performance code. The irradiation temperature and fuel structure of the high burn-up specimen were not given. The phonon model is expressed as:

$$\lambda_{phonon} = \frac{1}{A + B \bullet T + f(bu) + g(bu) \bullet h(T)}$$

where: A, B, C, D, K and Q are constants;

f(bu) = effect of fission products in crystal matrix (solution);

= C≅bu;

g(bu) = effect of irradiation defects, D ≅ bu$^{0.28}$;

h(T) = temperature dependence of annealing on irradiation defects;

$$= \frac{1}{1 + K \bullet e^{-Q/T}};$$

T = temperature, K.

Examination of the three model predictions at 1 GWd/MTU (Figure 1) demonstrates that the lack of the irradiation damage term in the Halden model results in a significantly higher thermal conductivity (16 to 20%) at temperatures below 900 K than calculated by the Lucuta and NFI models. Wiesenack has presented [3] some Halden centreline temperature data (< 670 K) from fresh fuel that demonstrates that fuel centreline temperatures do not increase in the first few hours of irradiation. This suggests that the thermal conductivity degradation at low temperatures and burn-up may not be as significant as predicted by the Lucuta or NFI models. However, this is contradictory to the earlier Daniel and Cohen [6] data and the qualitative theory that lattice vibrations and defects reduce thermal conductivity. Therefore, the dependence or lack of dependence of early-in-life irradiation on thermal conductivity needs to be confirmed by further experimentation. One experimental approach would be to irradiate UO$_2$ samples isothermally at low temperatures to low burn-ups, e.g. 1 GWd/MTU.

Examination of the three model predictions at 30 GWd/MTU (Figure 2) demonstrates that the NFI model predicts 12 to 20% lower thermal conductivity below 1000 K than the Lucuta and Halden model predictions and at 60 GWd/MTU the NFI model predictions are lower by 10 to 17%. This lower thermal conductivity of the NFI model can result in up to 100 K higher temperatures at 40 kW/m than predicted with the Lucuta or Halden models.

Evaluation of in-reactor data and comparison to model predictions

Evaluation of data

The in-reactor data are from rods instrumented with either centreline thermocouples or with expansion thermometers that measure fuel centre temperature as a function of time (burn-up) and linear heat generation rate (LHGR). These data are more relevant than ex-reactor data because they include the dynamics between temperature (anneals out Frenkel defects and promotes bubble precipitation/growth by gas atom diffusion) and irradiation effects (creates defects and resolves gas atoms from bubbles back into matrix solution as point defects). One of the major problems with in-reactor data is that measured centreline temperatures only provide a measure of the integrated (temperature) effects of burn-up degradation and do not give a measure of degradation at a given temperature level. Another major problem is the centreline temperature is also a function of fuel-cladding gap conductance that typically has a large uncertainty.

The in-reactor data presented here are from Halden and RISØ III programmes. The Halden data consists of the Halden Ultra High Burn-up (HUHB) experiment [3]; rod 4A from the FUMEX test cases [9]; and a section from a BWR Ringhals rod irradiated to a burn-up of 67 GWd/MTU that was re-instrumented by Halden with a centreline thermocouple and ramped in power and held for a few days. The HUHB experiment consisted of four rods with expansion thermometers but only one rod was selected for these analyses because all rods exhibited very similar thermal behaviour. The two RISØ III experimental rods were fabricated from two sections of commercial BWR rods with burn-ups of 24 and 42 GWd/MTU that were re-instrumented with centreline thermocouples by RISØ and ramped in power at DR-2 [8]. The details of each of the experimental rods are provided in Table 1.

Table 1. In-reactor data used to assess burn-up degradation of thermal conductivity

Reactor and type (and reactor from ramping), reference	Rod identification	Rod-average burn-up, GWd/MTU	As fabricated diametrical gap size (μm) [gap-to-diameter ratio (%)]	Initial fill gas type and room temperature pressure, psia (MPa)	Maximum rod-average LHGR, kW/ft (kW/m)
Halden Wiesenack 1997	Halden Ultra-High Burn-up (HUHB)	80	100 (1.7)	[145]He (1.0)	11.6 (37.9)
Ringhals BWR (Halden)[a]	Halden, Rod 2	67	265 (2.5)	[73]He (0.50)	7.6 (25)
Quad Cities BWR (DR-2), Knudsen et al. 1993	GE-2, GE-4 from RISØ III	42, 24	225 (2.1)	[73]He (0.5)	12.6 (41), 13.2 (43)
Halden (Halden), Chantoin et al. 1997 (FUMEX)	FUMEX-4A	39	220 (2.1)	[44]He (0.30)	15.7 (52)

[a] Halden Reactor Project, 1997. Personal communication with US NRC.

Comparison of models to data

In order to compare each of the thermal conductivity models to the in-reactor temperature data, each was programmed into the FRAPCON-3 fuel performance code [10] along with the input for each experimental rod. Because there are large variations in the LHGRs for the different experimental rods the difference between predicted and measured centreline temperature was ratioed to the difference between measured centreline and coolant temperature in order to normalise out LHGR differences. The resulting rations for the Lucuta, Halden and NFI models are provided in Figures 4, 5 and 6 respectively. As demonstrated in Figure 4, the Lucuta model has the largest underprediction of the in-reactor data above 40 GWd/MTU and the HUHB data are underpredicted by 12 to 18% for burn-up greater than 60 GWd/MTU. The Halden model (Figure 5) underpredicts the HUHB data by a similar amount (7 to 15% underprediction) but does significantly better for the Halden ramped Ringhals rod and the RISØ III rod (GE-2) at 42 GWd/MTU. The NFI model (Figure 6) does a significantly better prediction on all rods above 40 GWd/MTU but significantly overpredicts the FUMEX-4A rod at 39 GWd/MTU and underpredicts the RISØ III rod (GE-4) at 22 GWd/MTU. The reason for the underprediction of the RISØ III GE-4 rod and a relatively better prediction of the higher burn-up GE-2 rod from the same programme is unknown except that the fuel-cladding gap in the GE-4 rod may be larger than that calculated by the code at this low burn-up. The reverse may be true for the FUMEX-4A rod overprediction.

From the comparisons to the in-reactor data, it may be concluded that the NFI model provides the best prediction at burn-ups greater than 40 GWd/MTU, but additional in-reactor and out-of-reactor data are needed to better quantify burn-up degradation as a function of temperature. It can also be concluded that the greatest uncertainty in the lower burn-up rods is the gap size and, therefore, gap conductance. Gap conductance is less of an uncertainty for higher burn-up where the gap is closed.

Figure 4. FRAPCON-3 [Lucuta 1996 model] predicted-minus-measured centreline temperatures, normalised to measured-minus-coolant temperature difference, as a function of burn-up

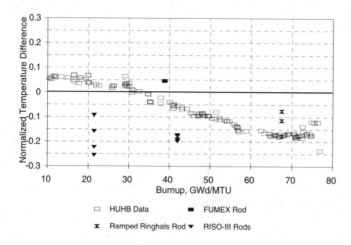

Figure 5. FRAPCON-3 [Halden 1997 model] predicted-minus-measured centreline temperatures normalised to measured-minus-coolant temperature difference, as a function of burn-up

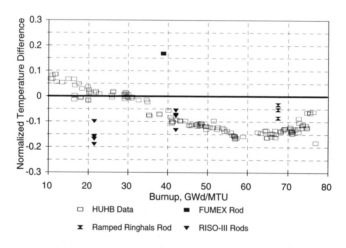

Figure 6. FRAPCON-3 [NFI model] predicted-minus-measured centreline temperatures normalised to measured-minus coolant temperature difference, as a function of burn-up

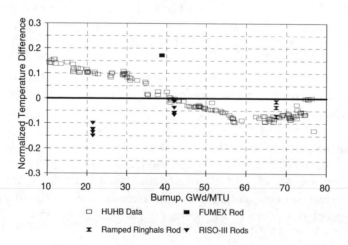

Evaluation of out-of-reactor diffusivity data and comparison to models

Background and data limitations

Thermal diffusivity is a parameter that occurs in the transient form of the general equation for heat transfer in a solid. It is defined as the ratio of the (local) thermal conductivity to the product of the density and the specific heat, and is analogous to the diffusion constant in the transient form of the diffusion equations in solids; it characterises the diffusion of heat throughout the solid as a function of time. The diffusivity of irradiated fuel pellet material is typically determined by the "laser flash" method. In this method, a small, disk-like sample is cut from a pellet and mounted in a tube furnace that has quartz windows. The sample is brought to a specified isothermal ("test") temperature in the furnace. Then the "front" side of the sample is subjected to a pulse of laser energy (through a quartz window) and the small resultant transient peak in the Abackside ≅ temperature is measured optically and recorded. The analysis of the time-temperature trace yields an estimate of the thermal diffusivity of the sample at the Atest ≅ temperature.

The thermal conductivity of the sample at the test temperature is derived by multiplying the diffusivity by the sample density and specific heat where each has been appropriately corrected to the test temperature. Only a small correction is made to the room-temperature estimate of sample density, to account for thermal expansion strain. The specific heat is a weak function of temperature, and also of burn-up. However, Lucuta *et al.* [5] and others (e.g. [11]) have demonstrated that the burn-up dependence is very small; thus correlations for specific heat of unirradiated fuel as a function of temperature will apply. To demonstrate this, the MATPRO [12] correlation for unirradiated UO_2 specific heat is shown in Figure 7 in comparison to Lucuta's results at various simulated burn-ups. In the comparisons to thermal conductivity models shown below, the MATPRO specific heat has been used to reduce the reported diffusivity data to conductivity values.

Figure 7. MATPRO specific heat and measured SIMFUEL specific heat from [5]

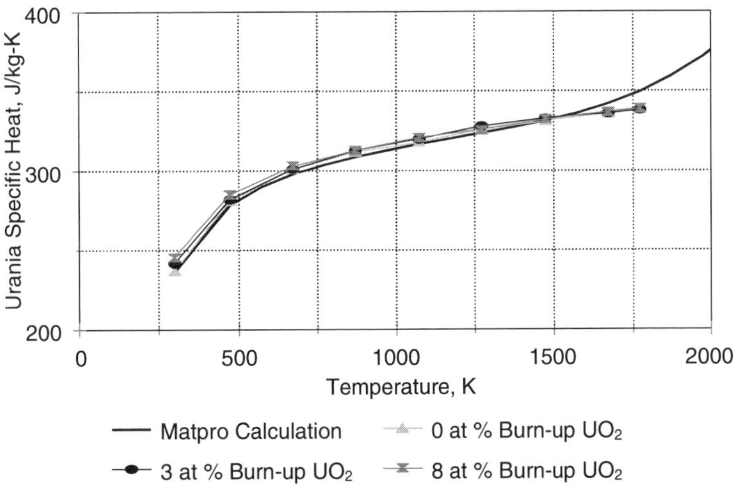

Two data sets from diffusivity measurements on irradiated pellet samples have been examined. The first data set measured by Carrol [1], listed here as Halden data, is a test on fuel irradiated to ~40 GWd/MTU in the Halden reactor, Norway, and diffusivity measured at Windscale Laboratory in the UK. The second data set [2] from the Japan Atomic Energy Research Institute (JAERI), listed here as JAERI data, is on fuel irradiated in a power reactor to 63 GWd/MTU and diffusivity measured in JAERI facilities. These two data sources were also used by Lucuta *et al.* [1] in verifying their thermal conductivity correlation.. In this comparison, alteration was made to the pore factor in the porosity correction factor for some samples, based on cermography results. The results presented here use the standard (spherical) pore factor of 1.5 from FRAPCON-3 for all model calculations.

It should be noted that the heating of the fuel sample for the laser-flash measurement of thermal diffusivity at various temperatures affects the measurements due to defect annealing. As the test temperature is stepped to levels above the sample irradiation temperature, damage annealing occurs, and the diffusivity values for the "cool-down" stage of a stepped diffusivity test are often greater than the values at corresponding temperatures during the "heat up" stage. Similarly, if multiple stepped tests (i.e. multiple "runs") are conducted on the same sample, the diffusivity and conductivity at corresponding test temperatures are likely to increase from one run to the next. For this reason, data used here are only from the reported "first heat up, first run" for comparison to the conductivity model predictions. In addition, only the measured diffusivity at the sample irradiation temperature should be used for comparison but this is not possible fore the Halden data because the irradiation temperatures are unknown. An example of the effect of irradiation temperatures is illustrated by hypothetically looking at three samples with three different irradiation temperatures. If Sample A is irradiated at 700 K, Sample B is irradiated at 900 K, and Sample C at 1100 K; Sample A will have the greatest number of point defects (and lower diffusivity) while Sample C the least point defects and highest diffusivity. Heating Sample A to 1100 K out-of-reactor for a given time period will not give the same number of point defects (a lower defect number will result) as Sample C in-reactor at 1100 K because there is no irradiation present and, therefore, Sample A diffusivity at 1100 K will be greater than Sample C at 1100 K. As noted above, the irradiation temperatures of the two Halden samples are unknown so their value is limited for evaluating in-reactor thermal conductivity. However, because these are the only data available they will be compared to the three models presented in this paper.

Comparisons of models to data

The comparison of model-predicted thermal conductivity to data-derived thermal conductivity from the Halden (first run) data is shown in Figure 8. Two samples were tested, both taken from the pellet mid-radius region. It is assumed that both these samples have equal nominal density and burn-up of ~42 GWd/MTU. The Lucuta and Halden models result in a small overprediction (5 to 6%) of Sample 1 and a greater overprediction for Sample 2, while the NFI model matches Sample 1 very well. The higher measured diffusivities (7 to 8%) of Sample 1 compared to Sample 2 below 1000 K may be due to a higher irradiation temperature for Sample 1. It should be remembered that power reactor fuel pellet operating temperatures, even at the outer surface, will be in excess of 600 K.

Figure 8. Comparison of Lucuta, Halden and NFI model predictions to out-of-reactor data from samples irradiated in Halden (see [1]); burn-up = 42 GWd/MTU

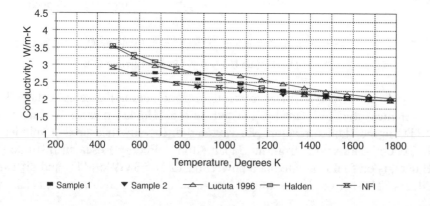

There are two separate JAERI samples, with differing estimated density at different radial locations in the pellet. The first sample, cut from the pellet central region, has an estimated as-irradiated density of 92 to 96% and would be at a higher irradiation temperature. The second sample, cut from the pellet outer region, has an estimated density of 83 to 89% and would be at a considerably lower irradiation temperature. The estimates of fuel density were based on optical ceramography of companion fuel by JAERI. The average burn-up in both samples is reported as approximately equal to 63 GWd/MTU by the authors but irradiation temperatures were significantly different. If irradiation temperature has a significant impact on thermal conductivity than the higher temperature sample should be underpredicted by the models at low temperatures (< 1000 K) even when model predictions are corrected for the higher density of this sample. The lower temperature sample (86% TD) should be predicted by the models much better at low temperatures than the higher temperature sample (92 to 96% TD). The comparison of the models to the higher density data (at 86% TD) is shown in Figure 10. All three models significantly underpredict the (higher temperature) 96% TD data, with the least underprediction (9 to 15% above 550 K) associated with the Lucuta and Halden models. The Lucuta and Halden models overpredict the 86% TD data (6 to 15% above 550 K) while the NFI model fits these data very well. These comparisons to these data that correct for density effects tend to corroborate the hypothesis that sample irradiation temperature has a significant impact on the measured out-of-reactor thermal conductivity and the measured values are only applicable to in-reactor thermal conductivity at the irradiation temperature.

Figure 9. Comparison of Lucuta, Halden and NFI model predictions to out-of-reactor JAERI data for an irradiated fuel sample[2]; estimated as-irradiated density = 96% TD, burn-up = 63 GWd/MTU

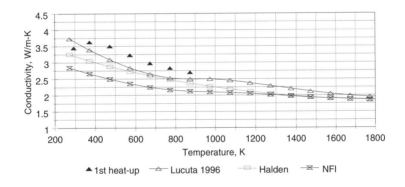

Figure 10. Comparison of Lucuta, Halden and NFI model predictions to out-of-reactor JAERI data for an irradiated fuel sample[2]; estimated as-irradiated density = 86% TD, burn-up = 63 GWd/MTU

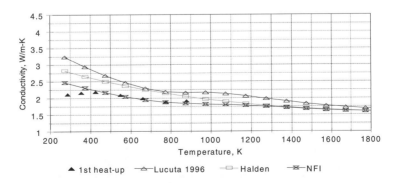

Thus, there is a significant variability among the examined out-of-reactor data that is probably due to differences in the in-reactor irradiation temperatures of the different fuel samples. The NFI thermal conductivity model appears to provide the best prediction of irradiated samples at low irradiation temperatures if it is assumed that the Halden samples were irradiated at low temperatures.

Conclusions

Based on an examination of the theory for thermal conductivity degradation of UO_2 fuel in an irradiation environment, including the effects of temperature, it is suggested that out-of-reactor thermal diffusivity data from irradiated samples are only applicable to in-reactor thermal conductivity when the measurement is performed at the recent irradiation temperature of the sample. The qualitative theory suggests that diffusivity measurements at temperatures below or above the sample irradiation temperature are not applicable to in-reactor thermal conductivity. This appears to be corroborated by comparison of model predictions at low temperature to conductivities derived from diffusivity measurements at low temperatures of the two JAERI samples taken from high and low temperature regions of the fuel.

In-reactor measures of thermal conductivity degradation by centreline temperature measurements also have some limitations. These data only measure the integrated effect of degradation and do not provide a measure of degradation as a function of temperature. There is also a large degree of uncertainty in degradation at low to moderate burn-up levels due to large uncertainties in gap conductance. However, these uncertainties are reduced at high fuel burn-ups (> 50 GWd/MTU) and this is corroborated by examination of the in-reactor data as a function of burn-up.

Of the three models examined, the NFI fuel thermal conductivity model has provided the best prediction of the in-reactor data at burn-ups greater than 40 GWd/MTU. This model also provides the best prediction of out-of-reactor conductivity data from samples believed to be irradiated at low temperatures. However, this model still underpredicts most of the in-reactor data and, because irradiation temperatures are unknown for the ex-reactor diffusivity data, the value of these data for quantifying burn-up degradation as a function of irradiation temperature is limited. Consequently, none of the three models presented has clear advantages over the others for use at high burn-ups.

Additional data are needed to verify or dispel the effects of early-in-life defects on in-reactor thermal conductivity because there is conflicting evidence. Also, ex-reactor thermal diffusivity data are needed from fuel samples at moderate-to-high burn-ups with known irradiation temperatures in order to quantify burn-up degradation as a function of irradiation temperature.

REFERENCES

[1]   Lucuta, P., Hj. Matske and I.J. Hastings, "A Pragmatic Approach to Modeling Thermal Conductivity of Irradiated Fuel: Review and Recommendations", *Journal of Nuclear Materials*, Vol. 232, pp. 166-180, 1996.

[2]   Nakamura, J. *et al.*, "Thermal Diffusivity Measurement of a High Burn-up UO_2 Pellet", Proceedings of the 1997 International Topical Meeting on Light Water Reactor Fuel Performance, Portland, Oregon, 2-6 March 1997 (published by the American Nuclear Society), 1997.

[3]   Wiesenack, W., "Assessment of UO_2 Conductivity Degradation Based on In-Pile Temperature Data", Proceedings of the 1997 International Topical Meeting on Light Water Reactor Fuel Performance, Portland, Oregon, 2-6 March 1997 (published by the American Nuclear Society), 1997.

[4]   Ohira, K. and N. Itagaki, "Thermal Conductivity Measurements of High Burn-up UO_2 Pellet and a Benchmark Calculation of Fuel Temperature", Proceedings of the 1997 International Topical Meeting on Light Water Reactor Fuel Performance, Portland, Oregon, 2-6 March 1997 (published by the American Nuclear Society), 1997.

[5]   Lucuta, P., *et al.*, "A Thermal Conductivity of SIMFUEL", *Journal of Nuclear Materials*, Vol. 188, pp. 198-204, 1992.

[6]   Daniel, R.C. and I. Cohen, "In-Pile Effective Thermal Conductivity of Oxide Fuel Elements to High Burn-up Depletion", WAPD-246, Bettis Atomic Power Report, Westinghouse Advanced Products Division, 1964.

[7]   Harding, J.H. and D.G. Martin, "A Recommendation for the Thermal Conductivity of UO_2", *Journal of Nuclear Materials*, Vol. 166, pp. 223-226, 1989.

[8]   Knudsen, P., C. Batter, M. Mogensen and H. Toftegaard, "Fission Gas Release and Fuel Temperature During Power Transients in Water Reactor Fuel at Extended Burn-up", Proceedings of a Technical Committee Meeting held in Pembroke, Ontario, Canada, 28 April-1 May 1992, IAEA-TECDOC-697, International Atomic Energy Agency, Vienna, Austria, 1993.

[9]   Chantoin, P.M., J.A. Turnbull, W. Wiesenack, "How Good is Fuel Modelling at Extended Burn-up? The IAEA's FUMEX Programme Provides Some Answers", *Nuclear Engineering International*, September 1997.

[10] Berna, G.A., C.E. Beyer, K.L. Davis and D.D. Lanning, "FRAPCON-3: A Computer Code for the Calculation of Steady-State, Thermal-Mechanical Behavior of Oxide Fuel Rods for High Burn-up", NUREG/CR-6534, Vol. 2 (PNNL-11513, Vol. 2), prepared for the US Nuclear Regulatory Commission by Pacific Northwest National Laboratory, Richland, Washington, 1997.

[11] Matsui, T. *et al.*, "High-Temeprature Heat Capacities and Electrical Conductivites of UO_2 Doped with Yttrium and Simulated Fission Products", *Journal of Nuclear Materials*, Vol. 188, pp. 205-209, 1992.

[12] Hagrman, D.L., G.A. Reymann and R.E. Mason, "MATPRO-Version 11 (Revision 2). A Handbook of Materials Properties for Use in the Analysis of Light Water Reactor Fuel Rod Behavior", NUREG/CR-0479 (TREE-1280, Rev. 2), prepared for the US Nuclear Regulatory Commission by EG&G Idaho, Inc., Idaho Falls, Idaho., 1981.

IRRADIATION AND MEASUREMENT TECHNIQUES AT THE HALDEN REACTOR PROJECT AND THEIR APPLICATION TO HIGH BURN-UP FUEL PERFORMANCE STUDIES

C. Vitanza

OECD Halden Reactor Project, Norway

Abstract

The paper is intended to review methods and techniques used in the Halden reactor for fuel performance assessments and in particular for determination of fuel thermal behaviour. Besides reviewing the Halden fuel rod instruments and their use, the paper presents applications of the refabrication/reinstrumentation techniques for high burn-up studies. Practical examples of experiments carried out or planned at Halden are also given.

Introduction

The characterisation of fuel thermal behaviour is fundamental for a proper understanding of the performance of the fuel itself and for the determination of the safety margins in a variety of conditions. During service, the fuel undergoes a number of processes which are strongly affected by temperature. For operation at extended burn-up it is important that the rod pressure does not increase beyond certain limits, which implies that the release of fission gas from the fuel should be kept under control. Unfortunately, both fuel restructuring and fuel thermal conductivity degradation contribute to enhancing the fission gas release at high burn-up. Reliable predictions of the fuel thermal behaviour are also important when the cladding temperature is to be calculated in coolant transients, since the heat stored in the fuel contributes significantly to overheating the cladding.

The fuel thermal performance is addressed at Halden, primarily by means of in-reactor tests involving measurements of the fuel temperature. The scope of these tests was originally limited to low or moderate burn-up and had the objective of forming the database for model developments as well as for optimising the fuel design. These tests cover a wide range of power and fuel design conditions. In particular, the effect of gap size and of the filler gap composition have been extensively addressed, such that the database could also be applied in presence of irradiation-induced phenomena, such as gap closure and fission gas release.

In-reactor fuel temperature data have been generated for mixed oxide fuels for fuel containing burnable poison, as well as in bilateral irradiations of fuel containing additives aimed at improving specific properties. Tests have also addressed the effect of fuel microstructure and of the fuel pellet surface conditions (roughness).

The increased emphasis on extended burn-up has posed new requirements to Halden experimentations. Irradiations techniques have been devised to produce data at high burn-up within acceptable time. These primarily consist of irradiating the fuel at high fission rates and at relevant temperatures. Isothermal fuel samples inserted in specially designed capsules have also been irradiated at Halden in order to facilitate ceramographic examinations and determinations of fuel thermal diffusivity in hot cells.

In parallel, techniques have been developed which enable to measure fuel temperatures in fuel segments pre-irradiated in commercial reactors to high burn-up and subsequently loaded, after refabrication, in the Halden reactor.

Fuel thermocouple measurements

In Halden experiments, the fuel temperature is measured at the centreline of the fuel column. For this purpose, a centre hole is drilled on the pellets at the end of the fuel column where the thermocouple is mounted. Calculations show that a thermocouple penetration of ~20 mm into the fuel stack would be sufficient to avoid end effects caused either by power end peaking or by axial heat transport along the thermocouple. Conservatively, depths from 40 to 70 mm are currently used in the Halden design.

The size of the pellet centre hold depends on the thermocouple thickness and may range from 1.2 to 1.8 mm. The centre hold size is a rather relevant parameter since it affects the extent to which the measured temperature differs from the real centre temperature in the actual (solid) pellet.

This difference arises from the fact that the power generated in the centre hole pellet is somewhat lower than in a solid pellet and, more importantly, by the smaller heat diffusion length for the hollow pellet. The latter is accounted for by the equation:

$$\int_{T_s}^{T_o} k\, dT = \frac{q_L \cdot f}{4\pi}$$

$f = 1$ for solid pellet.

$f = \left(1 - \frac{2R_I^2}{R_F^2 - R_I^2} \ln \frac{R_F}{R_I}\right)$ for annular pellet.

where: T_s, T_o are the fuel surface, centre temperature, °C
   $k(T)$ is the fuel thermal conductivity, W/°C cm
   q_L is the linear heat rating
   R_F, R_I are the pellet and centre hole diameter, cm

These equations (which are strictly valid for a flat power profile across the pellet radius) say that for a given heat rating, in presence of the centre hole the centre temperature is equal to the one in a solid pellet in which the power is reduced by a factor f. The value of f for various centre hole sizes and for three pellet diameters is given in Figure 1. The figure shows that f can be as high as 0.96 for 10.4 mm pellets and 1 mm hole and as low as 0.80 for 7.5 mm pellets and 2 mm hole. It goes without saying that the measured temperatures need in any case to be properly analysed in order to apply them correctly to the actual solid pellet geometry.

In Halden tests instrumented with fuel thermocouples, the thermocouple tip is positioned at mid-pellet position, in order to eliminate potential end pellet effects, which for instance can be of some relevance in the presence of pellet dishing.

The fuel thermocouples used at Halden are normally tungsten-rhenium types, insulated with compacted alumina powder and inserted into a refractory metal sheath. Over the years, depending on the availability and quality of suppliers, production time and experience, different types of such thermocouples have been used. The most used materials are as follows:

Thermocouple leads:	**W3%Re/W25%Re**
	W5%Re/W26%Re
Insulation:	**Al_2O_3**
	Mg O
	BeO
Sheath:	**Tungsten**
	Rhenium
	W-Re alloys
	Molybdenum

The fuel thermocouple design which is currently in use at Halden is as follows:

Sheath outer diameter:	**1.6 mm**
Sheath material:	**Mo - 50%Re**
Insulation material:	**Beryllium oxide**
Thermocouple leads:	**W5%Re/W26%Re**
Thermocouple leads size (Ø):	**0.20 mm**

Refractory metals cannot be welded in Zircaloy or stainless steel and cannot withstand a water environment for a long time. Thus, the fuel thermocouple must be connected to a suitable extension cable within the plenum of the fuel rod itself. In earlier tests, this connection was made at Halden and Chromel/Alumel cables were used. Based on this experience, we have gradually qualified suppliers able to deliver thermocouples already connected to the extension cables. A key requirement is that the thermocouple junction to the extension cable must be gas tight, have small size, contain nuclear inert materials and be reliable in terms of electrical continuity.

It is desirable that the extension cables do not introduce spurious emf due to the temperature difference between the plenum and the external reference temperature. To this end, compensation cables are used which reproduce the same emf output as the W-Re thermocouple, but in a lower temperature range.

The Halden design of the overall assembly of a fuel thermocouple inserted into a fuel rod is shown in Figure 2, containing all the relevant aspects which have been discussed above. It should be noted that the extension/compensation cables as fabricated by commercial suppliers are sheathed in an Inconel sheath (Inconel 600). Thus a transition Zircaloy-stainless steel must be realised at the rod end plug. Both co-extruded and swage-lock techniques have been used, with the swage-lock solution being the one that is currently preferred (see Figure 2). Other important details have gradually been refined, such as the welding of the extension cable onto the end plug and the "seal tube" solution to avoid that cable vibrations as consequence of the coolant flow induced fatigue-induced failure of the cable sheath.

Differently than at other establishments – where some fuel thermocouple measurements have been performed for short time duration (typically 24 hours) – Halden experiments involve tests of very long duration, that is, up to many years. It is apparent that this poses quite different challenges to the thermocouple performance. Although fuel temperature measurements up to 2400°C have been successfully carried out at Halden, the general experiences is that for long term operation (years) the thermocouples perform well below 1400-1500°C, not so well beyond 1600°C. It is believed that chemical interaction of fission products with the thermocouple sheath is the cause for decreased reliability at high temperature.

High temperature W-Re thermocouples decalibrate during service primarily as a consequence of rhenium conversion into osmium due to thermal neutron captures. The decalibration is irrelevant or not so important for short term irradiations (lasting up to 1-2 years), but it can become substantial for irradiations which last many years. In order to correct the measurements, an experimental campaign was carried out many years ago consisting of recalibrating thermocouples which had been in service at Halden for various periods of time. The outcome of these results, together with new experimental evidence, has recently been reviewed resulting in the decalibration correlation shown in Figure 3.

Based on this curve, which expresses the decalibration as a function of thermal fluence, one can derive that for most Halden irradiation the decalibration correction ranges between zero and 15%, but in some cases it is as high as 30-35%. The correlation shown in Figure 1 is now used on-line to correct all Halden W-Re thermocouple data.

Variants to the basic thermocouple design have been used for fuel temperature measurements in special cases. For studies of fuel thermal behaviour in the presence of water ingress and steam environment, for instance, a special coating with metal disilicide has been used. However, experience showed that normal W-Re thermocouples behaved in a manner comparable to the specially coated thermocouples.

For conditions below 1300°C, nicrosil-nisil thermocouples have been successfully used in many Halden tests. These thermocouples are easier to handle, do not require a transition junction in the plenum of the fuel rod and apparently have very little decalibration. They are also less expensive than W-Re thermocouples.

As a final remark concerning fuel temperature measurements at Halden, attempts were also made in the past to perform such measurements at the radial periphery of the pellet. However, the outcome was so poor that such attempts were quickly abandoned.

Fuel temperature measurements with expansion thermometers

High temperature, refractory metal thermocouples have been successfully employed for a long time and constitute the basis for most of the Halden fuel temperature data. At some point, however, a number of considerations suggested that it would have been desirable to have an alternative measuring technique. This was accomplished in 1989, after three years of development and qualification, by introducing the expansion thermometers (ET).

The main motivations for using this type of detector are that:

- It does not require cable penetration into the fuel rod, i.e. the sensor is totally enclosed into the rod itself. Apart from other considerations, this gives much more flexibility in test rig utilisation as the rod can be moved in and out of the test rig when it so desired.

- The expansion thermometer is a robust and repeatable sensor which is based on materials readily available on the market and which is easy to install.

- The sensor output is virtually unaffected by long term irradiation.

The principle of the ET measuring technique is shown in Figure 4. Basically it consists of a long, tiny cylinder placed along the centreline of the entire fuel pellet stack length. Three different materials were originally tested and all exhibited good performance. Of these, the material finally selected was 8% zirconia-stabilised tungsten, primarily because of its cubic structure, which prevents axial growth, and because it is readily available in repeatable and pure form.

The tungsten cylinder carries a Halden-standard magnetic core at one end. This rod end is designed for insertion into a differential transformer, such that an axial displacement of the magnetic core, induced by the thermal expansion of the tungsten cylinder, is converted into an AC signal change. Depending on the length of the fuel column, this set-up provides for a measurement sensitivity of ±2 to ±5°C, and for a validated measuring range up to 2400°C.

A basic difference between the fuel temperature determinations made with thermocouples and with expansion thermometers is that the former provide local measurements, whereas the latter give the average fuel centre temperature along the active length. In principle, there is full equivalence between the two as in both cases one has to use the measured value to infer the temperature distribution along the rod. A possible disadvantage of the ETs is however that the entire fuel column has pellets with centre hole.

As part of the ET validation, fuel centre measurements carried out with ETs have been compared with the ones obtained with thermocouples in a pair of rods having equal design and placed in the same test rig. The results are shown in Figure 5 and display the excellent comparability of the two types of measuring techniques.

Re-fabrication of commercial fuel

The experimental needs expressed by the Halden Project participants often require the use of instrumented commercial fuel previously irradiated in power reactors. This has been responded to by developing and implementing a refabrication technique at the Kjeller hot cells, a process which required two years of qualification work and which was ready for use in 1991. Since then, about 50 fuel rods have been refabricated, half of which were equipped with centreline thermocouples.

The details of the technique are many and complex and have been reported in the papers given in the reference list. Thus, they are not discussed here.

A schematic of the refabricated rod and thermocouple assembly is shown in Figure 6. The technical aspects described in the section entitled *Fuel thermocouple measurements* apply also for refabricated fuel rods. In particular, the same thermocouple, junction and extension cables are used for fresh and pre-irradiated fuel. The only difference is that for the latter applications a molybdenum tube is inserted after drilling in order to prevent that fuel fragments fall in the centre hole and to facilitate the thermocouple installation and positioning. The centre hole normally has a 2.5 mm diameter.

Irradiation technique for high burn-up applications

Besides the "normal" irradiations, which carry on for long periods of time, means have been devised to produce high burn-up data in shorter periods of time. These techniques are shown in Figure 7 and consist of:

- *Accelerated irradiations in small diameter rods.* In this case, fuel pellets having small diameter and high enrichment are used. This enables to produce prototypical linear heat rating the thus prototypical temperature across the pellet, but at considerably high burn-up accumulation rate. Typically, the pellets have a diameter of 5.5 mm and an enrichment between 10 and 13%. The burn-up accumulation rate is typically ~20 MWd/kg per year. These irradiations have been used for fuel thermal conductivity assessments and for comparative testing of the fission gas release and swelling behaviour for different types of fuel microstructures in the high burn-up range (60-80 MWd/kg).

- *Irradiations of commercial fuel rods.* In this case, the high burn-up has already been achieved during the base irradiation in power plants. The tests at Halden normally involve irradiations devoted to fuel thermal behaviour, fission gas release and PCI assessments. Refabricated and instrumented commercial fuel are also being used for determining the consequences of rod overpressure at high burn-up.

- *Isothermal disks.* The purpose of this type of irradiation is to provide well-characterised, uniform specimens for PIE determinations. In particular, it has been used for fuel diffusivity measurements carried out at specialised hot cells by means of the laser flash technique. Fuel diffusivity is proportional to fuel thermal conductivity and thus these PIE methods are complementary to the direct in-pile fuel thermocouple measurements, as both are used to infer the thermal conductivity decrease at increasing burn-up. Further, the isothermal specimens can be used for examination of the fuel structure at high exposure.

The isothermal conditions are achieved by placing the fuel disks between molybdenum disks, in the fashion shown in Figure 7. The heat produced in the fuel flows axially into the molybdenum disks and it is transferred through the molybdenum-to-cladding gap and to the outer coolant. The gas gap and composition and the molybdenum disk height determine the operating temperature at a given specific fuel heat rating. The recommended disk height is 1 mm, whereas disk diameters from 5 to 10 mm have been utilised. Irradiations have been carried out in the temperature range 500-700°C in most cases, however temperatures up to ~1000°C have been requested for some irradiations. In some cases, gas lines have been attached for the purpose of temperature control and/or for on-line fission gas release determinations.

Thermal conductivity degradation with burn-up

In Halden tests equipped with fuel centre temperature or with expansion thermometer, the degradation of the thermal conductivity at increasing burn-up is detected in terms of a corresponding increase of the temperature versus power curve. The data so far generated up to 80 MWd/kg have resulted in the following correlation of the thermal conductivity versus burn-up:

$$k = \frac{1}{A + \beta \cdot B_{up} + (C - \alpha \cdot B_{up}) \cdot T} + EXP(T)$$

where: k is the thermal conductivity in W/m °K
T is the fuel temperature in °C
B_{up} is the burn-up in MWd/kg UO_2
$A = 0.1148$ $C = 2.475 \cdot 10^{-4}$
$EXP(T) = 0.0132 \exp(0.00188 \cdot T)$
$\beta = 0.0035$ $\alpha = 8.0 \cdot 10^{-7}$

It should be noted that the above correlation is intended to provide a best fit with centre temperature data and as such may not provide the absolute truth on the detail dependence of k versus temperature. Said in other words, one may find that somewhat different correlations provide equally good centre temperature predictions.

As an example, Figure 8 shows that a different correlation (dotted line) giving a ~10% lower conductivity in the low temperature range (and a 5% higher conductivity in the high temperature range), virtually produce the same (within 40°C) centre temperature as the Halden correlation provided by the equation above.

REFERENCES

[1]   T. Johnsen, P. Dalene, H.P. Bøhnsdalen, "Fuel Thermocouples Re-Instrumentation of Irradiated Rods", EHPG Meeting Bolkesjø, HWR-314, June 1991.

[2]   T. Johnsen, H.P. Bøhnsdalen, H.A. Sleipnæs, "Re-Instrumentation of Irradiated Rods (Nos. 807 and 808) with Fuel Thermocouples", HWR-316, February 1992.

[3]   O. Aarrestad, "Instrumentation Capabilities at Halden", HWR-351, February 1993.

[4]   M. Vankeerberghen, "The Integral Fuel Rod Behaviour Test IFA-597.2: Precharacterisation and Analysis of the Measurements", EHPG Meeting Loen, HWR-422, May 1996.

[5]   R. Van Nieuwenhove, "Assessment of Fuel Temperature Sensor Decalibration Effects During In-Pile Service", EHPG Meeting Loen, HWR-443, May 1996.

[6]   H.K. Jenssen, "Re-Instrumentation and Re-Fabrication of Irradiated Fuel Rods", Enlarged HPG Meeting, Storefjell 1993, HPR-343/8.

Figure 1. Calculated values for the factor f for centre holes varying between 1 and 2 mm and for pellet diameters of 01.4, 8.1 and 7.5 mm

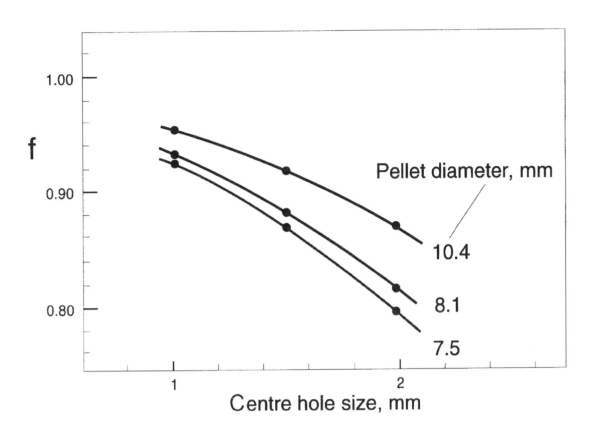

Figure 2. Thermocouple arrangement

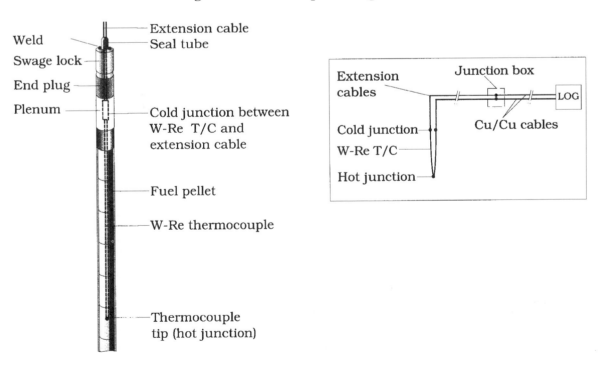

Figure 3. Decalibration correlation for W/Re thermocouples vs. thermal neutron fluence

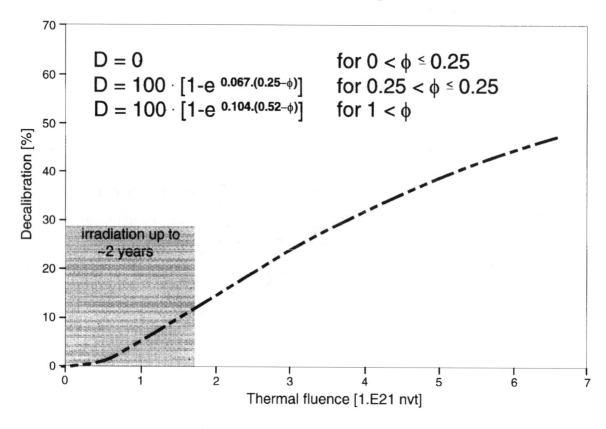

$$D = 0 \qquad\qquad \text{for } 0 < \phi \leq 0.25$$
$$D = 100 \cdot [1-e^{\,0.067 \cdot (0.25-\phi)}] \qquad \text{for } 0.25 < \phi \leq 0.25$$
$$D = 100 \cdot [1-e^{\,0.104 \cdot (0.52-\phi)}] \qquad \text{for } 1 < \phi$$

irradiation up to
~2 years

Decalibration [%]

Thermal fluence [1.E21 nvt]

Figure 4. Expansion thermometer arrangement

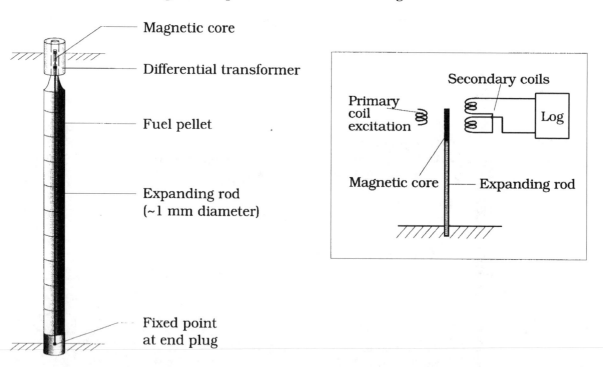

Magnetic core

Differential transformer

Fuel pellet

Expanding rod
(~1 mm diameter)

Fixed point
at end plug

Secondary coils

Primary
coil
excitation

Log

Magnetic core —— Expanding rod

Figure 5. Comparison between thermocouples and expansion thermometers (IFA-597.1)

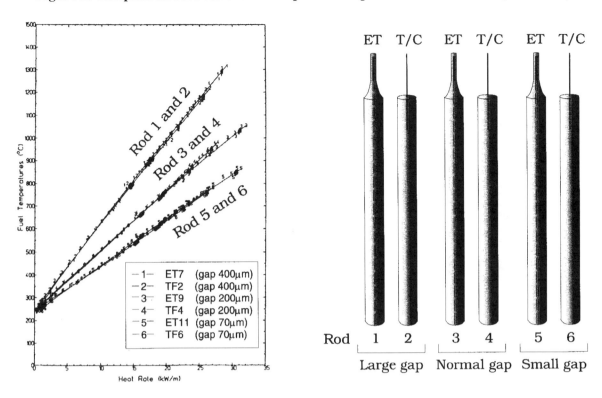

Figure 6. Schematic of the refabrication process

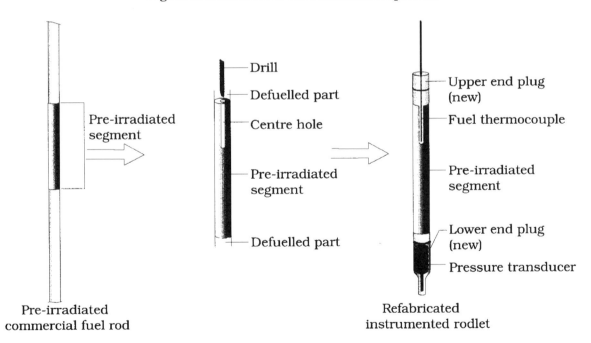

Figure 7. Irradiation techniques for high burn-up studies

Fuel /molybdenum disks
* high fission density
* Isothermal conditions

Small diameter rod
* high fission density
* representative temperature profile

LWR segments
* refabrication/reinstrumentation
* prototypical conditions

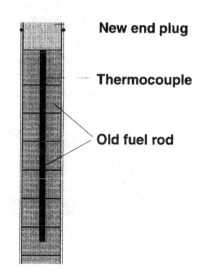

Measurements
* Gas flow lines
* Temperature monitoring

Measurements
* Fuel temperature
* Fission gas release

Measurements
* Fuel temperature
* Fission gas release

Figure 8. Fuel conductivity degradation at high burn-up, corresponding fuel temperature increase (PWR fuel)

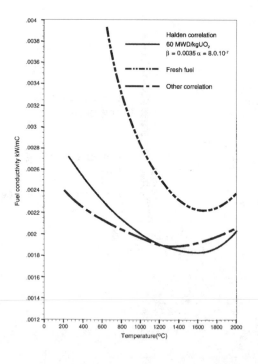

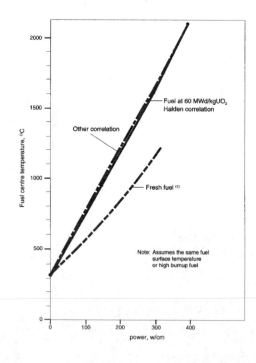

SESSION II

Thermal Conductivity Modelling

Chairs: F. Sontheimer, Y. Guérin

THERMAL PROPERTIES OF HETEROGENEOUS SOLIDS

M. Quintard
LEPT-ENSAM (URA CNRS), Esplanade des Arts et Métiers
33405 Talence Cedex – France

Abstract

This paper presents a summary of research concerning heat transfer in heterogeneous solids. In particular, effective properties and equations, as obtained from scaling up theories, are briefly reviewed.

Introduction: Macroscopic models

The macroscopic description of heat transfer in the heterogeneous medium represented in Figure 1 is of classical interest in many engineering applications. In this paper the two phases are denoted σ and β. The physical properties of the two phases are constant. Under these conditions, heat transfer is often described macroscopically by a single energy equation which we write as:

$$\left(\varepsilon_\beta \left(\rho c_p\right)_\beta + \varepsilon_\sigma \left(\rho c_p\right)_\sigma\right)\frac{\partial \langle T\rangle}{\partial t} = \nabla \cdot \left(\mathbf{K}_{\it eff} \cdot \nabla\langle T\rangle\right) \tag{1}$$

where $\langle T\rangle$ is the superficial average temperature, ε_β and $(\rho c_p)_\beta$ are the β-phase volume fraction and heat capacity respectively.

In Eq. 1, $\mathbf{K}_{\it eff}$ is the effective thermal conductivity (see [1] and [2] for a derivation of this equation using the method of volume averaging, and [3] and [4] for an analysis of the problem using the homogenisation method). The experimental determination of this parameter, or its direct estimation from the micro-scale description, is the key to macro-scale modelling.

Figure 1. Heterogeneous medium characteristic lengths

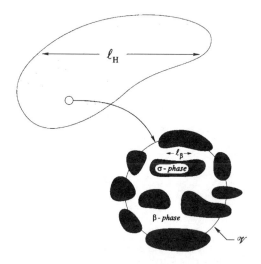

This one-equation model implies the assumption of *local thermal equilibrium* between the two phases, which corresponds to:

$$\left\langle T_\beta\right\rangle^\beta = \left\langle T_\sigma\right\rangle^\sigma = \langle T\rangle \tag{2}$$

where $\langle T_\beta\rangle^\beta$ and $\langle T_\sigma\rangle^\sigma$ are the intrinsic averaged temperature for the β-phase and the γ-phase respectively. The one-equation model expresses the fact that the time-scale associated to the diffusion process, at the micro-scale level, between the two phases is just fast enough to ensure that the two averaged temperatures are equal. Such a model is sufficient in many applications. However, highly transient phenomena, such as the one encountered when using transient thermal conductivity measurement methods like the flash method, or when a local source term leads to rapid increase of the temperature in one phase, call for local non-equilibrium models. This latter question is discussed below.

The hypothesis of local equilibrium has been investigated by several authors [5-11]. The assumption of local thermal equilibrium requires some estimate of the temperature difference $\langle T_\beta \rangle^\beta - \langle T_\sigma \rangle^\sigma$, and estimates have been obtained through some order of magnitude analysis [6,11]. They usually involve some combination of the relevant physical properties. However, it has been shown elsewhere [10,12] that the geometry has a tremendous influence on the thermal behaviour. As a consequence, assessing the validity of the assumption of local thermal equilibrium is not necessarily a simple task, and crude estimates might be useless. A possible solution to this problem is the use of two-temperature, or two-equation models. Their behaviour will provide conditions under which the local thermal equilibrium assumption is valid. Such models have been introduced heuristically in the literature [13,14] under the following form:

$$\varepsilon_\beta \left(\rho c_p \right)_\beta \frac{\partial \langle T_\beta \rangle^\beta}{\partial t} = \nabla \cdot \left(\mathbf{K}_\beta^* \cdot \nabla \langle T_\beta \rangle^\beta \right) - a_v h \left(\langle T_\beta \rangle^\beta - \langle T_\sigma \rangle^\sigma \right) \tag{3}$$

$$\varepsilon_\sigma \left(\rho c_p \right)_\sigma \frac{\partial \langle T_\sigma \rangle^\sigma}{\partial t} = \nabla \cdot \left(\mathbf{K}_\sigma^* \cdot \nabla \langle T_\sigma \rangle^\sigma \right) - a_v h \left(\langle T_\sigma \rangle^\sigma - \langle T_\beta \rangle^\beta \right) \tag{4}$$

Several questions need to be solved when using such models. What is the relevance of this model with respect to the micro-scale physical problem? How are the macroscopic properties related to the micro-scale physical properties of the system? Many of these questions have been clarified in a series of contributions during the last decade [1,12,15-17], and they are reviewed in this paper.

Micro-scale problem and generalised two-equation model

The micro-scale problem to be solved is written below in a very general manner including interfacial thermal barriers:

$$\left(\rho c_p \right)_\beta \frac{\partial T_\beta}{\partial t} = \nabla \cdot \left(k_\beta \nabla T_\beta \right) \tag{5}$$

in the β-phase.

$$\text{B.C.1} \qquad T_\beta = T_\sigma \tag{6}$$

at $A_{\beta\sigma}$.

$$\text{B.C.2} \qquad -\mathbf{n}_{\beta\sigma} \cdot k_\beta \nabla T_\beta = -\mathbf{n}_{\beta\sigma} \cdot k_\sigma \nabla T_\sigma = \frac{1}{R} \left(T_\beta - T_\sigma \right) \tag{7}$$

at $A_{\beta\sigma}$, where R is the thermal barrier resistance.

$$\left(\rho c_p \right)_\sigma \frac{\partial T_\sigma}{\partial t} = \nabla \cdot \left(k_\sigma \nabla T_\sigma \right) \tag{8}$$

in the σ-phase. We will follow the literature [1,12,15-17] to derive the macroscopic equations.

Averaged temperatures are defined as:

$$\langle T_\beta \rangle = \varepsilon_\beta \langle T_\beta \rangle^\beta = \frac{1}{V} \int_{V_\beta} T_\beta dV \tag{9}$$

These definitions are classical in the literature. However, the use of volume averages has been carefully investigated in a series of papers [12,18-23], and we refer the reader to this literature for a comprehensive analysis.

A solution of the problem is obtained in terms of the following Gray's decomposition [24]:

$$\tilde{T}_\beta = T_\beta - \langle T_\beta \rangle^\beta \, , \; \tilde{T}_\sigma = T_\sigma - \langle T_\sigma \rangle^\sigma \tag{10}$$

The method of volume averaging follows the steps outlined below:

- Take the average of the micro-scale equations.

- Express the micro-scale equations in terms of temperature deviations.

- Seek an approximate solution of the system of coupled equations, i.e. the micro-scale and averaged equations. This is obtained through a representations of the temperature deviations in terms of the macroscopic quantities such as averaged temperatures, their gradients, etc.

- Introduce these representations into the averaged equations to get the macroscopic equations in a closed form involving only averaged temperatures. This also provides an *explicit* link between the macroscopic effective properties and the micro-scale characteristics.

We briefly outline these different steps [1,12,15-17].

The averaged equation for the β-phase is obtained as:

$$\varepsilon_\beta (\rho c_p)_\beta \frac{\partial \langle T_\beta \rangle^\beta}{\partial t} = \nabla \cdot \left(k_\beta \left(\varepsilon_\beta \nabla \langle T_\beta \rangle^\beta + \frac{1}{V} \int_{A_{\beta\sigma}} \mathbf{n}_{\beta\sigma} \tilde{T}_\beta dA \right) \right) + k_\beta \int_{A_{\beta\sigma}} \mathbf{n}_{\beta\sigma} \cdot \nabla \tilde{T}_\beta dA \tag{11}$$

where the last term will account for the heat exchange between the two phases.

The micro-scale deviation equations for the β-phase is:

$$(\rho c_p)_\beta \frac{\partial \tilde{T}_\beta}{\partial t} = \nabla \cdot \left(k_\beta \nabla \tilde{T}_\beta \right) - \varepsilon_\beta^{-1} \nabla \cdot \left(k_\beta \frac{1}{V} \int_{A_{\beta\sigma}} \mathbf{n}_{\beta\sigma} \tilde{T}_\beta dA \right) - \varepsilon_\beta^{-1} k_\beta \int_{A_{\beta\sigma}} \mathbf{n}_{\beta\sigma} \cdot \nabla \tilde{T}_\beta dA \tag{12}$$

where the velocity deviation will be the source term for the dispersion mechanism, the first integral on the right hand side will account for tortuosity effects.

Looking at the system of equations involving the micro-scale boundary conditions, Eqs. 11 and 12, and the corresponding equations for the σ-phase, we can propose the following approximation for the micro-scale temperature field [1,12,15-17]:

$$\tilde{T}_\beta = T_\beta - \left\langle T_\beta \right\rangle^\beta = \mathbf{b}_{\beta\beta} \cdot \nabla \left\langle T_\beta \right\rangle^\beta + \mathbf{b}_{\beta\sigma} \cdot \nabla \left\langle T_\sigma \right\rangle^\sigma - s_\beta \left(\left\langle T_\beta \right\rangle^\beta - \left\langle T_\sigma \right\rangle^\sigma \right) + \dots \tag{13}$$

$$\tilde{T}_\sigma = T_\sigma - \left\langle T_\sigma \right\rangle^\sigma = \mathbf{b}_{\sigma\beta} \cdot \nabla \left\langle T_\beta \right\rangle^\beta + \mathbf{b}_{\sigma\sigma} \cdot \nabla \left\langle T_\sigma \right\rangle^\sigma - s_\sigma \left(\left\langle T_\sigma \right\rangle^\sigma - \left\langle T_\beta \right\rangle^\beta \right) + \dots \tag{14}$$

If these decompositions are introduced into the averaged equations one obtains the following macroscopic equations:

$$\varepsilon_\beta \left(\rho c_p \right)_\beta \frac{\partial \left\langle T_\beta \right\rangle^\beta}{\partial t} - \mathbf{u}_{\beta\beta} \cdot \nabla \left\langle T_\beta \right\rangle^\beta - \mathbf{u}_{\beta\sigma} \cdot \nabla \left\langle T_\sigma \right\rangle^\sigma$$
$$= \nabla \cdot \left(\mathbf{K}_{\beta\beta}^* \cdot \nabla \left\langle T_\beta \right\rangle^\beta + \mathbf{K}_{\beta\sigma}^* \cdot \nabla \left\langle T_\sigma \right\rangle^\sigma \right)$$
$$- a_v h \left(\left\langle T_\beta \right\rangle^\beta - \left\langle T_\sigma \right\rangle^\sigma \right) \tag{15}$$

$$\varepsilon_\sigma \left(\rho c_p \right)_\sigma \frac{\partial \left\langle T_\sigma \right\rangle^\sigma}{\partial t} - \mathbf{u}_{\sigma\beta} \cdot \nabla \left\langle T_\beta \right\rangle^\beta - \mathbf{u}_{\sigma\sigma} \cdot \nabla \left\langle T_\sigma \right\rangle^\sigma$$
$$= \nabla \cdot \left(\mathbf{K}_{\sigma\beta}^* \cdot \nabla \left\langle T_\beta \right\rangle^\beta + \mathbf{K}_{\sigma\sigma}^* \cdot \nabla \left\langle T_\sigma \right\rangle^\sigma \right)$$
$$- a_v h \left(\left\langle T_\sigma \right\rangle^\sigma - \left\langle T_\beta \right\rangle^\beta \right) \tag{16}$$

where the major parameters are the effective thermal dispersion tensors $\mathbf{K}_{\beta\beta}^*$, $\mathbf{K}_{\beta\sigma}^*$, $\mathbf{K}_{\sigma\beta}^*$, and $\mathbf{K}_{\sigma\sigma}^*$, and the heat exchange coefficient $a_v h$. These coefficients are given by a set of three closure problems involving the mapping fields found in Eqs. 13 and 14, which we write below.

Problem 1

$$0 = k_\beta \nabla^2 \mathbf{b}_{\beta\beta} - \varepsilon_\beta^{-1} \mathbf{c}_{\beta\beta} \tag{17}$$

in V_β.

$$\text{B.C.1.} \qquad \mathbf{n}_{\beta\sigma} \cdot k_\beta \nabla \mathbf{b}_{\beta\beta} + \mathbf{n}_{\beta\sigma} k_\beta = \mathbf{n}_{\beta\sigma} \cdot k_\sigma \nabla \mathbf{b}_{\sigma\beta} \tag{18}$$

at $A_{\beta\sigma}$.

$$\text{B.C.2.} \qquad -\mathbf{n}_{\beta\sigma} \cdot k_\beta \nabla \mathbf{b}_{\beta\beta} - \mathbf{n}_{\beta\sigma} k_\beta = \frac{1}{R} \left(\mathbf{b}_{\beta\beta} - \mathbf{b}_{\sigma\beta} \right) \tag{19}$$

at $A_{\beta\sigma}$.

$$0 = k_\sigma \nabla^2 \mathbf{b}_{\sigma\beta} - \varepsilon_\sigma^{-1} \mathbf{c}_{\sigma\beta} \tag{20}$$

in V_σ.

$$\mathbf{b}_{\beta\beta}(\mathbf{r}+\mathbf{l}_i)=\mathbf{b}_{\beta\beta}(\mathbf{r}) \ , \ \ \mathbf{b}_{\sigma\beta}(\mathbf{r}+\mathbf{l}_i)=\mathbf{b}_{\sigma\beta}(\mathbf{r}) \ , \ \ i=1,2,3 \tag{21}$$

$$\left\langle \mathbf{b}_{\beta\beta}\right\rangle=0 \ , \ \ \left\langle \mathbf{b}_{\sigma\beta}\right\rangle=0 \tag{22}$$

$$\mathbf{c}_{\beta\beta}=\left\langle \mathbf{n}_{\beta\sigma}\cdot k_{\beta}\nabla\mathbf{b}_{\beta\beta}\delta_{\beta\sigma}\right\rangle=-\mathbf{c}_{\sigma\beta} \tag{23}$$

where $\delta_{\beta\sigma}$ is the Dirac distribution associated with the interface $A_{\beta\sigma}$, and $\mathbf{l}_i$ are the unit lattice vectors.

Problem 2

$$0=k_{\beta}\nabla^2\mathbf{b}_{\beta\sigma}-\varepsilon_{\beta}^{-1}\mathbf{c}_{\beta\sigma} \tag{24}$$

in V_{β}.

B.C.1. $\quad \mathbf{n}_{\beta\sigma}\cdot k_{\beta}\nabla\mathbf{b}_{\beta\sigma}=\mathbf{n}_{\beta\sigma}\cdot k_{\sigma}\nabla\mathbf{b}_{\sigma\sigma}+\mathbf{n}_{\beta\sigma}k_{\sigma}$ \hfill (25)

at $A_{\beta\sigma}$.

B.C.2. $\quad -\mathbf{n}_{\beta\sigma}\cdot k_{\beta}\nabla\mathbf{b}_{\beta\sigma}=\dfrac{1}{R}\left(\mathbf{b}_{\beta\sigma}-\mathbf{b}_{\sigma\sigma}\right)$ \hfill (26)

at $A_{\beta\sigma}$.

$$0=k_{\sigma}\nabla^2\mathbf{b}_{\sigma\sigma}-\varepsilon_{\sigma}^{-1}\mathbf{c}_{\sigma\sigma} \tag{27}$$

in V_{σ}.

$$\mathbf{b}_{\beta\sigma}(\mathbf{r}+\mathbf{l}_i)=\mathbf{b}_{\beta\sigma}(\mathbf{r}) \ , \ \ \mathbf{b}_{\sigma\sigma}(\mathbf{r}+\mathbf{l}_i)=\mathbf{b}_{\sigma\sigma}(\mathbf{r}) \ , \ \ i=1,2,3 \tag{28}$$

$$\left\langle \mathbf{b}_{\beta\sigma}\right\rangle=0 \ , \ \ \left\langle \mathbf{b}_{\sigma\sigma}\right\rangle=0 \tag{29}$$

$$\mathbf{c}_{\beta\sigma}=\left\langle \mathbf{n}_{\beta\sigma}\cdot k_{\beta}\nabla\mathbf{b}_{\beta\sigma}\delta_{\beta\sigma}\right\rangle=-\mathbf{c}_{\sigma\sigma} \tag{30}$$

Problem 1 and 2 can be combined to obtain the micro-scale closure problem giving the effective conductivity in Eq. 1.

Problem 3

$$\left(\rho c_p\right)_{\beta}\mathbf{v}_{\beta}\cdot\nabla s_{\beta}=k_{\beta}\nabla^2 s_{\beta}-\varepsilon_{\beta}^{-1}h_{\beta} \tag{31}$$

in V_{β}.

B.C.1. $\quad \mathbf{n}_{\beta\sigma}\cdot k_{\beta}\nabla s_{\beta}=\mathbf{n}_{\beta\sigma}\cdot k_{\sigma}\nabla s_{\sigma}$ \hfill (32)

at $A_{\beta\sigma}$.

$$\text{B.C.2.} \qquad \mathbf{n}_{\beta\sigma} \cdot k_\beta \nabla s_\beta = \frac{1}{R}\left(-s_\beta + s_\sigma\right) + \frac{1}{R} \tag{33}$$

at $A_{\beta\sigma}$.

$$0 = k_\sigma \nabla^2 s_\sigma - \varepsilon_\sigma^{-1} h_\sigma \tag{34}$$

in V_σ.

$$s_\beta\left(\mathbf{r}+\mathbf{l}_i\right) = s_\beta\left(\mathbf{r}\right) , \quad s_\sigma\left(\mathbf{r}+\mathbf{l}_i\right) = s_\sigma\left(\mathbf{r}\right) , \quad i = 1,2,3 \tag{35}$$

$$\left\langle s_\beta \right\rangle = 0 , \quad \left\langle s_\sigma \right\rangle = 0 \tag{36}$$

$$h_\beta = \left\langle \mathbf{n}_{\beta\sigma} \cdot k_\beta \nabla s_\beta \delta_{\beta\sigma} \right\rangle = h_\sigma = a_v h \tag{37}$$

Looking at Problem 3, one obtains the important result that the volume heat exchange term is given by this integro-differential problem. Other macroscopic coefficients are given by:

$$\mathbf{K}_{\beta\beta} = k_\beta\left(\varepsilon_\beta \mathbf{I} + \left\langle \mathbf{n}_{\beta\sigma} \mathbf{b}_{\beta\beta} \delta_{\beta\sigma} \right\rangle\right) \tag{38}$$

$$\mathbf{K}_{\beta\sigma} = k_\beta\left\langle \mathbf{n}_{\beta\sigma} \mathbf{b}_{\beta\sigma} \delta_{\beta\sigma} \right\rangle \tag{39}$$

$$\mathbf{K}_{\sigma\beta} = k_\sigma\left\langle \mathbf{n}_{\sigma\beta} \mathbf{b}_{\sigma\beta} \delta_{\beta\sigma} \right\rangle \tag{40}$$

$$\mathbf{K}_{\sigma\sigma} = k_\sigma\left(\varepsilon_\sigma \mathbf{I} + \left\langle \mathbf{n}_{\sigma\beta} \mathbf{b}_{\sigma\sigma} \delta_{\beta\sigma} \right\rangle\right) \tag{41}$$

$$\mathbf{u}_{\beta\beta} = \mathbf{c}_{\beta\beta} - k_\beta\left\langle \mathbf{n}_{\beta\sigma} s_\beta \delta_{\beta\sigma} \right\rangle \tag{42}$$

$$\mathbf{u}_{\beta\sigma} = \mathbf{c}_{\beta\sigma} + k_\beta\left\langle \mathbf{n}_{\beta\sigma} s_\beta \delta_{\beta\sigma} \right\rangle \tag{43}$$

$$\mathbf{u}_{\sigma\beta} = \mathbf{c}_{\sigma\beta} - k_\sigma\left\langle \mathbf{n}_{\sigma\beta} s_\sigma \delta_{\beta\sigma} \right\rangle \tag{44}$$

$$\mathbf{u}_{\sigma\sigma} = \mathbf{c}_{\sigma\sigma} + k_\sigma\left\langle \mathbf{n}_{\sigma\beta} s_\sigma \delta_{\beta\sigma} \right\rangle \tag{45}$$

The validity domain of the proposed theory has been tested on the basis of numerical experiments. Examples of these comparisons are given in [11,12]. To understand the theoretical status of the proposed model it is best to look at a very simple problem, for which the relaxation time to reach local thermal equilibrium is essentially related to the solid phase behaviour.

The example corresponds to the limiting case:

$$k_\beta / k_\sigma \to +\infty \tag{46}$$

and the behaviour of such a system is discussed in [17,25].

It is shown that the field inside the low conductivity medium can be expressed as a time convolution integral involving the temperature at the boundary, i.e. $\langle T_\beta \rangle^\beta$. There is no particular reason that would yield a *general* representation of this flux in terms of a heat exchange term similar to the one in the macroscopic Eqs. 3 and 4, i.e.:

$$q = -a_v h \left(\langle T_\beta \rangle^\beta - \langle T_\sigma \rangle^\sigma \right) \tag{47}$$

However, it happens that, under some conditions, the time convolution integral can be *approximated* by expressions like Eq. 47.

A simple definition could be (see example in [26]):

$$a_v h = \lim_{t \to \infty} \left(q \Big/ \langle T_\beta \rangle^\beta - \langle T_\sigma \rangle^\sigma \right) \tag{48}$$

The heat transfer coefficient can be estimated as the best fit between an experiment and the solution of the macroscopic equation [27] assuming the theory is correct.

Other theories try to represent the micro-scale temperature field in terms of the macroscopic source terms, here ($\langle T_\beta \rangle^\beta$-$\langle T_\sigma \rangle^\sigma$), we have:

$$T_\sigma = \langle T_\sigma \rangle^\sigma + s_\sigma(x) \left(\langle T_\sigma \rangle^\sigma - \langle T_\beta \rangle^\beta \right) \tag{49}$$

For instance, Warren and Root [28] proposed the use of parabolas, and Kazemi *et al.* [29] linear interpolation schemes. A comparison between Eq. 49 and 14 shows that the proposed theory represents an extension of the approximation discussed for this simple case. Table 1 shows the values obtained from the different theories for a one-dimensional system. Indeed, differences may be important. The value 12 is the best fit, and is also the only one compatible with the asymptotic behaviour of the two-equation model [30].

Table 1. Values of $a_v h l_\sigma^2 / k_\sigma$ for different theories

Eq. 48	π^2
Best fit	12
Warren and Root [28]	12
Two-equation model	12
Kazemi *et al.* [29]	4

It is clear at this point that the two-equation model has limitations which correspond to the limitations discussed in this section. While the two-equation model cannot handle high frequency heat waves, several numerical experiments have proven its usefulness [11,12].

We have now at our disposal two models which can be used depending on the condition of local equilibrium or non-equilibrium. In addition, the closure problems can be used to calculate the effective thermal conductivity tensors, provided some information is known about the micro-scale structure. The use of such calculations is illustrated in the next section.

Calculation of effective properties

Problems 1-3 involve integrals of the solutions as source terms in the right-hand side of the bulk equations. Special numerical procedures to avoid such integro-differential equations are presented in Quintard *et al.* [16]. Numerical results have been obtained for different unit cells:

- One-dimensional unit cells corresponding to stratified systems [12,16]. These results are complementary to those obtained by Zanotti and Carbonell [31]. However, these unit cells are not very interesting for many applications since the thermal dispersion mechanism is represented through a square dependence of the longitudinal thermal dispersion coefficient, which is quite different from the coefficient close to unity observed on more dispersed systems.

- Two-dimensional in-line and staggered arrays of cylinders [12,16,32]: these results, especially those corresponding to staggered arrays, are closer to data for real systems.

- Three-dimensional unit cells [16,25].

Examples of these calculations can be found in the cited literature. Figure 2, extracted from Nozad *et al.* [2], illustrates two important behaviours:

- The effective conductivity K_{eff}/k_β increases with the ratio k_σ/k_β. If the σ-phase is discontinuous, the effective conductivity becomes almost constant for values of k_σ/k_β greater than 100.

- In the presence of contact points, the curve departs significantly from this discontinuous behaviour. This has been particularly investigated by several authors (see for instance Shonnard and Whitaker [33]).

Figure 2. Comparison between theoretical predictions and experimental values

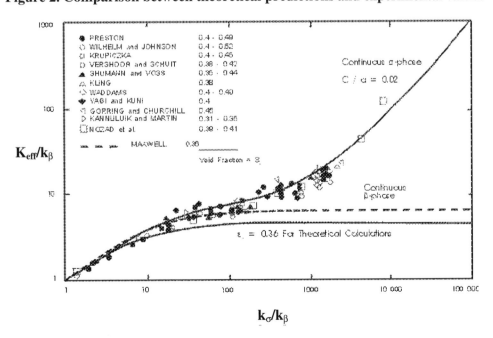

93

Properties associated with the two-equation model have also been calculated. For instance, Figure 3 shows the evolution of the heat transfer coefficient versus k_σ/k_β, for various β-phase volume fractions, and for the represented unit cell.

Figure 3. Heat transfer coefficient

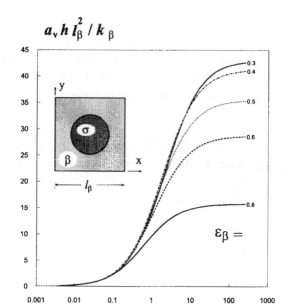

These numerical models can be used in different instances, for example, to check the experimental estimation of effective properties if some micro-scale information is available, or to investigate the possible modification of the effective thermal conductivity if one changes the micro-scale properties. It is beyond the scope of this summary to provide extensive examples.

Conclusion

In this paper, the introduction of macroscopic models representing heat transfer in heterogeneous media has been reviewed.

It is then explained how the closure problems that allow for the calculation of macroscopic effective properties can be solved numerically. Results for simple unit cells are given to illustrate this process.

Acknowledgement

Support from NSF/CNRS is gratefully acknowledged.

REFERENCES

[1]   R.G. Carbonell and S. Whitaker, "Heat and Mass Transfer in Porous Media", in *Fundamentals of Transport Phenomena in Porous Media*, J. Bear and M.Y. Corapcioglu, eds., Martinus Nijhof Publ., pp. 121-198, 1984.

[2]   I. Nozad, R.G. Carbonell, and S. Whitaker, "Heat Conduction in Multi-Phase Systems i: Theory and Experiment for Two-Phase Systems", *Chem. Eng. Sci.*, 40, pp. 843-855, 1985.

[3]   H.I. Ene, D. Polisevski, "Thermal Flow in Porous Media", D. Reidel Publishing Company, Boston, 1987.

[4]   J-L. Auriault, "Macroscopic Modelling of Heat Transfer in Composites with Interfacial Thermal Barrier", *Int. J. Heat and Mass Transfer*, 37 (18), pp. 2885-2892, 1994.

[5]   H. Truong and G. Zinsmeister, "Experimental Study of Heat Transfer in Layered Composites", *Int. J. Heat and Mass Transfer*, 21 (7), pp. 905-909, 1978.

[6]   S. Whitaker, "Local Thermal Equilibrium: An Application to Packed Bed Catalytic Reactor Design", *Chem. Engng. Sci.*, 41, pp. 2029-2039, 1986.

[7]   M. Sözen and K. Vafai, "Analysis of the Neo-Thermal Equilibrium Condensing Flow of a Gas Through a Packed Bed", *Ing. J. Heat Mass Transfer*, 33, pp. 1247-1261, 1990.

[8]   M. Kaviany, "Principles of Heat Transfer in Porous Media", Springer-Verlag, New York, 1996.

[9]   D.A. Nield and A. Bejan, "Convection in Porous Media", Springer-Verlag, New York, 1992.

[10]   C. Gobbé and M. Quintard, "Macroscopic Description of Unsteady Heat Transfer in Heterogeneous Media", *High Temperatures – High Pressures*, 26, pp. 1-14, 1994.

[11]   M. Quintard and S. Whitaker, "Local Thermal Equilibrium for Transient Heat Conduction: Theory and Comparison with Numerical Experiments", *Int. J. Heat Mass Transfer*, 38 (15), pp. 2779-2796, 1995.

[12]   M. Quintard and S. Whitaker, "One- and Two-Equation Models for Transient Diffusion Processes in Two-Phase Systems", *Advances in Heat Transfer*, Vol. 23, Academic Press, New York, pp. 369-464, 1993a.

[13]   D. Vortmeyer and R.J. Schaefer, "Equivalence of One- and Two-Phase Models for Heat Transfer Processes in Packed Beds, One-Dimensional Theory", *Chem. Eng. Sci.*, 29, pp. 485-491, 1974.

[14] E.U. Schlünder, "Equivalence of One- and Two-Phase Models for Heat Transfer Processes in Packed Beds: One-Dimensional Theory", *Chem. Eng. Sci.*, 30, pp. 449-452, 1975.

[15] F. Zanotti and R.G. Carbonell, "Development of Transport Equations for Multiphase Systems iii: Application to Heat Transfer in Packed Beds, *Chem. Eng. Sci.*, 39, pp. 299-311, 1984a.

[16] M. Quintard, M. Kaviany and S. Whitaker, "Two-Medium Treatment of Heat Transfer in Porous Media: Numerical Results for Effective Properties", *Advances in Water Resources*, 20 (2-3), pp. 77-94, 1997.

[17] J. Batsale, C. Gobbé and M. Quintard, "Local Non-Equilibrium Heat Transfer in Porous Media", *Recent Research Developments in Heat, Mass and Momentum Transfer*, Vol. 1, Research Signpost, India, pp. 1-24, 1996.

[18] M. Quintard and S. Whitaker, "Transport in Ordered and Disordered Porous Media: Volume-Averaged Equations, Closure Problems and Comparison with Experiment", *Chem. Eng. Science*, 48, pp. 2537-2564, 1993b.

[19] M. Quintard and S. Whitaker, "Transport in Ordered and Disordered Porous Media i: The Cellar Average and the Use of Weighting Functions", *Transport in Porous Media*, 14, pp. 163-177, 1994b.

[20] M. Quintard and S. Whitaker, "Transport in Ordered and Disordered Porous Media ii: Generalized Volume Averaging", *Transport in Porous Media*, 14, pp. 179-206, 1994c.

[21] M. Quintard and S. Whitaker, "Transport in Ordered and Disordered Porous Media iii: Closure and Comparison Between Theory and Experiment", *Transport in Porous Media*, 15, pp. 31-49, 1994d.

[22] M. Quintard and S. Whitaker, "Transport in Ordered and Disordered Porous Media iv: Computer Generated Porous Media for Three-Dimensional Systems", *Transport in Porous Media*, 15, pp. 51-70, 1994e.

[23] M. Quintard and S. Whitaker, "Transport in Ordered and Disordered Porous Media v: Geometrical Results for Two-Dimensional Systems", *Transport in Porous Media*, 15, pp. 183-196, 1994f.

[24] W. Gray, "A Derivation of the Equations for Multi-Phase Transport", *Chem. Eng. Sci.*, 3, pp. 229-233, 1975.

[25] G. Grangeot, M. Quintard and S. Whitaker, "Heat Transfer in Packed Beds: Interpretation of Experiments in Terms of One- and Two-Equation Models", 10th Int. Heat Transfer Conference, Vol. 5, pp. 291-296, 1994.

[26] J-L. Auriault and P. Royer, "Double Conductivity Media: A Comparison Between Phenomenological and Homogenization Approaches", *Int. J. Heat Mass Transfer*, 36, pp. 2613-2621, 1993.

[27] H.H. Gerke and M.T. Van Genuchten, "Macroscopic Representation of Structural Geometry for Simulating Water and Solute Movement in Dual-Porosity Media", *Advances in Water Resources*, 19 (6), pp. 343-357, 1996.

[28] J. Warren and P. Root, "The Behavior of Naturally Fractured Reservoirs", *Society of Petroleum Engineers Journal*, pp. 245-255, September 1963.

[29] H. Kazemi, L.S. Merril, K.L. Porterfield and P.R. Zeman, "Numerical Simulation of Water-Oil Flow in Naturally Fractured Reservoirs", *Society of Petroleum Engineers Journal*, pp. 317-326, December 1976.

[30] A. Ahmadi, M. Quintard and S. Whitaker, "Transport in Chemically and Mechanically Heterogenous Porous Media v: Two-Equation Model for Solute Transport with Adsorption", *Advances in Water Resources*, to be published, 1997.

[31] F. Zanotti and R.G. Carbonell, "Development of Transport Equations for Multiphase Systems i: General Development for Two-Phase Systems", *Chem. Eng. Sci.*, 39, pp. 263-278, 1984b.

[32] M. Quintard and S. Whitaker, "Convective and Diffusive Heat Transfer in Porous Media: 3-D Calculations of Macroscopic Transport Properties", EU-ROTHERM Seminar 36 on Advanced Concepts and Techniques in Thermal Modelling, Elsevier, Paris, pp. 301-307, 1994a.

[33] D.R. Shonnard and S. Whitaker, "The Effective Thermal Conductivity for a Point-Contact Porous Medium: An Experimental Study", *Int. J. Heat and Mass Transfer*, 32, pp. 503-512, 1989.

FUEL THERMAL CONDUCTIVITY: A REVIEW OF THE MODELLING AVAILABLE FOR UO_2, $(U\text{-}Gd)O_2$ AND MOX FUEL

Daniel Baron
EDF Études et Recherches, France

Abstract

A large amount of thermal conductivity data has been collected on UO_2 and MOX fuel over the last 30 years. Most are reported in the MATPRO reference manual. More recently, data have been provided on $(U\text{-}Gd)O_2$ fuels. Previously, the thermal conductivity data on fresh fuel were obtained by calorimetric measurements. With the improvement of laboratory methods, the LASER flash method allows to collect more accurate data via the diffusivity measurement on fresh fuel and irradiated material as well. The thermal conductivity dependence on temperature, deviation from stochiometry, porosity, or Pu and Gd contents, have been investigated. Concerning the burn-up degradation effect, data have recently been obtained from discs irradiated up to 80 GWd/tU within the NFIR international programme, enlightening the recovering processes with temperature.

This database has been used at the EDF to review and evaluate the modelling proposed in the literature to decide which model to use in the new design code CYRANO3. Most of the models are specific to a kind of fuel, accounting for stochiometry, Pu weight content or gadolinium content. Some are designed to account for all these parameters. The burn-up effect was accounted for, up to now, using only a correcting factor. The Wiesenack and the Lucuta models incorporate this parameter in the thermal conductivity formulation.

This review gives a comparison regarding the capability of these different models and enlightens upon the weak effect of the Pu as an element on the fuel conductivity but also the strong effect of the deviation from stochiometry. This confirms the importance of this last parameter in MOX fuel manufacturing.

A large amount of thermal conductivity data has been collected on UO_2 and MOX fuel over the last 30 years. Most are reported in the MATPRO reference manual [1]. More recently, data have been provided on $(U-Gd)O_2$ fuels. Previously, the thermal conductivity data on fresh fuel were obtained by calorimetric measurements. With the improvement of laboratory methods, the LASER flash method allows to collect more accurate data via the diffusivity measurement on fresh fuel and irradiated material as well [2-6]. The thermal conductivity dependence on temperature, deviation from stochiometry, porosity, or Pu and Gd contents, have been investigated. Concerning the burn-up degradation effect, data have recently been obtained from discs irradiated up to 80 GWd/tU within the NFIR international programme, enlightening the recovering processes with temperature [7]. In order to evaluate the thermal conductivity degradation due to fission products, out of pile thermal diffusivity measurements have been done on SIMFUEL [8].

The data collected by the calorimetric method show a strong discrepancy depending on the sources (Figures 1 and 2), certainly related to different manufacturing processes to elaborate the samples, or the relative control of the experimental thermal conditions or a lack in samples characterisation. Furthermore, it is often difficult to find a coherence on the influence of certain parameters such as the sample average porosity or the material deviation from stochiometry.

As far as this paper aims to evaluate a certain number of fuel conductivity models, it is preferable to refer to a relatively coherent database. Then, we selected the more recent data obtained by thermal diffusivity measurements when it was available or other data where the samples were accurately characterised. All the data have been reduced to a fuel material containing 5% of porosity (95% TD), using the porosity and densities provided by the laboratories, and more precisely before and after the thermal diffusivity test, in the case of the irradiated materials. The porosity correction was obtained using the LOEB and ROSS correlation [9]:

$$K(T) = K(T)_{95\%} (1 - por\ \alpha(T)) \quad (W/cm/°C)$$

where: T = the temperature (K)
       por = the relative porosity volume
      $\alpha(T) = 2,7384-0,58\ 10^{-3}T$ *(if T < 1273 K)*
      $\alpha(T) = 2,00$ *(if T > 1273 K)*

Because few data are yet available on the evolution of the heat capacity with burn-up, the heat capacity has been assumed weakly modified by the irradiation.

Presentation of the conductivity formulations

The models tested against the database are as follows.

The STORA model (CEA)

The STORA model [28] is only available for fresh UO_2 fuel 96,6% TD, and formulated:

$$K(T) = 5.8915\ 10^{-2} - 4.3554\ 10^{-5}\ T + 1.3324\ 10^{-8}\ T^2 \quad (W/cm/°C)$$

where: T = temperature (°C).

This model has been used for many years by the EDF in the CYRANO2 code and was fitted on the CYRANO experiments in Grenoble in the 60s.

The COMETHE formulation (Belgonucléaire)

The COMETHE formulation [11] is available for UO_2 and MOX fuel, 95% TD and temperatures from 20°C to 2000°C. This formulation was modified to account for Pu content on the basis of data from Gibby, Van Craeynest and Weilbacher and also Schmidt and Richter:

$$K(T) = \frac{A_0}{A_1 + A_2 x + (1 + B_0 q) T} + CT^3 \quad (W/cm/°C)$$

where: T = temperature (K)
x = absolute value of the stochiometry deviation (abs(2-O/M))
q = plutonium content (-)
$A_0 = 40.05 \quad A_1 = 129.4 \quad A_2 = 16020 \quad B_0 = 0.8 \quad C = 0,6416\ 10^{-12}$

NFIR formulation for gadolinium fuel

The NFIR formulation is proposed for $(U-Gd)O_2$ fuels up to 12%wt Gd, 95% TD and available from room temperature to 1800°C [13]:

$$K(T) = \frac{1}{A_0 + A_1 g + A_2 g^2 + (B_0 + B_1 g + B_3 g^2) T} \quad (W/cm/°C)$$

where: T = temperature (K)
g = gadolinium content (-)
$A_0 = 5.9089 \quad A_1 = 3.4009 \quad A_2 = 1564.7$
$B_0 = 0.021621 \quad B_1 = 0.066866 \quad B_2 = -1.3491$

Baron-Hervé – 1994 model (EDF)

The Baron-Hervé model version 1994 [14] is available for UO_2, UO_2 with gadolinium and MOX fuels, with a density of 95% TD and has been fitted to NFIR data for the gadolinium content effect:

$$K(T) = \frac{1}{A_0 + A_1 x + A_2 g + A_3 g^2 + (B_0 (1 + B_1 q) + B_2 g + B_3 g^2) T} + CT^3 \quad (W/m/K)$$

where: T = temperature (K), (0 to 2000 K)
x = absolute value of the stochiometry deviation (abs(2-O/M))
q = plutonium weight content (-)
g = gadolinium weight content (-), 0 to 12%
$A_0 = 5.24\ 10^{-2} \quad A_1 = 4 \quad A_2 = 0.3079 \quad A_3 = 12.2031$
$B_0 = 2.553\ 10^{-4} \quad B_1 = 8\ 10^{-3} \quad B_2 = 8.606\ 10^{-4} \quad B_3 = -0.0154$
$C = 40\ 10^{-12}$

Baron-Hervé – 1995 model (EDF)

The Baron-Hervé version 1995 is available for UO_2, UO_2 with gadolinium and MOX fuels, with a density of 95% TD. It is the same as the previous model, but is modified for the high temperature simulation on the basis of Delette and Charles' works [16]:

$$K(T) = \frac{1}{A_0 + A_1 x + A_2 g + A_3 g^2 + (B_0(1 + B_1 q) + B_2 g + B_3 g^2)T} + \frac{C + Dg}{T^2} \exp\left(-\frac{W}{kT}\right) \quad (W/m/K)$$

where: T = temperature (K), (0 to 2000 K)
     x = absolute value of the stochiometry deviation (abs(2-O/M))
     q = plutonium weight content (-)
     g = gadolinium weight content (-), 0 to 12%
     k = Boltzmann constant ($1,38.10^{-23}$ J/K)
     W = 1,41 eV = $1,41*1,6.10^{-19}$ J
     $A_0 = 4,4819.10^{-2}$ (m.K/W) $A_1 = 4$ (-) $A_2 = 0,611$ (m.K/W) $A_3 = 11,081$ (m.K/W)
     $B_0 = 2,4544.10^{-4}$ (m/W) $B_1 = 0,8$ (-) $B_2 = 9,603.10^{-4}$ (m/W) $B_3 = -1,768.10^{-2}$ (m/W)
     $C = 5,516.10^9$ (W.K/m)
     $D = -4,302.10^{10}$ (W.K/m)

This formulation is available for temperatures up to 2600 K.

Wiesenack model (HALDEN)

The Wiesenack law [17], available for UO_2 fuels (95% TD), accounts for the burn-up effect. The A and B parameters of the phonon term have been fitted on in-pile temperature data collected from centreline thermocouples in IFA Halden experiments:

$$K(T) = \frac{1}{(A_0 + B1 * BU) + (B_0 + B_1 * BU)T} + C * exp(D * T) \quad (W/m/K)$$

where: T   = temperature (°C)
     BU = fuel burn-up (GWd/tU)
     $A_0 = 0,1148$ $A_1 = 0,0035$
     $B_0 = 0,0002474$ $B_1 = -8 * 10^{-7}$
     $C = 0,0132$ $D = 0,00188$

The METEOR model (CEA)

The METEOR modelling accounts for the eventual gadolinium content and is available for UO_2 fuel, 95% TD:

$$K(T) = \frac{1}{A_0 + A_1 g + B_0 T} + \frac{C}{T^2} \exp\left(-\frac{D}{T}\right) \quad (W/m/K)$$

where: T = temperature (K)

g = gadolinium weight content (-)

A_0 = $5,3.10^{-2}$ (m.K/W) B_0 = $2,2.10^{-4}$ (m/W)

A_1 = 0,81 (-)   C = $4,715.10^9$ (W.K/m)   D = -16361 (K)

Philipponeau law (UO₂ and MOX), (CEA)

The Philipponeau law is considered available for UO_2, 95%TD and MOX fuel as well.

$$K(T) = \frac{1}{A_0(\sqrt{x - A_1} - A_2) + B_0 T} + C.T^3 \quad (W/m/K)$$

where: T = temperature (K)

x = absolute value of the stochiometry deviation (abs(2-O/M))

A_0 = 1,32 A_1 = 0,00931 A_2 = 0,091

B_0 = $2,493.10^{-4}$ (m/W)   C = $88,4.10^{-12}$ (W.K/m)

Harding and Martin law

Harding and Martin recommended the following law [19] available for UO₂ and expressed for a fuel 100% TD.

$$K(T) = \frac{1}{A_0 + B_0 T} + \frac{C}{T^2} \exp\left(-\frac{D}{T}\right) \quad (W/m/K)$$

where: T = temperature (K)

A_0 = $3,75.10^{-2}$ (m.K/W) B_0 = $2,165.10^{-4}$ (m/W)

C = $4,715.10^9$ (W.K/m)   D = -16361 (K)

Lucuta law (AECL)

Lucuta's law [20] is a modification of Harding and Martin's law to account for the burn-up effect. This modelling is the most descriptive and separates the different irradiation effects. Most of the data were provided from SIMFUEL simulations. The fuel thermal conductivity is then a combination of several corrective terms applied on the fresh fuel Martin's law:

$$K(T) = K_{1d} * K_{1p} * K_{2p} * K_{3x} * K_{4r} * K_0(T) \quad (W/m/K)$$

where: K_{1d} = quantifies the effect of dissolved fission products

$$K_{1d} = \left(\frac{1,09}{\beta^{3,265}} + \frac{0,0643}{\sqrt{\beta}} \sqrt{T} \right) \arctan\left(\frac{1}{\frac{1,09}{\beta^{3,265}} + \frac{0,0643}{\sqrt{\beta}} \sqrt{T}} \right)$$

103

K_{1p} = describes the effect of precipitated solid fission products

$$K_{1p} = 1 + \frac{0,019\beta}{3 - 0,019\beta} \cdot \frac{1}{1 + \exp\left[-\frac{T - 1200}{100}\right]}$$

K_{2p} = the porosity factor
K_{3x} = account for non stochiometric materials
K_{4r} = the effect of the radiation change

$$K_{4r} = \frac{0,2}{1 + \exp\left[\frac{T - 900}{80}\right]}$$

T $\quad$ = temperature (K)
β $\quad$ = burn-up (at %)

The MATPRO law

The MATPRO law has been fitted on a large database available in the open literature in the 70s, where abnormal values have been evicted. The law proposed is as follows [1]:

From room temperature to 1650°C:

$$K(T) = \frac{K_1}{K_2 + T} + K_3 \exp(K_4 * T) \quad (W/m/K)$$

and from 1650°C to 2840°C:

$$K(T) = K_5 + K_3 \exp(K_4 * T)$$

where: T $\quad$ = temperature (°C)

for UO_2, 95% TD	for $(U,Pu)O_2$, 96% TD
K_1 = 4040	K_1 = 3300
K_2 = 464	K_2 = 375
K_3 = $1.216*10^{-2}$	K_3 = $1.540*10^{-2}$
K_4 = $1.667*10^{-3}$	K_4 = $1.710*10^{-3}$
K_5 = 1.91	K_5 = 1.71

Lokken and Courtright model

Apart from the Wiesenack and the Lucuta models, the other models are available for fresh fuel and must be corrected to account for burn-up degradation of the fuel conductivity. Presently, the more frequently used correlation is the Lokken and Courtright expression [21] which proposes a correction of the fuel thermal conductivity as follows:

$$K(T)_{irr} = \frac{1}{\frac{1}{K(T)_{fresh}} + \xi \frac{BU}{T}} \quad (W/cm/°C)$$

where: T   = temperature in °C

    BU   = fuel burn-up in MWd/tU

    ξ = fitting coefficient (default = 0,124)

    $K(T)_{fresh}$ = fresh fuel thermal conductivity

Comparison modelling/experiments

UO_2 fuel

The different models presented above have been compared with the thermal conductivity database deduced from the diffusivity measurements reported in [12] and [13] on non irradiated pure UO_2. All the data have been reduced to a density of 95%, using the LOEB and ROSS correlation and the models were plotted for this density as well. Figures 3 and 4 show that all the formulations fit well with the database used, in a range of temperatures representative of the fuel normal and abnormal operating conditions (class 1 & 2: 500-2000 K). A few divergences occur above 2000 K depending on which data each model has been fitted.

Effect of a deviation from stochiometry

The fuel thermal conductivity is highly affected by a deviation from stochiometry. This is related to oxygen vacancies or interstitial in the oxygen sublattice which interact with the phonons. This interaction of phonon-defects is represented by the A term of the phonon component; the modelling then accounts for a modification of this term, as a linear function or as a square root function of the stochiometry deviation. Figure 5 shows the real effect of a deviation from stochiometry for two temperatures: 800 and 1200 K. Five models account for this parameter: COMETHE, Baron-Hervé (94 and 95), Philipponeau and Lucuta. For all modelling, the response is within the uncertainty of the measurements. Degradation is about 10-12% for a stochiometry of 2.010.

Effect of an addition of gadolinium

Gadolinium is an additive whose mass is very different from these of uranium and oxygen. Then its presence changes the parameter B of the phonon component as it creates additive interactions phonon-phonon. Because of its trivalence, it also creates defects in the oxygen sublattice and modifies the interaction phonon-defects, represented by the A term.

Figure 6 shows the effect of the gadolinium content again for two temperatures: 800 and 1200 K. Four models account for this parameter: Baron-Hervé (94 and 95), NFIR1 and METEOR. They give a good approximation of the fuel conductivity degradation with gadolinium content. The degradation is about 3% per 1% addition of gadolinium at 800 K and about 2% per 1% addition of gadolinium at 1200 K.

Mixed oxide fuels

The MOX fuel is different from the UO_2 fuel for several reasons. The plutonium isotopes have an atomic mass which is slightly different than that of uranium, the non-homogeneity of the manufactured fuel giving (U-Pu)O_2 clusters in a matrix of pure UO_2 and the different oxygen potential introduced by the plutonium. One can theoretically consider that the slight difference of mass between U and Pu will not drastically affect the fuel conductivity. On the other hand it has previously been shown that a slight modification of the oxygen potential considerably degrades the fuel conductivity. Concerning non-homogeneity, this is a benefit for the material as far as the fuel conductivity of a composite material is a ponderation of the matrix thermal conductivity and of the cluster thermal conductivity.

For a stochiometric fresh fuel (O/M = 2.000), the degradation observed is about 0.5% per 1% addition of plutonium (Figure 8), but the database shows a certain discrepancy in the values measured. Figure 7 shows the capability of the modelling to account for stochiometric MOX. This confirms the observation of Philipponeau, who proposes to use the UO_2 thermal conductivity law for such materials.

On the other hand, Figures 9, 10 and 11 show the strong effect of a deviation from stochiometry. To evaluate the weight of each parameter, we have plotted against the data sets the Baron-Hervé (95) law for stochiometric UO_2, for non-stochiometric UO_2 and finally for non-stochiometric MOX. This demonstrates the importance of the O/M parameter when manufacturing the fuel. Obviously, this presentation is relative to fresh fuel. Which kind of evolution one can expect for a stochiometric MOX fuel under irradiation. Certainly, the oxygen potential evolves differently between the clusters and the UO_2 matrix. The experimental data collected with in-pile thermocouple exhibits a difference of about 8% between the UO_2 and MOX average thermal conductivity, even at BOL; one can then expect a modification of the fuel stochiometry under irradiation in the range 1.990 to 1.995 as soon as irradiation starts.

Burn-up effect

The burn-up increase obviously has an impact on the fuel thermal conductivity. The fissions produce a great number of fission products within the fuel matrix which can no longer be considered as a pure material. Its physical chemical composition is modified continuously with the apparition of fission products and an important disturbance of the matrix is induced by fission spikes which produce a great number of point defects, mainly vacancies in the oxygen sublattice. In consequence, the fuel conductivity is progressively degraded. However, the solid fission products in substitution in the UO_2 lattice lead to a permanent degradation while point defects and gaseous or volatile fission products are able to move if the local temperature is high enough, above the diffusion thresholds. Subsequently, it is observed during a thermal diffusivity measurement that a part of the diffusivity degradation can be recovered if the fuel material sustains an increase of temperature during a sufficient period.

Figures 12, 13, 14 and 15 compare the data obtained in the NFIR2 programme at respectively 26.8, 38.8, 63.6 and 80 GWd/tU, with the three models presented above and accounting for the burn-up effect: Lokken and Courtright associated with the Baron-Hervé (95) fuel conductivity model, Lucuta and Wiesenack. The Baron-Hervé (95) at zero burn-up is plotted as a reference. The data were obtained on fuel wafers, 1 mm thick and 10 mm in diameter, irradiated in the Halden reactor at average temperature starting from 800°C at BOL to about 600°C at EOL. All data are reduced to 95% TD using the measured porosity before and after the thermal diffusivity measurement (see [15]).

The Wiesenack model gives as expected a good average value within the data sets, as far as it has been fitted on instrumented rods in Halden. This kind of experiment gives a radial integrated value of the fuel thermal conductivity. It must be noticed that the in-pile fitting of the Wiesenack model is in a good agreement with the out-pile measurements. The Lokken and Courtright model is good value in the range 600-1800 K. Nevertheless it is not usable for temperature lower than 600 K, as it drastically overpredicts the thermal conductivity degradation at low temperatures. The Lucuta model is the more descriptive formulation and accounts for the recovering processes with temperature. However its predictions capabilities are equivalent to the Wiesenack model.

Conclusions

A large amount of thermal conductivity data have been collected on UO_2 and MOX fuel over the last 30 years. More recently, data have been provided on $(U-Gd)O_2$ fuels. Previously, the thermal conductivity data on fresh fuel were obtained by calorimetric measurements. With the improvement of laboratory methods, the LASER flash method allows to collect more accurate data via the diffusivity measurement on fresh fuel and irradiated material as well. The thermal conductivity dependence on temperature, deviation from stochiometry, porosity, or Pu and Gd contents, have been investigated.

The thermal conductivity database has been used by the EDF to review and evaluate modelling proposed in the literature. Most of the models are specific to a kind of fuel, accounting for stochiometry, Pu weight content or gadolinium content. Some are designed to account for all these parameters. This review gives a comparison on the capability of these different models and enlightens upon the weak effect of the Pu as an element on the fuel conductivity but also the strong effect of the deviation from stochiometry. This confirms the importance of this last parameter in MOX fuel manufacturing. Some questions stay open on the evolution of the fuel stochiometry during irradiation, mainly for the MOX. This review shows nevertheless that modelling is able to account for this parameter.

Concerning the burn-up degradation effect, data have been recently obtained from discs irradiated up to 80 GWd/tU within the NFIR2 international programme, enlightening the recovering processes with temperature. Presently the modelling (Lokken and Courtright, Wiesenack and Lucuta) gives a good average value of the burn-up degradation. However, the fuel sample wafers used in the NFIR2 programme have been irradiated with an average temperature higher than 800°C for part of the life. This means that recovering processes were activated during the samples irradiation. Then, one can assume that the fuel conductivity degradation should have been higher at rim temperature (about 500°C). This assumption will be confirmed in the near future by the data set expected on the low temperature wafer provided in the High Burn-up Rim Project.

REFERENCES

[1]   D.L. Hagrman, G.A. Reymann, "MATPRO-Version 11, A Handbook of Materials Properties for use in the Analysis of Light Water Reactor Fuel Rod Behavior", February 1979, NUREG/CR0497/TREE/1280/R3.

[2]   M. Hirai, M. Amaya et al., "Thermal Diffusivity Measurements of Irradiateds UO_2 Pellets", paper 3/3, IAEA TCM on Advances in Pellet Technology for Improved Performance at High Burn-up, TOKYO, Japan 28 Oct-1 Nov 1996.

[3]   M. Amaya, Hirai, "Thermal Diffusivity Measurements on Irradiated $(U,M)O_{2+x}$ Pellets", paper 4/2, IAEA TCM on Advances in Pellet Technology for Improved Performance at High Burn-up, TOKYO, Japan 28 Oct-1 Nov 1996.

[4]   J. Nakamura et al., "Thermal Diffusivity of High Burn-up UO_2 Pellet", IAEA TCM on Advances in Pellet Technology for Improved Performance at High Burn-up, TOKYO, Japan, 28 Oct-1 Nov 1996.

[5]   T.L. Shaw, J.C. Carrol, R.A. Gomme, "Thermal Conductivity Determinations for Irradiated Fuels", IAEA TCM on Advances in Pellet Technology for Improved Performance at High Burn-up, TOKYO, Japan, 28 Oct-1 Nov 1996.

[6]   K. Bakker and R.J.M. Konings, "The Effect of Irradiation on the Thermal Conductivity of High Burn-up UO_2 Fuel", paper 3/4, IAEA TCM on Advances in Pellet Technology for Improved Performance at High Burn-up, TOKYO, Japan, 28 Oct-1 Nov 1996.

[7]   H.G. Morgan, AEA Technology, NFIR-II, Update on Fuel Properties Measurements, 25th NFIR Steering Committee Meeting, TORONTO, Canada, 13-14 April 1994.

[8]   Lucuta, H.J. Matzke, R.A. Verral, "Modelling of UO_2-based SIMFUEL Thermal Conductivity: The Effect of the Burn-up", Journal of Nuclear Materials, 217 (1994), pp. 279-286.

[9]   A.L. Loeb, Journal of Amercican Ceramic Society, 37:96 (1954).

[10]   D.R. Olander, "Fundamental Aspect of Nuclear Reactor Fuel Element 1976", Department of Nuclear Engineering, Berkeley, CA, US Department of Commerce, National Technical Information Dervice, TID-26711-P1.

[11]   N. Hoppe, M. Billaux, J. Van Vliet, "COMETHE 3L, General Description", BN8201-04, January 1982.

[12] R.A. Bush, "NFIR-Properties of the Urania-Gadolinia System (Part 1)", NFIR-RP03 December 1986 (Final Report).

[13] R.H. Watson, "NFIR-Properties of the Urania-Gadolinia System (Part 2)", NFIR-RP04 March 1987 (Final Report).

[14] D. Baron and J-C. Couty, "A Proposal for a Unified Fuel Thermal Conductivity Model Available for UO_2, $(U-Pu)O_2$ and UO_2-Gd_2O_3 PWR Fuel", IAEA TCM on Water Reactor Fuel Element Modelling at High Burn-up and its Experimental Support, Windermere, UK, 19-23 September 1994.

[15] D. Baron, "A Proposal to Simulate the Fuel Thermal Conductivity Degradation at High Burn-up and the Recovering Processes with Temperature", HBRP 4th Steering Committee, Karlsruhe, November 1997.

[16] G. Delette, M. Charles, "Thermal Conductivity of Fully Dense Unirradiated UO_2: A New Formulation from Experimental Results Between 100°C and 2500°C, and Associated Fundamental Properties", IAEA TCM on Water Reactor Fuel Element Modelling at High Burn-up and its Experimental Support, Windermere, UK, 19-23 September 1994.

[17] W. Wiesenack, "Assessment of UO_2 Conductivity Degradation Based on In-Pile Temperature Data", OECD Halden Reactor Project, Report HWR 469, 1996.

[18] Y. Philipponneau, "Thermal Conductivity of Mixed Oxide Fuel", Note Technique SEFCA n°157, Janvier 1990.

[19] J.H. Harding, D.G. Martin, "A Recommendation for the Thermal Conductivity of UO_2", *Journal of Nuclear Materials*, 166 (1989), pp. 223 -226.

[20] P.G. Lucuta, H.J. Matzke, I.J. Hastings, "A Pragmatic Approach to Modelling Thermal Conductivity of Irradiated UO_2 Fuel, Review and Recommendations", *Journal of Nuclear Materials*, 232 (1996), pp. 166-180.

[21] R.O. Lokken and E.L. Courtright, "A Review of the Effects of Burn-up on the Thermal Conductivity of UO_2", Battelle Pacific Northwest Laboratory, BNWL-2270, Document prepared for the NRC.

[22] R.L. Gibby, "The Effect of Plutonium Content on the Thermal Conductivity of $(U-Pu)O_2$ Solid Solutions", *Journal of Nuclear Materials*, 38, p. 163-177, 1971.

[23] R.L. Gibby, "The Effect of Oxygen Stoichiometry on Thermal Diffusivity and Conductivity", USAEC Report BNWL 927, Battelle Northwest Laboratory, January 1969.

[24] S. Fukushima and T. Ohmishi, "The Effect of Gadolinium Content on the Thermal Conductivity of Near Stoichiometric $(U-Gd)O_2$ Solid Solutions", *Journal of Nuclear Materials*, 105, p. 201-210, 1982.

[25] J.C. Van Craynes and Weilbacher, *Journal of Nuclear Materials*, 26, p. 132, 1968.

[26] J-P. Alessandri, Y. Philipponneau, "Etude expérimentale de la diffusivité thermique des combustibles MOX", Note Technique LPCA n°36, Novembre 1994.

[27] Y. Philipponneau, "Thermal Conductivity of Mixed Oxide Fuel, Review and Recommendation", Note Technique SEFCA n°117, January 1990.

[28] J. Stora *et al.*, "Thermal Conductivity Sintered Uranium Oxide Under In-Pile Conditions", EURAEC 1095 (CEA-R 2586), August 1964.

Figure 1. UO$_2$ data from MATPRO

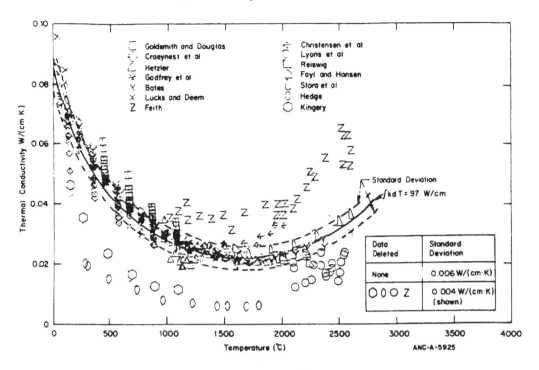

Figure 2. (U,Pu)O$_2$ data from MATPRO

Comparison of measured and predicted values of the thermal conductivity of (U,Pu)O$_2$
for materials corrected to 96% TD and standard deviation of data from the theoretical curve.

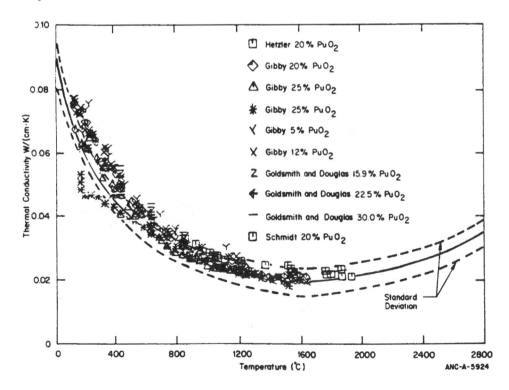

Figure 3. UO₂ thermal conductivity versus temperature

Density = 95% TD, O/M = 2, BU = 0, models available in the CYRANO3 code

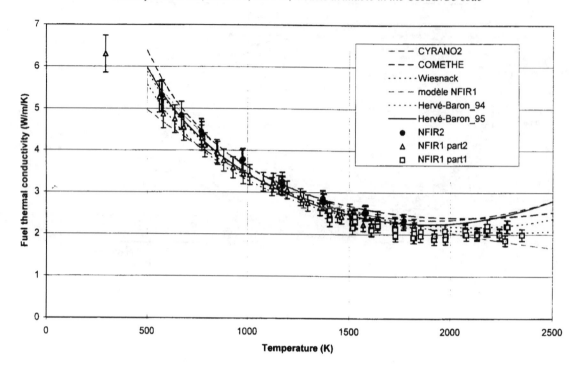

Figure 4. UO₂ thermal conductivity versus temperature

Density = 95% TD, O/M = 2, BU = 0, comparison with other models

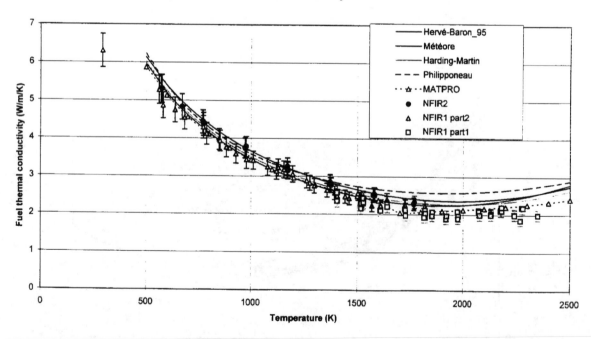

Figure 5. Influence of O/M ratio on oxide fuel thermal conductivity

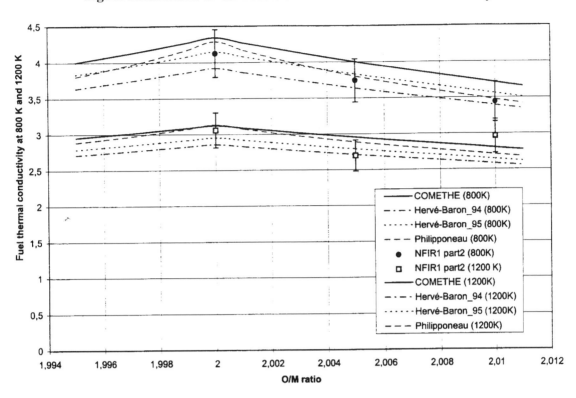

Figure 6. Influence of gadolinium content on the fuel oxide thermal conductivity

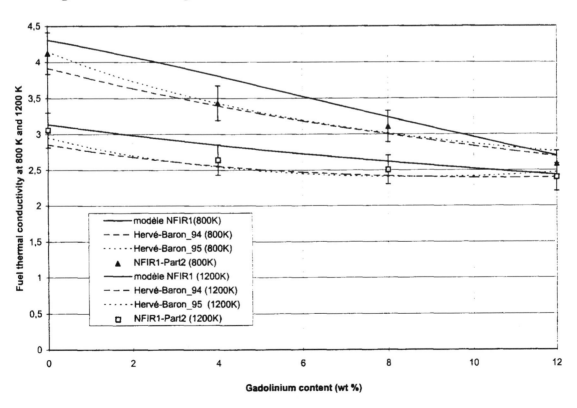

Figure 7. MOX fuel thermal conductivity for 20% Pu

95% TD, O/M = 2.000, BU = 0

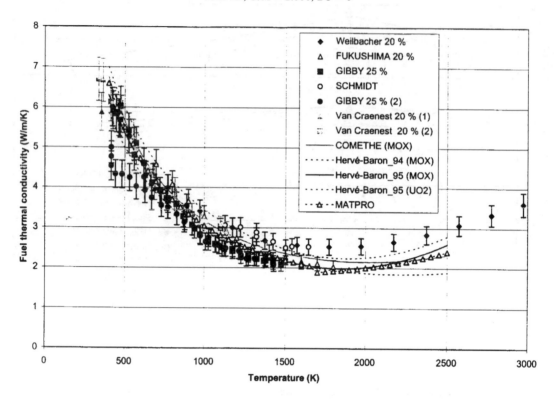

Figure 8. Influence of Pu content on fuel thermal conductivity

95% TD, O/M = 2.000, BU = 0

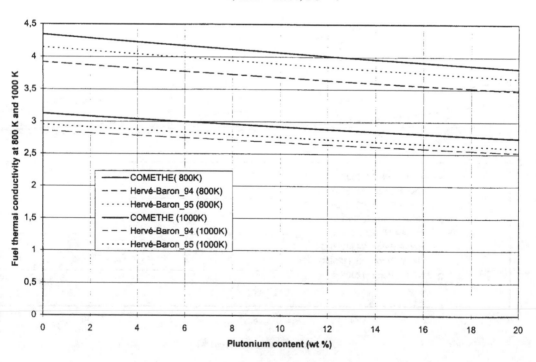

Figure 9. MOX fuel thermal conductivity

O/M = 1.98, 95% TD

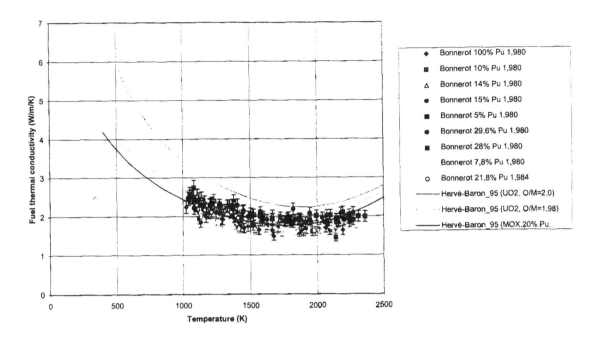

Figure 10. MOX fuel thermal conductivity

O/M = 1.98, 95% TD

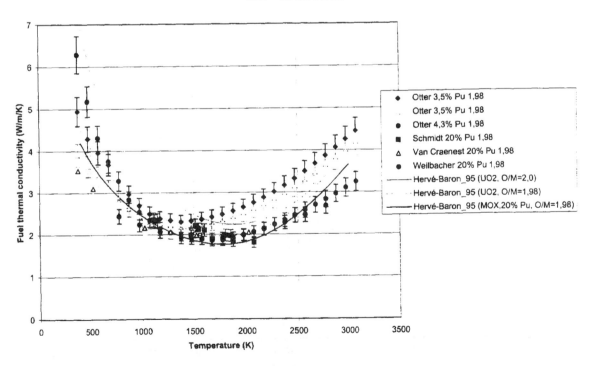

Figure 11. MOX fuel thermal conductivity

O/M = 1.975, 95% TD

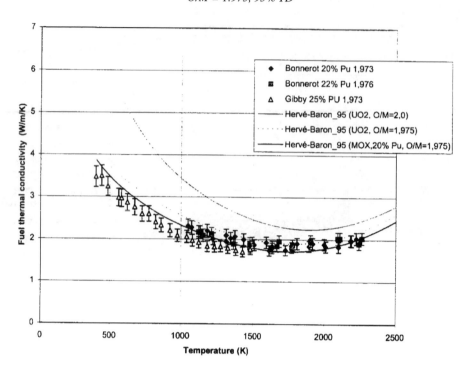

Figure 12. UO₂ fuel thermal conductivity at 26,8 MWd/KgU and reduced at 95% TD

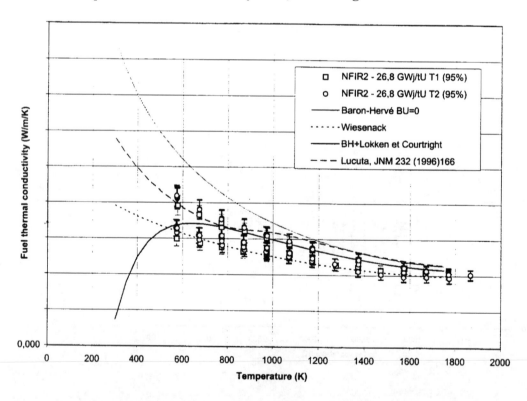

Figure 13. UO$_2$ fuel thermal conductivity at 38,8 MWd/KgU and reduced at 95% TD

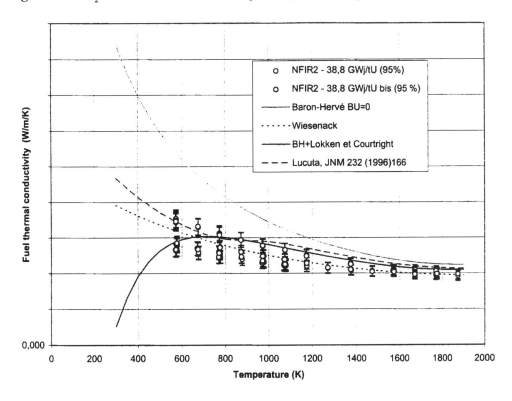

Figure 14. UO$_2$ fuel thermal conductivity at 63,6 MWd/KgU and reduced at 95% TD

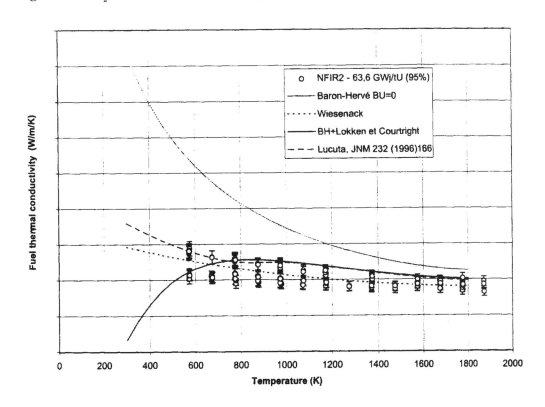

Figure 15. UO₂ fuel thermal conductivity at 80 MWd/KgU and reduced at 95% TD

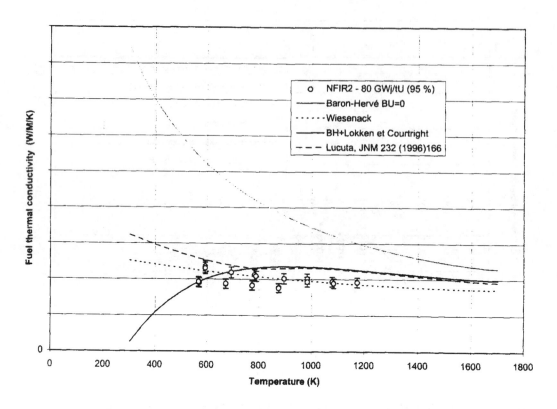

A FUEL THERMAL CONDUCTIVITY CORRELATION BASED ON THE LATEST EXPERIMENTAL RESULTS

F. Sontheimer, H. Landskron
Siemens AG, KWU
P. O. Box 3220, 91050 Erlangen, Germany

M. R. Billaux
Siemens Power Corporation,
P. O. Box 130, Richland, WA 99352, USA

Abstract

A new fuel thermal conductivity (ftc) correlation for UO_2 and $(U,Gd)O_2$ is presented, which is based on Klemens' relaxation-time theory [1]. The correlation is chosen because of its validity in a wide range of defect concentrations, as for instance encountered in fuel with a wide range of burn-up and gadolinia additions, as has been shown by Ishimoto [2].

The phonon term of the new correlation has the form $\frac{1}{x} \cdot \arctan(x)$, where x is a measure of the defect concentration introduced by burn-up and gadolinia additions.

For low defect concentrations, this term is identical with the classical form for the phonon term $\frac{1}{A + B \cdot T}$. At high defect concentrations, however, when phonon-point defect scattering starts dominating over phonon-phonon scattering, the new correlation deviates from the classical formulation and has a distinctly weaker dependence on temperature and defect concentration than the classical form.

The new arctan correlation in combination with an appropriate electronic ftc term is fitted to the Halden data base of fuel centre-line temperature measurements (represented by the "Halden ftc correlation recommendation"). Agreement is very good up to a burn-up of about 60 MWd/kgU; beyond, the arctan form has a saturating burn-up degradation.

The new arctan correlation in combination with an appropriate electronic ftc term is also shown to describe very well our latest ftc measurements [3] on unirradiated gadolinia fuel up to 9% gadolinia content.

Application to Halden measurements up to very high burn-up is successful, when combined with the so-called "rim-effect", which counteracts the saturation tendency of the new correlation at high burn-up.

The latest laser thermal diffusivity measurements on irradiated gadolinia fuel in the frame of the NFIR programme, although not yet open for literature and not discussed in the paper, indicate very good agreement with the new arctan correlation.

Introduction

The fuel thermal conductivity (ftc) of the nuclear fuel UO_2 is affected during irradiation not only due to dissolving and precipitation of fission products in the fuel matrix, but also to microstructural changes such as an O/M ratio change, bubble formation and irradiation damage accumulation; doping of UO_2 fuel with gadolinia has a similar effect as fission products accumulated with burn-up.

Up to now, many studies in the low and medium burn-up range have shown that the decrease in thermal conductivity of UO_2 fuel is proportional to the defect concentration introduced by irradiation and doping of the fuel and that the phonon term of the ftc is well described by the classical form $\dfrac{1}{A + A_{irr} + A_{dope} + B \cdot T}$, where the terms A_{irr} and A_{dope} describe the influence of the defects introduced by irradiation and doping; the term B describes the phonon-phonon interaction (Umklapp prozess).

However, at very high burn-ups and/or gadolinia additions, it has been shown [2] that there is a change in the physical processes causing ftc degradation from mainly phonon-phonon scattering to mainly phonon-point defect scattering, which is accompanied by a reduction in ftc degradation dependence on temperature and defect concentration. In this region, the above classical form of the phonon term no longer seems valid.

Based on literature and our own measurements, it is shown below that a correlation based on Klemens' relaxation-time theory [1], which has a phonon term of the form $\dfrac{1}{x} \cdot \arctan(x)$, where x is a measure of the defect concentration introduced by burn-up and doping (e.g. gadolinia additions), describes ftc degradation at large defect concentrations very well. Moreover, it is appropriate for the whole range of defect concentrations from low to high values, because it contains the classical form as a limit.

Properties of the arctan form of the ftc phonon term in different regions of defect concentrations

Ishimoto [2] modified the Klemens ftc theory for arbitrary defect concentrations and provides a phonon contribution λ_{Ph} to ftc, which is expressed by the following equations:

$$\lambda_{Ph} = \frac{\lambda_0}{x} \cdot \arctan(x) \qquad\qquad x = D \cdot \sqrt{\Gamma} \cdot \sqrt{\lambda_0}$$

Whereas in the Klemens theory D is a constant, it is slightly temperature dependent in Ishimoto. $\lambda_0 = \dfrac{1}{B \cdot T}$ is the thermal conductivity for a defect-free sample at Klemens (phonon contribution), whereas $\lambda_0 = \dfrac{1}{(A + B \cdot T)}$ is the ftc for fresh, undoped UO_2 at Ishimoto. For both Klemens and Ishimoto, Γ is the scattering cross-section parameter for all the point defects. For different contributions to scattering we have:

$$x = \sqrt{\sum x_i^2}, \text{ with } \quad x_i = D_i \cdot \sqrt{\Gamma_i} \cdot \sqrt{\lambda_0}$$

The index i stands for the different contributions to scattering as e.g. fission products, dopants, stoichiometry deviations, irradiation defects, etc.

Fresh UO₂, x=0

The term $\dfrac{\arctan(x)}{x}$ is equal to unity, giving $\lambda_{Ph} = \lambda_0 = \dfrac{1}{A + B \cdot T}$, the ftc phonon term for as fabricated, undoped UO₂.

Medium burn-up UO₂ and/or doping, x<1

The term $\dfrac{\arctan(x)}{x}$ can be expanded to $1 - \dfrac{x^2}{3}$, yielding $\lambda_{Ph} = \dfrac{1}{A + B \cdot T + \Gamma \cdot \dfrac{D^2}{3}}$, which

is exactly the classical form commonly used to describe the burn-up influence on ftc, when Γ is identified as burn-up (see also classical form of phonon term at low to medium defect concentration shown in the introduction). The linear temperature dependence in the denominator is still the same as for fresh UO₂; also the dependence on Γ is linear.

A value for the burn-up proportionality factor often found in literature (e.g. [4]) is 0.003 (mK)/W/(MWd/kgU). The condition x<1 then leads to a surprisingly low burn-up (<15 MWd/kgU) in the temperature range of interest, to which the classical form of the phonon term should be restricted; above, the arctan formulation appears to be more appropriate.

As mentioned, the burn-up proportionality factor $\dfrac{D^2}{3}$ has a slight temperature dependence in Ishimoto's formulation (see also description below). This is in agreement with a slight temperature dependence found in the Halden measurements [4].

High burn-up UO₂ and/or doping, x>>1

The term arctan(x) approaches $\pi/2$, giving:

$$\lambda_{Ph} = \frac{\lambda_0 \cdot \pi}{2 \cdot x} = \frac{\pi}{2} \cdot \frac{1}{\sqrt{A + B \cdot T}} \cdot \frac{1}{D \cdot \sqrt{\Gamma}}$$

In contrast to the region of low to medium defect concentrations x<1, the temperature dependence and the dependence on Γ in the denominator are now no longer linear for x>>1; instead, a square root dependence is observed. For high burn-up and gadolinia content, this results in saturation-like effects.

Confirmation of arctan form by Ishimoto

Figures 1, 2a and 2b, which are selected to demonstrate the applicability of the arctan formulation of the ftc phonon term in a wide range of defect concentrations, are taken from Ishimoto's paper. The measurements performed were laser flash determinations of ftc on pure UO₂ and 6% gadolinia fuel with simulated burn-up (SIM fuel) of 30, 60 and 90 MWd/kgU.

The relevance of the SIM fuel laboratory measurements for the selection of an appropriate ftc correlation is high, because the samples are unirradiated and can be precisely characterised and easily measured. An adequate representation of the real burn-up effect of the fission products on ftc by SIM fuel methods has been proven earlier [5].

Figure 1 shows that already at a simulated burn-up of 30 MWd/kgU, the temperature dependence of the ftc phonon term (denominator) becomes markedly weaker than linear (slope<1). This is in agreement with the above statement that at burn-ups above about 15 MWd/kgU (real burn-up !) already, the classical form with the linear temperature dependence in the denominator, becomes inappropriate.

Figure 1. Logarithmic plot of phonon contribution to normalised
thermal conductivities in UO$_2$ with simulated burn-up vs. temperature [2]

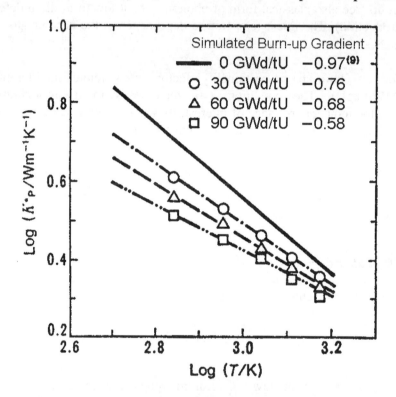

At a simulated burn-up of 90 MWd/kgU, the slope of the data in Figure 1 approaches 0.5, confirming the square root temperature dependence found in the high defect concentration approximation of the arctan formulation above.

Because SIM fuel only covers the effect of real burn-up on ftc concerning the fission products, a transition of the ftc correlation from the classical form with a linear temperature and burn-up dependence in the phonon term denominator to the high defect concentration approximation of the arctan formulation with square root dependencies, should occur already well below 90 MWd/kgU.

Comparison of Figures 2a and b shows the different effect of simulated burn-up on the ftc of pure UO$_2$ and 6% gadolinia fuel. Clearly, the effect on ftc of the gadolinia fuel is much smaller, especially in the low to medium temperature range. This is interpreted as the "saturation effect" due to the transition from a linear to a square root dependence on burn-up (Γ) described above.

Figures 2a, 2b. Thermal conductivities of UO_2 and $UO_2/6\%Gd_2O_3$ with simulated burn-up up to 90 MWd/kgU (normalised to 95% TD, from [2])

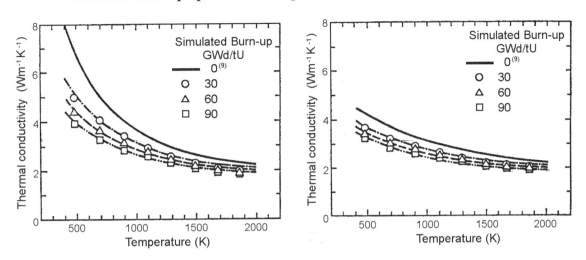

The curves drawn are arctan forms of the ftc phonon term (supplemented by a common, burn-up independent electronic term), fitting well through the data points.

Fitting of the arctan form to measurements on rods irradiated in Halden and to unirradiated UO_2/Gd_2O_3 fuel

The full expression for the ftc: $\lambda = \dfrac{\lambda_0}{x} \cdot \arctan(x) + \lambda_{electronic}$ is fitted to the data. Here λ_0 and $\lambda_{electronic}$ are the ftc phonon term for as fabricated UO_2 and the electronic term, used at Siemens, respectively and x is a superposition of a burn-up term and a gadolinia term:

$$x^2 = x_{bu}^2 + x_{Gd}^2 \text{ with } x_{bu}^2 = D_{bu}^2 \cdot burnup \cdot \lambda_0 \text{ and } x_{Gd}^2 = D_{Gd}^2 \cdot Gd\% \cdot \lambda_0$$

Just like Ishimoto, we assume D to have a moderate temperature dependence of the form $\exp(-const\cdot T)$, with the constant in the exponent of the order of $10^{-4}K^{-1}$.

Measurements in Halden

In [4], a large database of in-pile fuel centre-line temperature measurements in Halden and laser flash measurements of thermal diffusivity on irradiated fuel samples is discussed. There, a formula is also recommended which can be regarded as a consensus description of this database. This Halden recommendation formula was taken as a fitting basis for the new correlation (determining D_{bu}). The fitting range with respect to burn-up is limited to a burn-up of about 60 MWd/kgU because of the saturation behaviour of the new correlation.

In Figure 3 the Halden recommendation (points) and the fitted new correlation (curves) are compared. The Halden recommendation is used only in the temperature range covered in the experiments (up to about 1600°C). Agreement between the new model and the Halden recommendation is very good up to 40 MWd/kgU. At 60 MWd/kgU, the saturation tendency of the new model becomes apparent at low fuel temperature.

Figure 3. Arctan correlation fitted to the Halden recommendation formula (Wiesenack [4], ftc at 95% TD)

Fit of new correlation (curves) to Halden data (points)

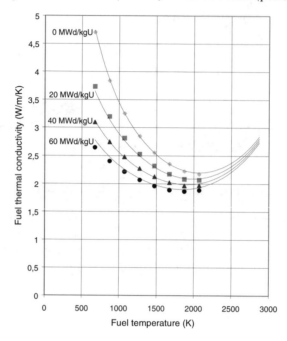

Unirradiated UO_2/Gd_2O_3 fuel with up to 9 w% Gd_2O_3

The new Siemens laboratory laser flash measurements [3] were taken to determine the Gadolinia fitting parameter D_{Gd} for the new correlation. It is shown in Figure 4 that the new correlation describes the measurements very well. Slight deviations occur for 9% Gd_2O_3, where the measurements (triangles) display a somewhat less pronounced temperature decrease.

Especially, the trend of vanishing Gd influence on ftc at high temperatures above about 1900 K and consequently approaching ftc values for pure UO_2 and UO_2/Gd_2O_3, independent of gadolinia content, is excellently described by the new correlation. Attempts to fit this trend with the old, classical correlation failed; it led to overlapping of the ftc for fresh UO_2 and gadolinia fuel in the high temperature range. The gadolinia data thus also show that the arctan form is more appropriate than the classical form.

By setting $D_{bu}^2 \cdot burn-up \cdot \lambda_0 = D_{Gd}^2 \cdot Gd\% \cdot \lambda_0$, the effect of the Gd dopant on ftc can be compared to the burn-up effect. The result in the temperature range of interest is that 1w/o Gd_2O_3 addition corresponds to 4 MWd/kgU.

Saturation of ftc degradation at high burn-up and Gd_2O_3 content

In Figure 5 the relative ftc as a function of burn-up is compared for the classical form $\dfrac{1}{\left(A + A_{irr} + A_{dope} + B \cdot T\right)}$ (points) and the new arctan correlation (curve) at 7w/o Gd_2O_3.

Figure 4. arctan correlation fitted to the latest Siemens fuel thermal conductivity measurements (laser flash method) on unirradiated UO₂/Gd₂O₃ fuel (at 95% TD, [3])

New correlation (curves) fitted to Gd measurements (points)

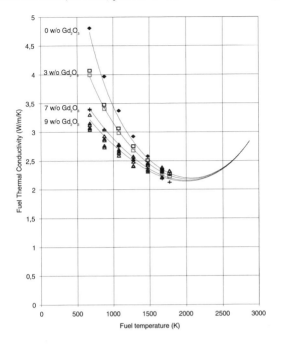

Figure 5. Relative fuel thermal conductivity degradation for high burn-up. Const x 0.0027 is the additive burn-up term in the classical phonon term.

New (curve) versus classical correlation (points)

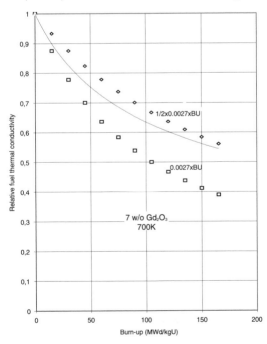

For $A_{irr}=0.0027 \cdot BU$ (a value used in our fuel rod design code) the degradation of the classical form is much steeper than the arctan function; a reduced burn-up dependence in the classical form of $A_{irr}=0.0014 \cdot BU$ comes closer. At very high burn-up well above 100 MWd/kgU, however, which becomes relevant in the rim region of the fuel pellet with high burn-up, the saturation trend of the arctan form is progressively stronger than in the classical form.

Application of the arctan form to Halden measurements up to very high burn-up

In the Halden Ultra High Burn-up (UHB) rig IFA-562, fuel centre-line temperatures are measured up to burn-ups above 80 MWd/kgU. At such high burn-ups, the pellet rim effect is known to have a strong influence on fuel temperature (increase).

The UHB data are central to the Halden recommended ftc correlation described by Wiesenack in [4], which fits the UHB data very well and thus can be used as a representation of the UHB data.

Comparing the UHB "data" (Wiesenacks correlation) with the classical form and the arctan form, taking the pellet rim effect into account, yields very good agreement for the arctan form and a fuel centre-line overprediction of nearly 100 K for the classical form at 80 MWd/kgU, as shown in Figure 6. The overprediction of the classical form at burn-ups above about 50 MWd/kgU is due to the rim effect; in the arctan form, the rim effect is balanced by the saturation in this burn-up region.

Figure 6. Pellet centre-line temperature vs. burn-up at 250 W/cm calculated with different fuel conductivity correlations with degradation of ftc in the rim region

Simplified calculation with flat power profile

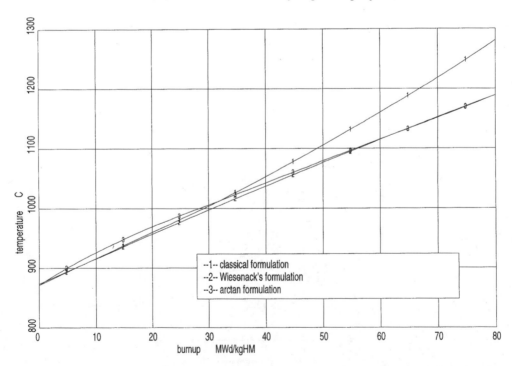

REFERENCES

[1]   Klemens, P.G., *Phys. Rev.*, 119, 507 (1960).

[2]   Ishimoto *et al.*, "Effects of Soluble Fission Products on Thermal Conductivities of Nuclear Fuel Pellets", *J. Nucl. Sci. Technol.*, 31 (8), pp. 796-802 (August 1994).

[3]   JTK 1998 G. Gradel, W. Doerr, H. Gross, to be published.

[4]   W. Wiesenack *et al.*, "Fuel Conductivity Degradation Assessment Based on HBWR Data", HWR-469, 96-05.

[5]   P.G. Lucuta *et al.*, "Proc. 2nd Int. Conf. CANDU Fuel", October 1989, Pembroke, Ontario, Canada, ed. I.J. Hastings, CNS, Toronto (1989) 132.

ABOUT THE MODELLING OF FUEL THERMAL CONDUCTIVITY DEGRADATION AT HIGH BURN-UP ACCOUNTING FOR RECOVERING PROCESSES WITH TEMPERATURE

Daniel Baron
EDF Etudes et Recherches, France

Abstract

In BWR and PWR commercial nuclear reactors, the average fuel discharge burn-up has been increased from 35 to 52 GWd/tM over the last 20 years. Furthermore, a large scale of experimental irradiation programmes have shown the capability of the actual design to achieve more than 72 GWd/tM if more appropriate zircaloy alloys allow to limit the corrosion process and hydrogen pick-up of the cladding.

One must account for the consequence of the burn-up increase on the material properties. In particular, the fuel material which is obviously the place for the fission process can no more be considered as pure UO_2. Its physical chemical composition is continuously modified with the apparition of fission products and a strong disturbance of the matrix is induced by fission spikes which produce a great number of point defects, mainly vacancies in the oxygen sublattice. Consequently, the fuel conductivity is progressively degraded. However, the solid fission products in substitution in the UO_2 lattice lead to a permanent degradation while point defects and gaseous or volatile fission products are able to move if the local temperature is high enough. Subsequently, a part of the fuel conductivity degradation can be recovered if the fuel material sustains an increase of temperature during a period of its history.

A modelisation is proposed to account for the burn-up degradation, very similar to the model developed previously by Dr. Wiesnack in Halden but with a recovery term based on the lattice parameter considered as a marker for the fuel disorder.

Introduction

In the BWR and PWR commercial nuclear reactors, the average fuel discharge burn-up has increased over the last 20 years from 35 to 52 GWd/tM. Furthermore, a large scale of experimental irradiation programmes have shown the capability of the actual fuel rod design to achieve more than 72 GWd/tM. The only limitation is indeed related to the cladding corrosion and the excessive hydrogen pick-up, but not to the fuel itself. The new zircaloy alloys, with a low tin content and other additives, would allow to extend the discharge burn-up in the next future.

This burn-up extension obviously has an impact on the fuel rod behaviour, then one must account for the consequence on the material properties in the code simulations. In particular, the fuel material which is the place for the fission process can no longer be considered as pure UO_2. Its physical chemical composition is modified continuously with the apparition of fission products [1,2] and a strong disturbance of the matrix is induced by fission spikes which produce a great number of point defects, mainly vacancies in the oxygen sublattice. As a consequence, the fuel conductivity is progressively degraded. However, the solid fission products in substitution in the UO_2 lattice leads to a permanent degradation while point defects and gaseous or volatile fission products are able to move if the local temperature is high enough. Subsequently, a part of the fuel conductivity degradation can be recovered if the fuel material sustains an increase of temperature during a period of its power history.

A modelisation is proposed to account for the burn-up degradation, very similar to the model developed previously by Dr. Wiesnack in Halden but with a recovery term based on the lattice parameter considered as a marker for the fuel disorder.

Theoretical aspects

Lattice perturbation

The overall phenomena induced during irradiation tend to degrade the fuel thermal conductivity. Nevertheless, the thermal agitation or all kinds of energy brought to the matrix can thwart this evolution.

All perturbations or defects in the fuel matrix increase the phonon dispersion and then decrease the heat transport. During irradiation, the lattice perturbations have many origins, including:

- The fission fragments which produce a great number of point defects, colliding along their track before they totally loose their initial kinetic energy. This points defects correspond mainly to oxygen atom ejected from their site but also to uranium atom displacement. Most of the time, this leads to a creation of pairs (intersticials/vacancies).

- When their energy is low enough, the fission fragments become fission products and stop in the matrix. Some are located in the place of an uranium atom (substitution) resulting in a deformation of the lattice due to their smaller size compared to uranium. Others can stop in an interstitial site, resulting again in a lattice deformation. For the gaseous forms such as xenon they tend to locate in a two or three vacancy sites.

- The substituting fission products have various affinities for oxygen. The main trend is strong modification of the oxygen sublattice with an increase in oxygen potential. This oxygen in excess is buffered by metallic fission products such as molybdenum, zirconium, etc.

- Many fission products are able to recombine with others elements to create intermediate phases which contribute to the fuel change in thermal conductivity as far as the fuel becomes a material multi-compounds multi-phases.

- Dislocation loops are formed provided with a flow of point defects rejected by the nearby matrix.

Gaseous fission products and porosities

The gaseous fission products which are non-soluble in the uranium matrix are strongly related to the fuel local temperature. At low temperature (< 600°C) the xenon and krypton atoms' free path is between 10-100 manometers. The gas atoms are able to come together in the location of a vacancy cluster and form nanobubbles containing a few gas atoms. The nanobubbles exist at such temperatures in large quantities. Their number is somehow limited by a resolution phenomena due to fission spikes. During irradiation an equilibrium in the nanobubble population is reached between creation and resolution processes.

If the temperature increases, the gas atom free path increases and the nanobubbles are able to move and grow, changing in the way the porosity aspect of the fuel. The effect of the bubbles on the fuel thermal conductivity is related to their spatial distribution, their number, their shape and the gases contained. For example, at high temperature (> 1600°C) most of the porosities are large lenticular porosities located on grain boundaries and at triple boundaries. This degrades more or less the heat transfer at the grain boundaries, depending on the gas mixture contained in the pores. On the other hand, the thermal conductivity is improved within the grain.

Reversibility

Depending on the local temperature the thermal agitation can allow the mobility of more or less large defects. This mobility can be increased by two factors: the fission kinetic energy and the lattice stored energy. All the causes described above to explain the fuel conductivity degradation can be considered as defects in the UO_2 lattice: oxygen or uranium vacancies, interstitial fission products, vacancies or interstitial clusters or dislocations. Even the nanobubbles could be considered as lattice defects.

When burn-up proceeds, the fuel matrix is progressively distorted by the presence of these defects. The average lattice parameter evolves with this distortion. This corresponds to an increase of the energy stored in the lattice which tends on another hand to eliminate the defects to reduce its energy. Most of the fission products in interstitial position are in a metastable situation. This results in local stresses field (nano scale) which reduces potential barriers around atom locations. These defects are able to move if – at the same time – the thermal agitation level is high enough for the atom to cross the potential barrier.

The point defects, interstitial and vacancies are able to move for temperatures around 800 K [3]. This threshold certainly decreases with burn-up. At higher temperatures (1000-1100 K), dislocations and larger defects such as xenon or nanobubbles are able to move at long distances. This results in an apparent purification of the lattice and then to a partial recovering of thermal properties. The change in the porosity distribution, due to fission gas diffusion, also contributes to the recovering of the thermal properties with a kinetic nevertheless different (bubble diffusion and coalescence).

Rim specificity

The rim is characterised by two factors: a low irradiation temperature (< 800 K) and a high density of fission (twice the average) due to the ^{238}U self-shielding effect which results in a strong accumulation of plutonium participating in the fission process.

Due to the low temperature the recovering processes are quasi non-operant during the overall irradiation. Nevertheless, the stored energy due to the accumulation of defects in the matrix creates such high stresses that above 55 GWd/tU local burn-up, the defects seem able to move at short distances, allowing a fuel restructuration. From the observation available today, the process could be as follows: accumulation of dislocations in preferential crystallographic planes [1,1,1] or [1,0,0]. Their intersections are preferential locations to pin and stabilise the nanobubbles in perpetual creation and resolution. Then, when stabilised, these bubbles can grow and form cavities, provided with available vacancies in the close neighbourhood. The sliding planes can act as drains around this cavities for vacancies and gaseous fission products (a type of short cut).

H.J. Matzke [11] suggests that these bubbles are overpressurised. This overpressurisation is then the origin of the fragmentation around the pores. The round shape of the grains around the pore suggests another mechanism proposed by N. Lozanno [12] and based on diffusion processes: the flow of fission products toward the free surface inside the cavities can induce a concentration difference between the bulk and the open surface. As a consequence, it could produce a surface rearrangement and the apparition of the so observed fine round grains around the cavities. Further from the bubbles, the polygonal shape of the subgrains formed with a greater size suggests a non-diffusionnal mechanism with a mechanical process induced by stress gradients.

Figure 1. Rim structure formation

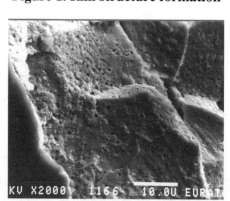

In relation with the description here above, the fuel matrix conductivity is strongly degraded due to lattice defects, the numerous fission products, the decrease of the grain size (increase in the grain boundaries length) and the increase of the fuel porosity. One question is left open to discussion about the distribution of the fission gases between the matrix, the bubbles or the free volumes. Are the bubbles overpressurised or in equilibrium with the hydrostatic pressure?

The lattice parameter

In the cold fuel region, the lattice parameter increases with burn-up as shown by Une and Nogita [4]. A saturation of this parameter likely occurs when the rim formation starts around 60 GWd/tU. For higher temperature regions (greater than 1100 K) the lattice parameter decreases with burn-up. The results are consistent with the measurement on SIMFUEL where fission products are artificially implanted in a fuel matrix. The difference between these two trends is the material damage as far as at high temperature irradiation damage is recovered.

Figure 2. Burn-up dependence of lattice parameter at the
fuel periphery ($r/r_0 = 0.8$-1.0) of irradiated UO_2 (from [4])

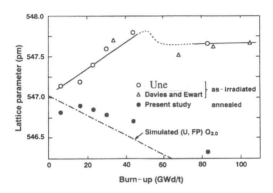

Then, the average lattice parameter can be considered as a marker of the damage state of the material. During a thermal diffusivity test, this parameter evolves depending on the temperature history of the sample, meaning that the first set of diffusivity obtained during such a test when temperature is increasing is representative of the behaviour of a damaged material. On the other hand, the set of measurements during the last temperature decrease is representative of a material where all irradiation damages able to be recover are recovered.

We propose in the following a utilisation of the lattice parameter to represent the recovery status in the fuel conductivity modelling.

Fuel conductivity formulation

General formulation

Two phenomena are responsible for the thermal transfer through a solid: the propagation of atoms' vibrations around their lattice site and the electronic transport, corresponding to the kinetic energy carried by free electrons at high temperatures. The solid thermal conductivity can therefore be written as the sum of two components:

$$K(T) = K_S(T) + K_e(T)$$

The phonon component is as follows:

$$Ks(T) = \frac{1}{A + BT}$$

The first term, A, depends on the number of crystal defects, whatever the local temperature. However, at very low temperature, in a solid free of defects, the mean free path can be equivalent to the sample width. The thermal conductivity becomes in such a case a function of the sample dimensions.

If interatomic strength were purely harmonic, phonon-phonon scattering should be enabled and the mean particles' free path would be limited only by the crystal boundaries or by the crystal defects. In the uranium dioxide phonon-phonon scattering results of the enharmonic components of the crystal vibrations.

In an ionic crystal, lattice enharmonicity indeed increases with the mass difference between anions and cations. In UO_2 or PuO_2 oxides, this mass difference is much greater than in all other common oxides. For that reason, the thermal conductivity of such actinide oxides are considerably lower than most of the crystalline oxides. This effect, thermal dependent is represented by the term BT.

Then, Ks(T) is a continuous decreasing function of the local temperature. However, it can be observed at high temperatures (> 2000°C), that the fuel thermal conductivity is improving again. This next component is representative of an improvement of the energy transportation by free electrons at high temperatures. The electron-vacancies pairs, created by the reaction $U^{4+} \rightarrow U^{5+} + U^{3+}$ can be considered as small polarons. The heat transfer conductivity is similar to an electrical charge transfer in a dielectric matrix, and can be formulated as follows [7]:

$$Ke(T) = \frac{C}{T^2} \exp\left(-\frac{W}{kT}\right)$$

The total formulation for the fuel thermal conductivity for a stochiometric UO_2 is:

$$K(T) = \frac{1}{A + BT} + \frac{C}{T^2} \exp\left(-\frac{W}{kT}\right)$$

where: K(T) = the fuel conductivity in W/m/K
      T   = the fuel local temperature in K
      k   = the Boltzmann constant = $(1.38\ 10^{-23}\ J/K)$
      A   = $4.4819\ 10^{-2}$ (m.K/W) (m.K/W)
      B   = $2.4544\ 10^{-4}$ (m/W)
      C   = $6.157\ 10^{9}$ (W.K/m)
      W   = 1.41 eV = $1.41 * 1.6\ 10^{-19}$ J

In order to account for non-stochiometric fuel, the formulation used is more precisely [6]:

$$A = 4.4819\ 10^{-2} + 4 *(2 - O/M)$$
$$B = 2.4544\ 10^{-4} * (1 + 0.8\ q)$$

where: q   = the plutonium content
    O/M = the oxygen to metal ratio

NFIR database reduction

In the framework of the NFIR programme, UO_2 fuel wafer samples have been irradiated to different burn-ups in the Halden reactor: 27, 39, 64 and 80 GWd/tU. Thermal diffusivity measurements have then been conducted on these different samples, well characterised before and after the test (density, porosity and lattice parameter). These TD measurements were conducted in AEA using a LASER flash method.

The database obtained has been reduced and analysed by the EDF to obtain a fuel conductivity database at 95% of the theoretical density, by the relation:

$$K = \rho \, Cp \, \lambda$$

where ρ is the fuel density, Cp the heat capacity and λ the diffusivity measured. The heat capacity was considered as not modified by irradiation. The MATPRO formulation was used [8]:

$$Cp = \frac{a_1 \theta^2 \exp\left(\dfrac{\theta}{T}\right)}{T^2 \left[\exp\left(\dfrac{\theta}{T}\right)-1\right]^2} + a_2 T + \left(\frac{1}{2} O/M\right) \frac{a_3 E_d}{RT^2} \exp\left(-\frac{E_d}{RT}\right)$$

with: a_1 = 296.7 J/Kg/K
a_2 = 2.43 * 10^{-2} J/Kg/K
a_3 = 8.745 * 10^{107} J/Kg/K
E_d = 1.577 * 10^{105} J/mol
R = 8.3143 J/mol/K

where: T = the fuel temperature (K)
θ = the Debye temperature = 535.3 K
O/M = the oxygen to metal ratio
E_d = the activation energy of Frankel defects (J/mol)

The porosity correction is the Loeb and Ross formulation [6,9]:

$$\text{Correction} = \frac{1 - 0.05 * \varepsilon(T)}{1 - \varepsilon(T) * \text{por}}$$

where: por = the relative porosity volume
$\varepsilon(T)$ = 2.7384 - 0.58 * 10^{-3} * T   (for T < 1273 K)
$\varepsilon(T)$ = 2   (for T $\geq$ 1273 K)

Search for the A and B coefficients

For each sample, the A + BT value have been plotted versus temperature for the batch of data during the first rise of temperature (before defect recovery) and the batch during the last temperature decrease (after recovery) (Figures 3a and 3b). This was obtained as follows:

$$A + BT = \cfrac{1}{K - \cfrac{C}{T^2} \exp\left(-\cfrac{W}{kT}\right)}$$

where K is the fuel conductivity obtained from TD measurements. The values plotted are aligned for each sample before and after TD measurements so it has been easy to fit straight lines in order to obtain A and B from their equations.

A and B coefficient have then been plotted versus burn-up (Figures 4a and 4b). The trend obtained has suggested a linear dependence with burn-up, but also another linear dependence with the recovery status to explain the shift between the two set of points, before and after TD measurements. As proposed before, we used the measured lattice parameter values to fit this second term in the A and B formulations. This gives A and B the following formulation:

$$A = 0.044819 + 0.005 * BU + 20 * (\xi - 5.4702)$$
$$B = 2.4544 - 0.0125 * BU - 70 * (\xi - 5.4702)$$

where: BU = the burn-up in GWd/tU
ξ = the lattice parameter in Å

The final formulation has been compared against the NFIR fuel conductivity database and the other modelling available in the literature [12,13,14]. The graphs are presented in Figures 5a, 5b, 5c and 5d. The measured lattice parameters were used for each sample except for the highest burn-up (80 GWd/tU) for which measurements were not available at the time. Then, for this particular sample, we assumed a value of 5.4688 before recovery (consistent with the values obtained on the other samples) and a value 5.4630 after recovery (consistent with the values reported by Une and Nogita): see Figure 6.

The fuel sample wafers used in the NFIR programme were irradiated in the Halden reactor with an average temperature higher than 800°C for part of the life. This means that recovering processes were activated during the samples irradiation. This can explain why the lattice parameters measured are lower than those reported by Une and Nogita. Then, one can assume that the fuel conductivity degradation should have been higher at rim temperature (about 500°C). This assumption will be confirmed in the near future by the data set expected on the low temperature wafer provided in the High Burn-up Rim Project.

Final formulation of the Baron-Hervé – 97 model

The final overall formulation of the unified Baron-Hervé (97) correlation used by the EDF is:

$$K(T) = \cfrac{1}{A_0 + A_1 x + A_2 g + A_3 g^2 + (B_0(1 + B_1 q) + B_2 g + B_3 g^2)T} + \cfrac{C + Dg}{T^2} \exp\left(-\cfrac{W}{kT}\right) \quad (W/m/K)$$

where: T = temperature (K)
x = absolute value of the stochiometric deviation (abs(2-O/M))
q = plutonium weight content (-)
g = gadolinium weight content (-), 0-12%
k = Boltzmann constant ($1.38.10^{-23}$ J/K)
W = 1.41 eV = $1.41 * 1.6 \, 10^{-19}$ J

$A_0 = 0.044819 + 0.005 * BU + 20 * (\xi - 5.4702)$ (m.K/W)

$B_0 = (2.4544 - 0.0125 * BU - 70 * (\xi - 5.4702))10^{-4}$ (m/W)

$A_1 = 4$ (-) $B_1 = 0.8$ (-)

$A_2 = 0.611$ (m.K/W) $B_2 = 9.603.10^{-4}$ (m/W)

$A_3 = 11.081$ (m.K/W) $B_3 = -1.768.10^{-2}$ (m/W)

$C = 5.516.10^{9}$ (W.K/m)

$D = -4.302.10^{10}$ (W.K/m)

BU = burn-up in GWd/tU

ξ = the lattice parameter in Å

Conclusions

The fuel thermal conductivity is progressively degraded during irradiation. This degradation is variable depending on the fuel local temperature. When the thermal diffusion processes are very low, the accumulation of irradiation defects can be so high that it leads in the rim area to a non-thermal restructuration process beyond 55 GWd/tU. As soon as the irradiation temperature increases, the point defects recovery can be first thermally activated above 800 K and then the recovery of much bigger defects above 1100 K.

The fuel thermal conductivity is affected by these defects as well as by the porosity which can also evolve depending on the local thermal condition. Fuel diffusivity measurements have been conducted up to 80 GWd/tU in the framework of the NFIR high burn-up programme. The data obtained enlighten the degradation effect with the burn-up and the fuel diffusivity partial recovery with temperature.

A model is proposed within this paper to account for the burn-up degradation with burn-up on the fuel thermal conductivity. It is obviously not the first model of this type. The proposal here is to account for the thermal recovery using the lattice parameter as a marker of the matrix damage evolution. In this work, the evolution of the fuel porosity was assumed to be properly modelled with the Loeb and Ross formulation.

REFERENCES

[1] D.R. Olander, "Fundamental Aspect of Nuclear Fuel Elements", TID-26711-P1, 1976.

[2] H. Kleykamp, "The Chemical State of Fission Products in Oxide Fuels", JNM 131 (1985), pp. 221-246.

[3] Hirai et al., "Thermal Diffusivity Measurements of Irradiated UO$_2$ Pellets", IAEA TCM on Advances in Pellet Technology for Improved Performance at High Burn-up, TOKYO, Japan, 28 Oct-1 Nov 1996.

[4]   K. Une, K. Nogita, S. Kashibe, M. Imamura, "Microstructural Change and its Influence on Fission Gas Release in High Burn-up Fuel", 188 (1992), pp. 65-72.

[5]   J. Nakamura *et al.*, "Thermal Diffusivity of High Burn-up UO_2 Pellet", IAEA TCM on Advances in Pellet Technology for Improved Performance at High Burn-up, TOKYO, Japan, 28 Oct-1 Nov 1996.

[6]   D. Baron and J-C. Couty, "A Proposal for a Unified Fuel Thermal Conductivity Model Available for UO_2, $(U-Pu)O_2$ and UO_2-Gd_2O_3 PWR Fuel", IAEA TCM on Water Reactor Fuel Element Modelling at High Burn-up and its Experimental Support, Windermere, UK, 19-23 September 1994.

[7]   G. Delette, M. Charles, "Thermal Conductivity of Fully Dense Unirradiated UO_2: A New Formulation from Experimental Results Between 100°C and 2500°C, and Associated Fundamental Properties", IAEA TCM on Water Reactor Fuel Element Modelling at High Burn-up and its Experimental Support, Windermere, UK, 19-23 September 1994.

[8]   D.L. Hagrman, G.A. Reymann, "MATPRO-Version 11, A Handbook of Materials Properties for Use in the Analysis of Light Water Reactor Fuel Rod Behavior", February 1979, NUREG/CR0497/TREE/1280/R3.

[9]   A.L. Loeb, *Journal of American Ceramic Society*, 37:96 (1954).

[10]   H.J. Matzke, A. Turos and G. Linker, *Nucl. Instr. and Meth. in Physic*, Res B91 (1994).

[11]   N. Lozanno, L. Desgranges, "High Magnification SEM Observations for Two Types of Granularity in a High Burn-up PWR Fuel Rim", to be published in the *Journal of Nuclear Materials*.

[12]   R.O. Lokken and E.L. Courtright, "A Review of the Effects of Burn-up on the Thermal Conductivity of UO_2", Battelle Pacific Northwest Laboratory, BNWL-2270, Document prepared for the NRC.

[13]   P.G. Lucuta, H.J. Matzke, I.J. Hastings, "A Pragmatic Approach to Modelling Thermal Conductivity of Irradiated UO_2 Fuel, Review and Recommendations, *Journal of Nuclear Materials*, 232 (1996), pp. 166-180.

[14]   W. Wiesenack, "Assessment of UO_2 Conductivity Degradation Based on In-Pile Temperature Data", OECD Halden Reactor Project – Report HWR 469, 1996.

Figure 3a. A + BT curves before recovering

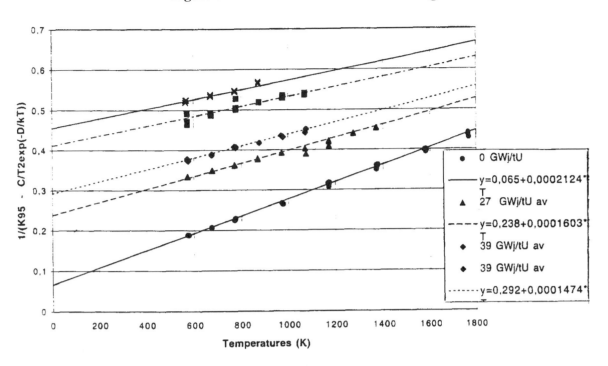

Figure 3b. A + BT curves after recovering

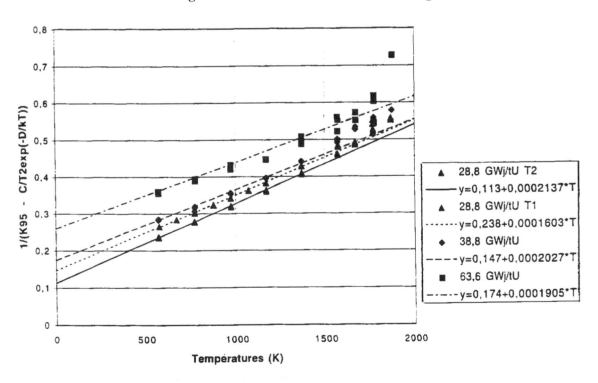

Figure 4a. Variation of the A coefficient with burn-up

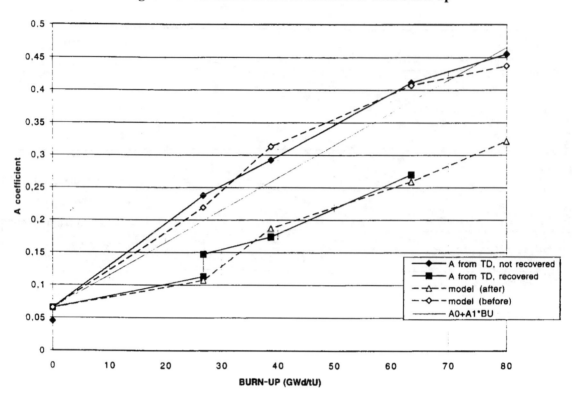

Figure 4b. Variation of the B coefficient with burn-up

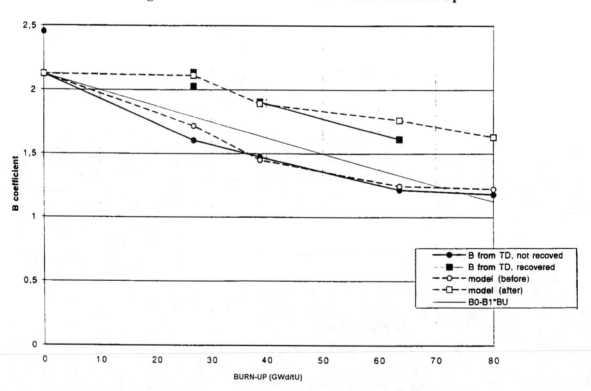

140

Figure 5a. UO$_2$ fuel thermal conductivity at 26.8 MWD/KgU and reduced at 95% TD

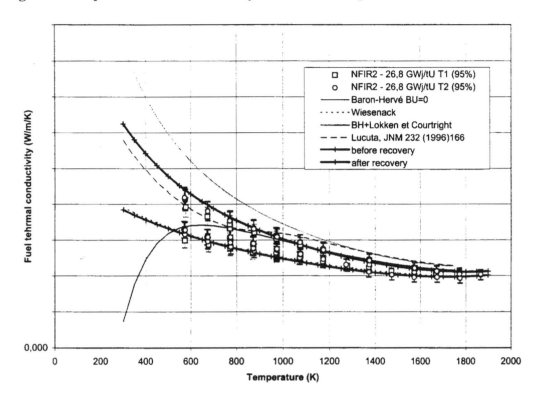

Figure 5b. UO$_2$ fuel thermal conductivity at 38.8 MWd/KgU and reduced at 95% TD

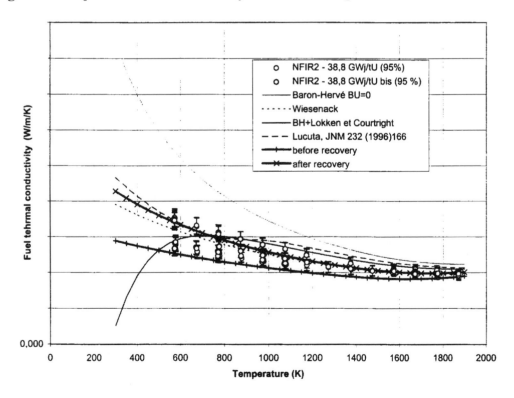

Figure 5c. UO$_2$ fuel thermal conductivity at 63.6 MWd/KgU and reduced at 95% TD

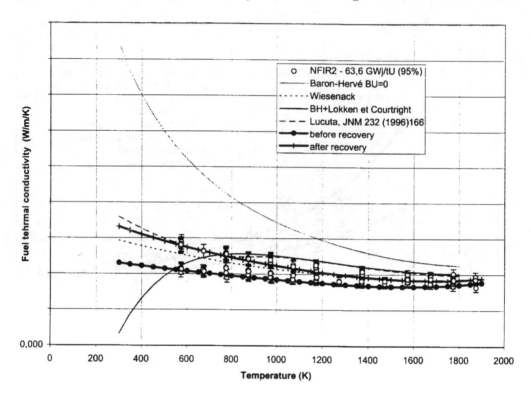

Figure 5d. UO$_2$ fuel thermal conductivity at 80 MWd/KgU and reduced at 95% TD

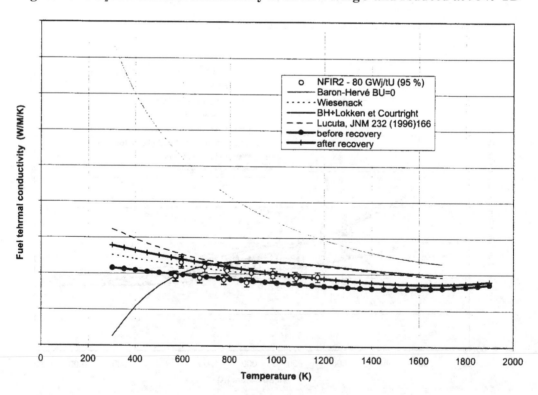

Figure 6

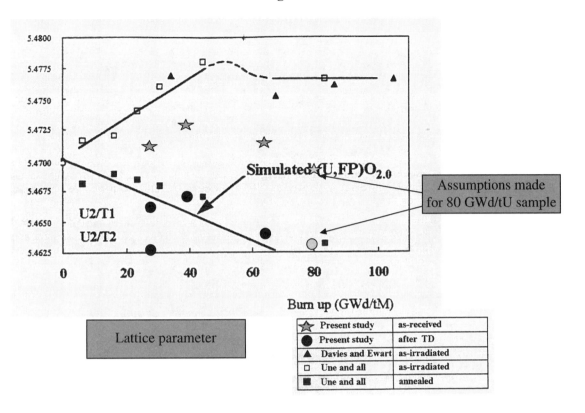

Burn up (GWd/tM)

Lattice parameter

	Present study	as-received
●	Present study	after TD
▲	Davies and Ewart	as-irradiated
□	Une and all	as-irradiated
■	Une and all	annealed

THERMAL PROPERTIES OF HETEROGENEOUS FUELS

D. Staicu[1,2], M. Beauvy[1], M. Laurent[2], M. Lostie[2]
[1]Commissariat à l'Energie Atomique, Direction des Réacteurs Nucléaires,
Département d'Etudes des Combustibles, Centre de Cadarache, 13108 St. Paul Lez Durance Cedex
[2]Centre de Thermique de Lyon, INSA de Lyon
Bât. 404, 20 Av. A. Einstein, 69621 Villeurbanne Cedex

Abstract

Fresh or irradiated nuclear fuels are composites or solid solutions more or less heterogeneous, and their thermal conductivities are strongly dependent on the microstructure. The effective thermal conductivities of these heterogeneous solids must be determined for the modelling of the behaviour under irradiation. Different methods (analytical or numerical) published in the literature can be used for the calculation of this effective thermal conductivity. They are analysed and discussed, but finally only few of them are really useful because the assumptions selected are often not compatible with the complex microstructures observed in the fuels. Numerical calculations of the effective thermal conductivity of various fuels based on the microstructure information provided in our laboratory by optical microscopy or electron micro-probe analysis images, have been done for the validation of these methods. The conditions necessary for accurate results on effective thermal conductivity through these numerical calculations will be discussed.

Introduction

The nuclear fuels are found to be heterogeneous or multi-phased when they are observed on a small scale. Their microstructure may be of the solid solution type (in which case the conductivity varies continuously in accordance with the composition), or else it may be of the composite type in the form of a matrix with conductivity k_m containing inclusions of conductivity k_f or in the form of interconnected phases, the phase conductivities will then be denoted k_i. On a large scale they behave as homogeneous materials whose equivalent thermal conductivity k_{eq} can be evaluated using the conductivities and the distributions of the phases, based on the following assumptions:

- if the equivalent conductivity is evaluated over a volume V, there exists a value of V beyond which the value of the conductivity no longer varies if V is increased, this volume is referred to as the "Elementary Representative Volume" (ERV);

- the medium is statistically homogeneous: the statistical distribution of the phases does not depend on the position within the material, the equivalent conductivity is the same irrespective of the position of the ERV;

- the dimensions of the material are large compared with the dimensions of ERV;

- steady-state conditions are used;

- the medium is opaque;

- Fourier's law applies within the ERV and in the entire medium;

- there is no mass transfer.

The different scales which may be considered are shown in Figure 1, m is the microscale, L the mesoscale and M the macroscale. The relationship which they must verify is: $M \gg L \gg m$, this being the notion of separation of scales [1,15].

Figure 1. Notion of separation of scales

| Heterogeneous solid | Elementary Representative Volume | Equivalent medium |

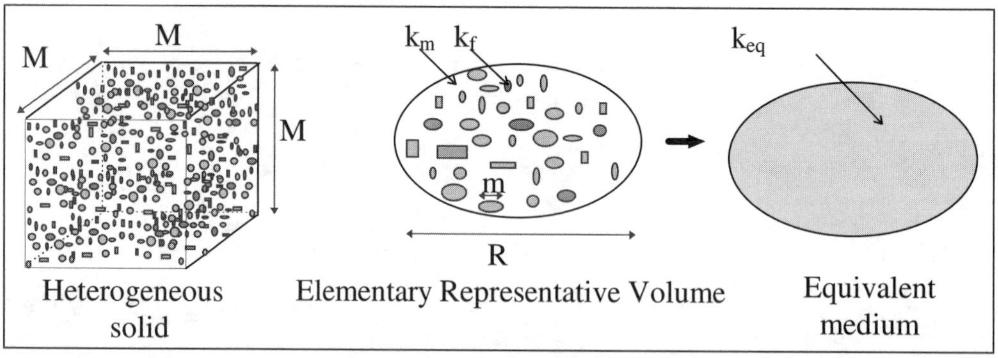

The different methods for calculating the equivalent thermal conductivity as well as their conditions of application are analysed in relation to the nature of the material and the information available concerning its morphology. An example is given of the determination of the equivalent conductivity of a mixed oxide of uranium and plutonium.

Definition of equivalent conductivity

General definition

The equivalent thermal conductivity is defined by the mean temperature gradient within the material $<\nabla T>$ and the mean density of the heat flux within the material $<\varphi>$: $<\nabla T> = \dfrac{1}{V}\int_V \nabla T dV$

and $<\varphi> = \dfrac{1}{V}\int_V \varphi dV$, where the volume V is the ERV, $<\nabla T>$ and $<\varphi>$ are proportional: $<\varphi> = -k_{eq}<\nabla T>$.

Application

In order to evaluate the equivalent thermal conductivity of an heterogeneous solid, it is first necessary to choose a cell representative of the material, of a given shape and where the boundary conditions are such as to allow a thermal flux to circulate. For a parallelepiped shaped cell in Cartesian geometry, one possible choice is to consider a temperature difference between two isothermal faces corresponding to two opposite faces of the material, the other faces being insulated, and then to evaluate the flux Φ circulating between the two faces (Figure 2).

Figure 2. Chosen configuration for evaluating the equivalent conductivity

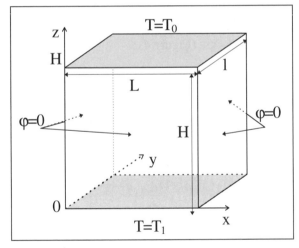

In this cell:

1. The temperature field T(x,y,z) is calculated using the boundary conditions specified in the figure by resolving:
$$\nabla.(k(x,y,z)\nabla T(x,y,z)) = 0$$

2. The total flux passing through the isothermal faces is given by:
$$\Phi = \int_{z=0} \varphi(x,y,z).dx.dy = \int_{z=H} \varphi(x,y,z).dx.dy$$
where: $\varphi(x,y,z) = -k(x,y,z)\nabla T(x,y,z)$

3. The equivalent conductivity along the z axis is:
$$k_{eq} = \frac{\Phi.H}{(T_1 - T_0)L.l}$$

The objective is to determine the intrinsic equivalent conductivity, i.e. which is independent of the size of the cell considered, and of the boundary conditions.

Influence of the size of the cell considered

In the first instance, the calculated equivalent conductivity varies when the size of the cell increases, then it stabilises at a value k_{eq} from a certain value (Figure 3).

Figure 3. Influence of the size of the cell considered on the value of the equivalent conductivity

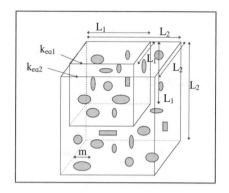 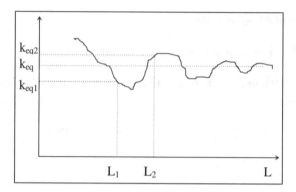

In practice, if m is the maximum dimension of the inclusions, L must be >> m. This condition can sometimes be difficult to satisfy numerically as it would require using too large a cell size. The appropriate value for L can be determined from the graph $k_{eq} = f(L)$ plotted for increasing values of L.

Influence of the boundary conditions

The isothermal surfaces T_0 and T_1, as well as the adiabatic surfaces ($\varphi = 0$) do not exist in the case of an actual transfer (except if the material is periodic and if the flow at the cell boundaries is parallel or perpendicular to the faces of the cell). Consequently, the heat transfer in the vicinity of the boundaries is disturbed. The temperature field calculation must therefore be performed over a domain larger than the ERV and keeping away from these boundaries to evaluate the equivalent thermal conductivity using a sub-domain in order to evaluate $<\nabla T>$ and $<\varphi>$ (Figure 4).

Figure 4. Influence of boundary conditions on the equivalent conductivity value

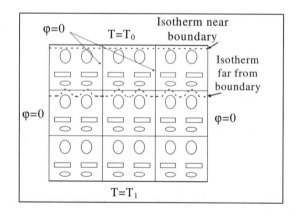 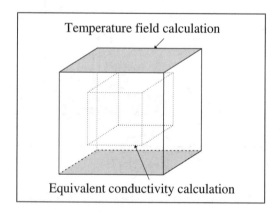

The case of nuclear fuels

The study of the thermal conductivity of the fuel in a reactor involves taking into account two apparently contradictory characteristics relative to the assumptions made for determining the equivalent conductivity: the thermal conductivity variation with temperature and the presence of internal heat sources.

Effect of the conductivity variation with temperature

The thermal conductivity k_{eq}, as previously defined, remains valid if the temperature difference at the cell boundaries used for calculating the equivalent conductivity is small enough for the constituent conductivities not to vary significantly within the cell. However, this equivalent conductivity is dependant on temperature and must therefore be considered as a local value, in the sense that it is not constant throughout the medium, as the temperature is not constant, although it is a homogenised value, i.e. there is no need to distinguish locally between the various constituents (Figure 5). In practice, the notion of separation of scales is sufficient for this condition to be satisfied.

Figure 5. Configuration used for studying the effect of conductivity variation with T

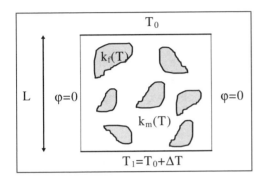

In practice, for k_m and k_f to vary little within the cell, then:

$$L\overline{\nabla T}\frac{\partial k(T)}{\partial T}\bigg|_{T_0} \ll k(T_0)$$

for $k(T)=k_m(T)$ and $k(T)=k_f(T)$

Influence of internal heat sources

When the medium contains heat sources with variable intensity in the space $g_0(x,y,z)$ the conventional formula $\langle \varphi \rangle = -k_{eq}(T_1-T_0)$ is no longer applicable. The calculation with heat sources no longer involves equivalent thermal conductivity, but becomes a calculation of heat transfer in a given configuration. This calculation would be much simpler if it were possible to 'homogenise' the heat sources, i.e. consider that the sources are no longer localised in certain zones of the material, but distributed homogeneously throughout it. In practice, the notion of separation of scales implies that the local fluctuations of the temperature field introduced by the heat sources are of low amplitude so that the temperature field is not significantly different from that which would be obtained with homogenised sources. In fact, the quantity of heat generated within the cell is small by comparison with the quantity of heat flowing through the cell and the heat sources can therefore be homogenised.

Equivalent conductivity calculation methods

Classification of the different approaches

When the morphology and the thermophysical properties of the phases and the interfaces are known, the equivalent thermal conductivity can be directly calculated using a numerical method. As this information is rarely exhaustive, the determination of the equivalent properties for a heterogeneous medium most often requires the use of a certain number of assumptions concerning the morphology, the heat transfer, the positions of the isotherms and the adiabatic surfaces. Additionally, mathematical simplifications are sometimes introduced during the calculation. The results obtained using the various methods published are not therefore all equivalent: their fields of application are dependant on the assumptions made and the simplifications adopted.

Three different types of approach can be distinguished for determining the equivalent thermal conductivity of a medium:

- For perfectly characterised heterogeneous media, the numerical calculation is in principle the best method. Numerical calculations have been performed on various periodic systems [2] and on other systems of known statistical description [3]. This type of approach will be presented in the chapter dedicated to numerical application;

- For well characterised media:

  If the morphology is simple, an exact analytical calculation is possible.

  If the morphology is complex, approximate analytical solutions can be obtained by modelling the geometry or by making calculation assumptions, these two types of approximation often being linked.

- For insufficiently characterised media, for example when the distribution of the constituents is not known, methods are used which provide maximum and minimum values of thermal conductivity or approximation methods are used which involve the use of adjustable parameters.

Exact calculation of the conductivity over a unit cell for well characterised morphology

On a microscopic scale, the material is assumed to be represented by a unit cell which repeats in the three space dimensions.

Exact result for a simple cubic distribution of spheres by direct analytical calculation

McPhedran and McKenzie [4] resolved the heat equation for a cell, this is the "conventional" analytical method. The cell must be limited by a flux tube and two isotherms. The simple morphology chosen is a simple cubic distribution of spheres (Figure 6). There are no other assumptions.

Figure 6. Geometry considered by McKenzie and McPhedran

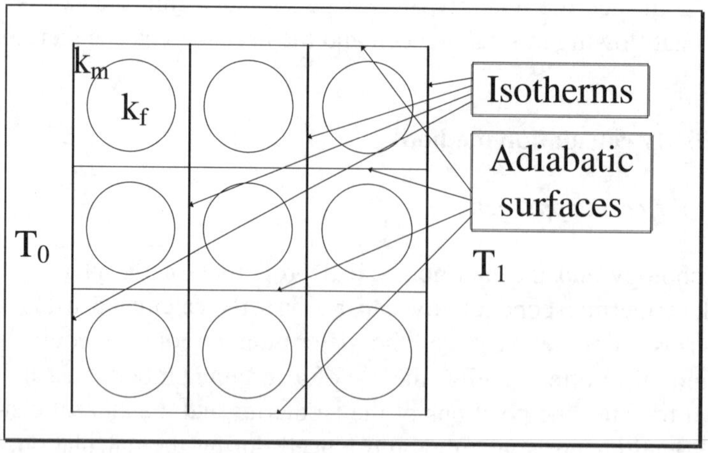

The advantages of this approach are as follows:

- the shape and distribution of the inclusions are taken into account;
- the result is exact for all values of k_m and k_f and for all admissible values of the voluminal fraction of inclusions.

The disadvantages are:

- the results are only available for very simple morphologies;
- the boundary conditions at the edge of the cell must be known;
- the calculations are complex.

In the case of morphologies for which the analytical solution is not known, a numerical calculation may be performed. In this case the calculation will only provide the conductivity for particular conductivity values of the constituents and of the voluminal fraction.

The averaging method and the homogenisation method for more complex morphology

These two approaches can be used to process more general cases than the previous method. They also apply to periodic media but do not demand that the cell be limited by a flux tube and two isotherms (Figure 7). However they often lead to the numerical resolution of an associated problem: this approach does not appear justified for cases where the boundary conditions at the edges of the cell are known, a numerical calculation can then be performed directly in order to resolve the heat equation.

Figure 7. Cells with non-trivial boundary conditions

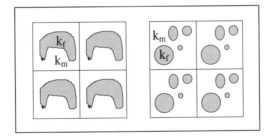

Averaging method

The local heat transfer equations are averaged over a large volume expressing the temperatures as a function of their mean value and of their spatial fluctuation: $T_f = <T_f> + \tilde{T}_f$. This provides expressions involving mean values of the different magnitudes as well as their local fluctuations. The number of equations obtained is less than the number of unknowns, and it is necessary to make some assumptions concerning the shape of the fluctuations in order to contain the problem (for example: temperature fluctuations proportional to the mean temperature gradient). The closure results in a local system whose solution is used for calculating k_{eq} [4]. In the case of a periodic and isotropic medium of period V:

$$k_{eq.} = \frac{1}{V} \int_V \left(k(x, y, z) + k(x, y, z) \frac{\partial \omega}{\partial x_i} \right) dx \quad \text{where } \omega \text{ is the solution to the local problem}$$

$$\frac{\partial}{\partial x_i} \left[k(x, y, z) \frac{\partial \omega}{\partial x_i} \right] = -\frac{\partial k(x, y, z)}{\partial x_i}$$

Homogenisation

The various magnitudes are expressed in the form of an asymptotic development: the temperature is given by the expression $T(x,y) = T_0(x,y) + \varepsilon T_1(x,y) + \varepsilon^2 T_2(x,y) + ... + \varepsilon^n T_n(x,y)$ where ε is a small parameter representing the ratio of the dimension of a period over the dimension of the material, x is the overall co-ordinate, used to describe the macroscopic behaviour of the material and y is the local co-ordinate, used within the microscopic cell. If only the first two terms of the development are considered, the problem is split into two: a macroscopic problem which involves the equivalent conductivity, and a local problem whose solution enables k_{eq} to be calculated [5]. This method gives similar results to those obtained by the averaging method and the comments made previously remain valid.

Methods involving assumptions or approximations for well-characterised morphology

Analytical calculation using approximations: Maxwell's model

The physical significance of the calculation approximations involved in certain models is difficult to ascertain. For instance Maxwell [6] considers spherical inclusions and assumes that there are no interactions, thus the arrangement and the radius of the spheres can therefore be random. The case of an isolated sphere can be resolved exactly (Figure 8) and the result for heterogeneous material consisting of a large number of spheres is obtained by assuming that there is no interaction between the spheres and by summing the effects of each sphere.

Figure 8. Maxwell's model

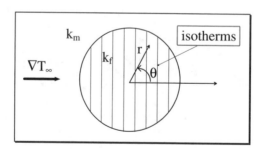

Models based on the assumption of one-dimensional heat transfer

The heat transfer is assumed to be one-dimensional. An equivalent conducting network is constructed and the thermal conductivity of the system is obtained by defining a simplified geometry for the isotherms or for the lines of flux. Two types of assumptions are possible, corresponding to maximum and minimum limit values of the equivalent conductivity (Figures 9 and 10).

Figure 9. Equivalent conducting diagram (series assembly)

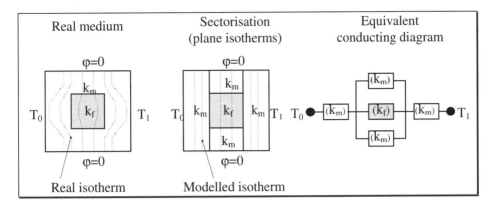

Figure 10. Equivalent conducting diagram (parallel assembly)

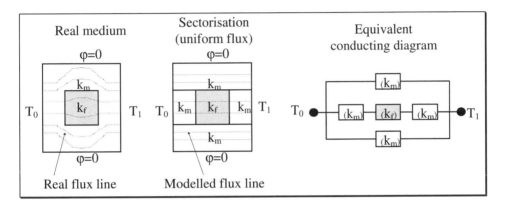

The geometry can be explicitly taken into account. Jefferson *et al.* [7] consider a cubic cell containing a sphere in its centre. The conductivity is evaluated by integration by cutting the medium into very thin slices parallel to the lines of flux and consisting of series mounted resistors, the various slices being assembled in parallel (Figure 11).

Figure 11. Jefferson's model, position of the tubes of flux and equivalent conducting diagram

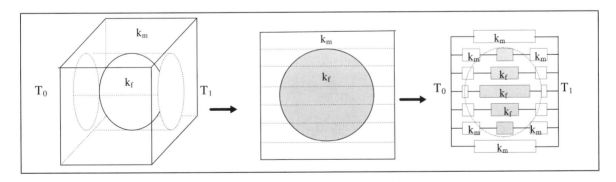

The advantages of this method are the ease of calculation, irrespective of the geometry and the clarity of the assumptions. Whereas many models are limited to two constituents (particles in a matrix), this method can take into account any number of constituents, without any restriction regarding their distribution. It also enables the use of contact resistance. The user can therefore easily adapt it to his particular case. The main limitation of this model is that the assumption of one dimensional heat transfer is not realistic if the values of the conductivities of the constituents are very different. Moreover the way in which the shape and the distribution of the constituents is taken into account is partly fictitious. This can be seen by noting that once the cube has been sliced into planar isotherms or into uniform tubes of flux, the slices can be switched around at will in order to achieve very different geometries from the actual geometry, without affecting the final result.

Non- or little-characterised media

Models involving an adjustable parameter for media whose morphology is not known

Certain models make use of an arbitrary parameter, which integrates the effect of the morphology. The value of this parameter must be evaluated experimentally. It is most often of no value to search for a very fine model and the assumption of one-dimensional heat transfer can in some cases be adequate (Figure 12).

Figure 12. Series, parallel and series/parallel models

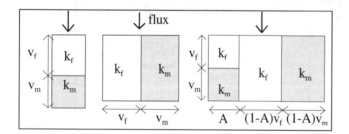

$$k_{ser.} = \frac{1}{\dfrac{v_f}{k_f} + \dfrac{v_m}{k_m}}$$

$$k_{par.} = v_f k_f + v_m k_m$$

$$k_{eq} = A k_{ser.} + (1 - A) k_{par.}$$

It should be noted that this type of parameter can also be introduced into the more elaborate models.

Calculation of limits using the variational method for media with partly characterised morphologies

This method uses a variational formulation of the heat equation and takes into account statistical parameters on the distribution of phases. It is not necessary to have a perfect knowledge of the geometry, as limits can be obtained by making use of partial information such as:

- general information on the material: homogeneity and isotropy;

- information on the shape of the particles;

- information on the layout of the particles;

- information on the relative sizes of the particles.

Hashin and Shtrikman [8] for example obtained limits by considering any homogenous and isotropic medium in which the temperature gradient in each of the grains was uniform. They showed that their limits were the best possible in view of these assumptions, and that these limits are achieved for a material consisting of spheres of the first material coated with a spherical shell of the second material and arranged in such a way that these structures completely fill the space (Figure 13).

Figure 13. Hashin and Shtrickman's composite assembly of spheres

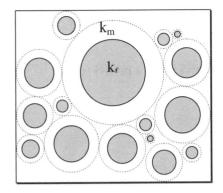

Supposing that $k_1 > k_2$:

The upper limit is achieved if
$$k_m = k_1$$
$$k_f = k_2$$

The lower limit is achieved if
$$k_m = k_2$$
$$k_f = k_1$$

This method provides good results, i.e. tight limits, when the conductivities of the constituents have similar values. In other cases, the limits are not very restrictive.

Influence of contact resistance at the interfaces

When crossing a solid-solid interface, the heat transfer and the temperature field are disturbed. An approximate modelling using thermal contact resistance can be established from the following assumptions:

- The width of the disturbed zone is zero.

- The complex but continuous variation in the temperature profile when crossing this zone is replaced by a temperature discontinuity at the interface. This step in temperature is proportional to the flux density crossing the interface as well as to a parameter called the thermal contact resistance R_c which characterises the overall effect of the imperfect contact (Figure 14).

Figure 14. Lines of flux and temperature field in the vicinity of the contact zone

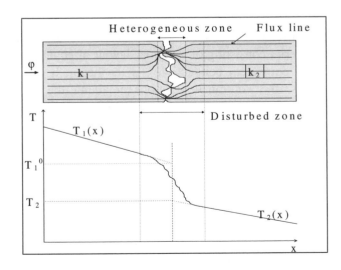

$$R_c = \frac{T_1^0 - T_2^0}{\varphi}$$

Hasselman [9] modified Maxwell's model to take into account the contact resistance at the interface between the particles and the matrix and Lu and Song [10] considered spheres of conductivity k_2 and radius a, surrounded by a shell of conductivity k_3 and radius b, in a matrix of conductivity k_1 (Figure 15).

Figure 15. Hasselman's model (left) and Lu and Song's model (right)

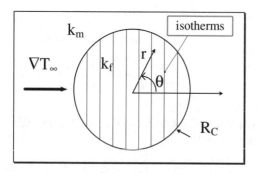 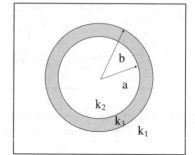

Influence of porosity

Not all the results obtained for thermal conductivity of solids containing inclusions are applicable to porous solids, due to the large difference which exists between the conductivity of the solid and the conductivity of the pore:

- The results obtained using the one-dimensional transfer assumption are not valid, since this assumption is not acceptable.

- The limits obtained using variation methods are insufficiently restrictive.

- The most suitable results are those obtained using an exact calculation over a unit cell.

Influence of aperiodicity

Most of the models consider equal sized particles perfectly aligned whereas in fact such a configuration is rare. We use as a reference a material consisting of a regular layout of equal sized particles, and we wish to know the change in the equivalent conductivity, if the layout becomes random or if the particles are of different sizes, the voluminal fraction remaining the same (Figure 16).

Figure 16. Configurations considered for the study of the influence of heterogeneity

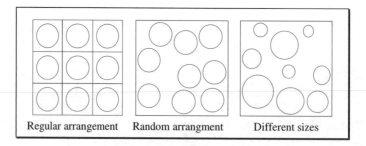

156

The results applicable for this type of study are those which in fact take into account the layout and the size of the particles. A bibliographical study shows that:

- For a conductivity ratio between the inclusions and the matrix close to 1, the thermal conductivity is not significantly affected by the size and distribution of the particles (Kou and Yu [11], Behrens [12], Hashin [13]).

- A random distribution of the particles results in a higher conductivity than a regular or periodic distribution (Gavioli [14], Ke-Da et al. [15], Sareni et al. [16]).

- A medium containing different sized particles also has a higher equivalent conductivity than one containing equal sized particles (Torquato et al. [17], Ke-Da et al. [15]).

Effect of morphology on conductivity

In the absence of contact resistance

The value of the conductivity is not dependant on the magnitude of the heterogeneities, but only on their sizes and their relative layout within the material The three media represented in Figure 17 have the same equivalent conductivity value:

Figure 17. Examples of media having the same equivalent conductivity

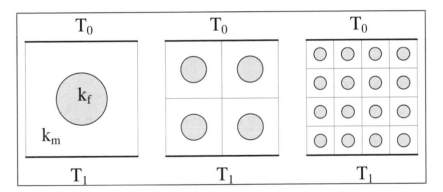

Hashin and Shtrikman [8] have defined the statistically homogeneous and isotropic material for maximum conductivity: the matrix consists of the best conductor and the geometry is the configuration of composite spheres as already described [8]. This result is commonly used for improving the equivalent conductivity of heterogeneous materials [8].

Effect of contact resistance

The size of the inclusions affects the conductivity value: the equivalent conductivity is all the higher if their dimension is large. The approach suggested by Davis [19] takes into account different sized particles in order to determine whether it is best to use particles of the same size or particles of different sizes.

Conclusion on conductivity models used for fuels

- An exact solution corresponding to a complex geometry is difficult to obtain. An approximate solution valid under certain conditions, can be obtained by using an approximation based on one-dimensional heat transfer.

- As far as we are aware, no exact calculation has yet been performed for an aperiodic medium.

The choice of model depends on the quantity of information available concerning the morphology and the thermophysical properties:

- When the morphology is known and periodic:
  - numerical methods, homogenisation and averaging technique (use not justified), or direct analytical calculations can be used if the boundary conditions at the edge of the cell are known;
  - homogenisation and averaging technique can be used if the boundary conditions at the edge of the cell are not known.

- When the morphology is known but aperiodic, the useful approaches are: numerical methods over a large dimension cell, analytical or numerical methods for an idealised morphology and calculation of limits.

- When the morphology is not known, arbitrary parameters or calculation of limits can be used.

This choice is also dependant on the value of k_f/k_m.

Application to mixed oxides of uranium and plutonium

MOX nuclear fuels consist of a mixture of uranium oxide UO_2 and plutonium oxide PuO_2 and take the form of a mixed oxide $(U_{1-y}Pu_y)O_{2-x}$. This heterogeneous solid solution exhibits zones of varying plutonium content. A quantitative microprobe picture of mixed oxide $(U_{0.79}Pu_{0.21})O_{2-x}$ shows this heterogeneity and the distribution of the Pu content present (Figure 18). The picture shows the local content, resolved into 256 values. It also enables modelling of the content distribution function $Fx(\%Pu)$ by the value of its mean and its standard deviation.

Figure 18. Microprobe picture (1 pixel = 0.5 μm) and characterisation of the Pu content

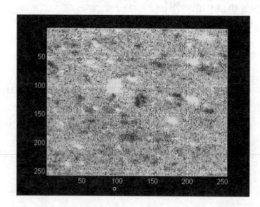
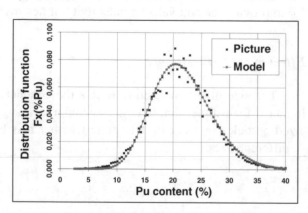

158

Choice of a conductivity law in accordance with the Pu content

Calculation of the equivalent conductivity of the heterogeneous solid solution requires knowledge of the local conductivity value. This value for a homogeneous solid solution is not well established because most of the measurements concern heterogeneous fuels and they already include the overall effect of heterogeneity, and are therefore unusable as soon as the microstructure is modified. As an example for this study, a published law covering the dependence of the thermal conductivity at 1000°C according to the Pu content, for a virgin homogeneous fuel, O/M = 2-x = 1.98 and free from porosities will be used as a basis for estimating the equivalent conductivity [20] (Figure 19). The results obtained will be compared to the values proposed [21] for a heterogeneous fuel when the influence of the plutonium content does not appear.

Figure 19. Variation of the conductivity k$_{homogeneous}$(%Pu) with Pu content at T = 1000°C for a homogeneous sample, 100% dense with a ration O/M = 2-x = 1.98

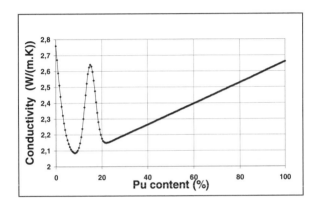

Determination of the equivalent conductivity

The thermal conductivity for the mixed oxide $U_{1-y}Pu_yO_{2-x}$ is not very dependant on the value of y; we therefore elected to calculate, in the first instance, the series and parallel limit values valid for all morphologies. We then calculated the Hashin and Shtrinkman limit values, assuming that the material was macroscopically homogeneous and isotropic. As the composition varies continuously, these limits will be calculated by integration over the different contents.

The series limit is given by: $\dfrac{1}{k_{series}} = \displaystyle\int_{\%Pu=0}^{\%Pu=100} \dfrac{Fx(\%Pu)}{k_{homogeneous}(\%Pu)} d(\%Pu)$ and the parallel limit

is given by: $k_{parallel} = \displaystyle\int_{\%Pu=0}^{\%Pu=100} k_{homogeneous}(\%Pu)Fx(\%Pu)d(\%Pu)$.

Numerical calculation and comparison with the theoretical and experimental values

Table 1 shows that the series and parallel limits are extremely close, they therefore enable a very good approximation of the equivalent conductivity to be estimated. A 2-D numerical calculation will be performed. The chosen resolution method is the finite differences method, as it is the most suitable

Table 1. Conductivities calculated using the theoretical formulae

Content (% Pu)	Standard deviation of the distribution of content (% Pu)	series conductivity W/(m.K)	parallel conductivity W/(m.K)	Hashin & Shtrikman W/(m.K) (2-D transfer)	Hashin & Shtrikman W/(m.K) (2-D transfer)
21.75	5.31	2.2517	2.2613	2.2583	2.2591

for the calculation based on a matrix obtained from a picture. The outline for resolving the heat equation used is that suggested by Ozisik [22] for the case of variable thermal conductivity. This calculation was performed on the picture presented in Figure 18, the mean Pu content was 21.75% and we obtained an equivalent conductivity value of 2.2591 W/(m.K), this value being compatible with the limits presented in Table 1.

Moreover the measured conductivity value [23] is: 2.3 ± 0.1 W/(m.K), this value coincides, within the error of uncertainty, with the values calculated numerically and theoretically.

Finally, the value published for a heterogeneous fuel, without any information on its microstructure, is 2.3 W/(m.K) [21], this value is consistent with the previous set of results.

The conductivity of the mixed non-irradiated oxide $U_{1-y}Pu_yO_{2-x}$ is therefore little dependant on its microstructure, for a given distribution function of Pu content. This is due to the low thermal conductivity variation with plutonium content.

The two-dimensional numerical calculations performed on the pictures lead to an under-estimation of the actual equivalent conductivity, which corresponds to a three-dimensional heat transfer. The numerical calculations must therefore in principle be performed on a 3-D morphology. This was not necessary in our case in view of the small difference between the series and parallel limits and the low error of uncertainty on the resulting equivalent conductivity value.

Conclusions

The choice of an equivalent thermal conductivity model is influenced by the relative values of the conductivities of the different phases and by the quantity of information available concerning the morphology. In the particular case of mixed oxide $U_{1-y}Pu_yO_{2-x}$, as the variation in conductivity with the plutonium content is low, the equivalent conductivity limit values calculated using the series and parallel models provide a good estimation. The morphology therefore has little influence on the value of the equivalent conductivity. As the conductivity variation in a perfectly homogeneous material is not monotonous with the plutonium content, the value is however influenced more significantly by the standard deviation of the plutonium content distribution, the latter can vary for a given overall content.

A study of irradiated materials, in the first instance segregating the different effects, will be performed by generalising the results of this study.

REFERENCES

[1]   Quintard, M., Whitaker, S., *Chem. Eng. Sci.*, Vol. 48, N° 14, pp. 2537-2564.

[2]   Helsing, J., *Phys. Rev. B*, 1991, Vol. 44, p. 11677.

[3]   Schwartz, L.M., Banavar, J.R., *Phys. Rev. B.*, 1989, Vol. 39, p. 11965.

[4]   McKenzie, D.R., McPhedran, R.C., *Nature*, 1977, Vol. 265, pp. 128-129.

[5]   Auriault, L., Royer, P., *Int. J. Heat Mass Transfer*, 1993, Vol. 36, N° 10, pp. 2613-2621.

[6]   Maxwell, J.C., "A Treatise on Electricity and Magnetism", 1, p. 440 (Clarendon, Oxford, 1892).

[7]   Jefferson, T.B., Witzell, O.W., Sibbitt, W.L., *Ind. Engng. Chem.*, 1958, Vol. 50, N° 10, pp. 1589-1592.

[8]   Hashin, Z., Shtrinkman, S., *J. Appl. Phys.*, 1962, Vol. 33, N° 10, p. 3125, 3131.

[9]   Hasselman, D.P.H., *J. Comp. Mat.*, 1986, Vol. 21, pp. 508-515.

[10]   Lu, S.Y., Song, J.L., *J. Appl. Phys.*, 1996, Vol. 79, N° 2, pp. 609-618.

[11]   Kou, H.S., Yu, K.T., *Computers & Structures*, 1994, Vol. 53, N° 3, pp. 569-577.

[12]   Behrens, E., *J. Comp. Mater*, 1968, Vol. 2, N° 1, pp. 2-17.

[13]   Hashin, Z., *J. Appl. Mech.*, 1983, Vol. 50, pp. 481-505.

[14]   Gavioli, A. J., *Math. Models Methods Appl. Sci.*, 1992, Vol. 2, N° 3, pp. 283-294.

[15]   Ke-Da, B., Hui, L., Grimwall, G., *Int. J. Heat Mass Transfer*, 1993, Vol. 36, N° 16, pp. 4033-4038.

[16]   Sareni, B., Krahenbuhl, L., Beroual A. *et al.*, *J. Appl. Phys.*, 1997, Vol. 81, N° 5, pp. 2375-2383.

[17]   Torquato, S., Kim, I.C., Thovezrt, J.F. *et al.*, *J. Appl. Phys.*, 1990, Vol. 67, N° 10, pp. 6088-6098.

[18]   Lurie, K.A., Cherkaev, A.V., *J. Optim. Th. Appl.*, 1985, Vol. 46, N° 4, pp. 571-580.

[19] Davis, L.C., Artz, B.E., *J. Appl. Phys.*, 1995, Vol. 77, N° 10, pp. 4954-4960.

[20] Beauvy, M., *High Temp. - High. Press.*, 1990, Vol. 22, pp. 631-639.

[21] Philliponneau, Y., Personal memo.

[22] Ozisik, N., Heat Conduction, p. 511, Willey, New York (1980).

[23] Duriez, C., Personal memo.

SESSION III

Fuel-Clad Gap Evolution Modelling

Chairs: K. Lassmann, C.E. Beyer

TEMPERATURE CALCULATIONS AND THE EFFECT OF MODELLING THE FUEL MECHANICAL BEHAVIOUR

P. Garcia, C. Struzik, N. Veyrier
Commissariat à l'Énergie Atomique, Centre de Cadarache, France
(DRN/DEC/SDC)

Abstract

The aim of this paper is to review some of the aspects pertaining to the thermal modelling included in the METEOR/TRANSURANUS fuel behaviour code, developed by CEA. An overview of the basic modelling options of the code is given. Experiments for which the fuel centreline temperature is monitored in-pile are interpreted and used to illustrate the range of operating conditions and fuel types which the modelling covers. These include high burn-ups, MOX and gadolinium doped fuels.

Although the expressions used for both the fuel thermal conductivity and the gap conductance are of major importance in accurately predicting the thermal response of a fuel rod, the effect of the fuel and cladding mechanical behaviour is also significant, albeit to a lesser degree.

Its influence is mainly felt through the fuel-cladding gap size and in cases where the gap is closed, through the fuel cladding contact pressure.

The fuel-cladding gap size is itself essentially the result of three phenomena that proceed simultaneously, i.e. fuel and cladding thermal expansion, cladding creep down and pellet fragment relocation. Only the latter is difficult to quantify independently or ascribe to a particular mechanism. This multidimensional effect is modelled in the METEOR/TRANSURANUS code through the use of an additional strain, which is empirically adjusted to fit fuel centreline temperature measurements and post irradiation cladding diameter changes. Three-dimensional calculations are shown to accurately reproduce previously adjusted relocation strains, which could pave the way for a model based on parametric studies.

Introduction

The METEOR/TRANSURANUS fuel performance code models the behaviour of LWR fuel rods [1]. Its current version is validated for UO_2 fuels up to 60 GWd/t and up to 45 GWd/t for MOX fuels in PWR conditions. The essential models aimed at describing the thermal behaviour of gadolinium doped fuels are also available.

The paper gives an overview of the main features pertaining to the thermal modelling included in the code. They are alluded to in detail in part one.

In part two of the paper, the main phenomena that lead to a change in the fuel-cladding gap size are reviewed. A strong emphasis is placed on fuel relocation and associated models. An attempt at elucidating the nature and quantifying the effect of the phenomenon is made through the critical analysis of multi-dimensional parametric calculations and available experimental data.

The last section is devoted to applications that serve to illustrate the breadth of operating conditions and fuel types the code is capable of covering. Applications include a validation against in-pile fuel centreline temperature measurements such as the IFA432 experiment, carried out in the Halden test reactor and the GDGRIFF experiment carried out in the Siloe test reactor. Calculations pertaining to a high burn-up base irradiated fuel rod are also presented to illustrate how the code can be used to quantify the effect on thermal behaviour of a porosity build-up in the peripheral part of the pellet or of the fuel and cladding mechanical model.

Overview of main model

Models pertinent to the thermal behaviour of a fuel rod can be thought of as belonging to one of two distinct groups according to whether or not they interact with the overall fuel rod modelling. The radial power distribution and external cladding corrosion models fit into the latter category whereas the fuel thermal conductivity and the fuel cladding gap heat transfer models fit into the former. In the following paragraphs, we endeavour to analyse these models and an attempt is made at highlighting the main interacting phenomena.

Radial power distribution

In the METEOR/TRANSURANUS fuel performance code, the radial power distribution can be introduced in the data file if it is known sufficiently well or a Radar type model (e.g. [2] or [3]), if applicable, can be used. The relevant cross-sections have been calculated by the reactor physics code APOLLO and fitted as a function of burn-up and initial enrichment. These calculations and fits were performed for UO_2, MOX and gadolinium doped fuels under PWR conditions, to which the model therefore applies.

Thermal conductivity

As regards the fuel thermal conductivity, the Harding-Martin [4] formulations are used for UO_2 type fuels whereas the European recommendations formulated within the framework of the European Fast Breeder Reactor Project are applied to MOX fuels. These formulations are used along with the

Lucuta recommendations for fission product effects [5]. The latest Lucuta recommendation [6], pertaining to the effect of radiation damage is not yet applied by default but has been successfully tested in the case of the high burn-up thermal experiment Extrafort carried out in the OSIRIS test reactor [7] (CEA/Saclay).

Corrective terms are added to account for the effect of cracks, deviation from stoichiometry, gadolinium in solid solution (for Gd doped fuels) and porosity. Porosity can either result from fission gas precipitation, including that produced in the RIM area, or have been left over from manufacturing. This gives an indication of how the thermal response of the code can interact with other models.

Heat transfer model through the fuel-cladding gap

The heat transfer model is very similar to the TRANSURANUS URGAP model [8], at least in formulation. The heat flow across the fuel cladding gap is modelled as the sum of the following three terms:

- The radiative transfer term which is always relatively low.

- The heat conduction term through the filling gas which is a function of gas conductivity, gap size and oxide and cladding conductivity. This term includes a solid(fuel or cladding)-gas heat transfer term along with a contribution relative to the transfer across the gas itself.

- The solid-solid contact term which is a function of pressure contact and surface roughness.

Two cases can then be considered:

- *Case 1*: *Open gap or low contact pressure*: In this case the heat transfer is strongly dependant on the gap size and the nature of the filling gas. For large gaps (above 10 microns), gaseous transfer predominates.

- *Case 2*: *Closed gap with high contact pressure*: The contribution of the contact term becomes substantial and can even become predominant if the filling gas is polluted by rare gas release, in which case it is essential that the contact pressure be adequately computed. However, the heat transfer coefficient is rather good whether the gap be filled with helium or xenon and available experimental data point to the fact that the influence of the gas composition is of the order of magnitude of the power uncertainty.

Phenomena affecting gap-size and gap closure kinetics

In PWR fuel rods, many phenomena such as fuel and cladding thermal expansion, cladding creep down, fuel swelling and densification contribute to varying degrees to closing the fuel cladding gap. The aim here is to define and quantify when possible the contribution to gap closure of each of these phenomena. As regards fuel and cladding thermal expansion or cladding creep down, accurate out-of-pile or in-pile measurements are available thus eliminating any uncertainty in assessing the associated fuel rod dimensional changes. However, modelling only the above mentioned phenomena leads to an over estimation of time to closure. This naturally introduces fuel relocation.

Fuel relocation-experimental evidence

Experimental evidence of fuel relocation is essentially indirect. In-pile experiments equipped with centreline temperature monitoring thermocouples help assess the amplitude of fuel relocation through the analysis performed using a fuel behaviour code. Other experimental evidence includes cladding diameter measurements that reveal primary ridging after a two reactor cycle base irradiation. The RECOR experiment [9], designed to quantify the effect of fuel relocation is possibly the closest one can get to direct experimental evidence of the phenomenon, although it is observed through in-pile cladding diameter measurements that obviously do not make it possible to dissociate cladding from pellet behaviour.

Fuel relocation covers a range of effects related to fuel cracking. As the linear power of a fuel rod reaches around 50 W/cm, the circumferential thermal stresses generated in the pellet periphery are of the order of magnitude of the material fracture stress and therefore are sufficient to induce radial pellet cracking and stress relief. As the power is increased a typical network of cracks appears for base irradiated rods. The overall result is that the apparent volume occupied by the fuel pellet is greater than that of the un-cracked pellet at the same temperature. Roughly speaking, for a standard PWR fuel rod and a linear power of approximately 200 W/cm, fuel thermal expansion, cladding creep down and pellet fragment relocation contribute to gap closure in similar proportions.

An approach to fuel relocation modelling

Existing models

The model available in the METEOR/TRANSURANUS fuel performance code is based on the GAPCON model [10]. The extent to which the fuel fragments relocate outwards is a function of the linear power, the initial gap size and the pellet radius and has been adjusted to fit fuel centreline temperature measurements. For sake of completeness, a word must also be said of pellet repositioning. The phenomenon which is thought to occur once the fuel cladding gap is closed is associated with the propensity of pellet fragments to return, to some extent, to their initial position as the contact pressure between fuel and cladding increases.

The common denominator to all relocation models is that they are essentially empirical. Although there is nothing wrong with that in principle, and these models serve their engineering purpose, the danger is of misunderstanding the underlying mechanisms. In the following paragraphs we show how a multi-dimensional finite element model can be used to shed light on these mechanisms.

3-D calculations

We describe here the finite element model used to determine gap size changes in relation to pellet relocation. The TOUTATIS code [11], within the framework of which the model is implemented is developed at CEA/Saclay. It is a fuel performance application which uses the CASTEM 2000 finite element package.

Hypotheses

The model describes thermo-mechanical effects on the scale of a fuel pellet. The latter is represented here as a radially cracked structure containing eight pre-existing, equally spaced fuel fragments. No axial cracks are assumed to appear in the course of the simulation. Both these hypotheses will be discussed later. Because of the symmetry of the problem, only one thirty second of the actual pellet is modelled. The mid-pellet plane along with the plane bisecting the fragment are planes of symmetry which for the latter implies that the displacement in the tangential direction is nought. Unilateral contact between fuel fragments and adjacent pellets is also assumed. There is no displacement condition on the points lying along the axis of symmetry.

Power histories consist of a fast rise to 400 W/cm in 50 W/cm increments and are the same for all cases. The initial gap size was chosen so that it does not close during the simulation. Several calculations pertaining to different pellet sizes (varying inner and outer fuel radii) were performed. Table 1 lists the cases run.

Table 1

	Case 1	Case 2	Case 2
Outer radius (mm)	5.335	5.335	4.1
Inner radius (mm)	0.9	0	0
Pellet height (mm)	13	13	13

Results and discussion

Figure 1 shows a scaled representation of the displacement field. It clearly shows how the pellet fragments deform. The result is that the displacement of the outer surface of the pellet exceeds that of an infinite expanding circular cylinder at an identical linear power.

- The consequence of wheatsheafing is an axial variation in the gap size and hence temperature. At 300 W/cm for Case 1, the displacement difference between a point lying in the inter pellet plane on the outer surface and a corresponding point in the mid pellet plane is approximately 30 μm. This corresponds to an axial temperature variation of approximately 40 degrees (for a 1 bar helium gap).

- There is no evidence of any temperature or displacement variation with angular position.

- The main calculation results are presented in Figure 2. It shows the displacement of a point on the outer surface of the fuel is linearly dependent on the temperature drop across the pellet albeit with distinct slopes according to the axial position of the point. This could possibly explain the power and gap size dependence (a reflection of the temperature drop across the pellet) in the original GAPCON model. Figure 2 also shows the axially averaged displacement as a function of the axially averaged temperature drop across the pellet which is useful when extrapolating to a simpler mono dimensional situation. The inner fuel radius only marginally affects the slope of the line.

Figure 1. Radial displacement field at 200 W/cm

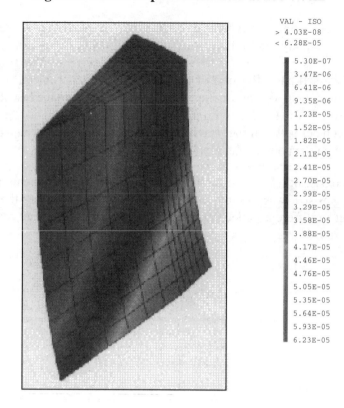

```
VAL - ISO
> 4.03E-08
< 6.28E-05

5.30E-07
3.47E-06
6.41E-06
9.35E-06
1.23E-05
1.52E-05
1.82E-05
2.11E-05
2.41E-05
2.70E-05
2.99E-05
3.29E-05
3.58E-05
3.88E-05
4.17E-05
4.46E-05
4.76E-05
5.05E-05
5.35E-05
5.64E-05
5.93E-05
6.23E-05
```

Figure 2. Case 1 and 2: Initial pellet radius 5.335 mm

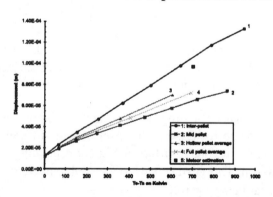

Figure 3. Case 3: Initial pellet radius 4.1 mm

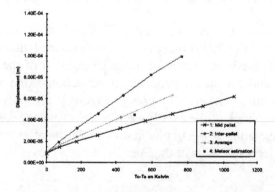

- The behaviour corresponding to a typical PWR fuel rod appears in Figure 3. Comparison with Figure 2 shows the effect the pellet radius has on fuel relocation. Possibly the most important lesson to be drawn from these figures is that the axially averaged three-dimensional displacement is very close to the mono dimensional displacement which includes both simple fuel expansion and fuel relocation.

It is interesting to note that the circumferential stresses go through a maximum along the plane bisecting a fragment but lie below the reported value for fuel fracture stresses, which justifies a posteriori the number of fragments initially surmised. At 300 W/cm, however, the axial stress lying in the mid pellet plane is of the same order of magnitude as the fuel fracture stress which suggests that the pellet would naturally tend to crack along that plane. The effect of such cracking should be the object of future investigations. We do not believe however that considering axial cracks would significantly alter the conclusions of this study.

Pellet repositioning has not been included in this study and is an area we intend to focus on in the near future. The aim would be to investigate the relationship between fragment repositioning and fuel creep.

The effect of axial cracks is still open to discussion. There also seems to be scope for developing models such as those developed specifically for modelling the behaviour of concrete structures. These could apply to oxide fuels and prove practical in so far as they do not require pre-defining the location of cracks in the pellet.

Model applications

Relocation, gap size and fission gas release

The IFA432 experiment carried out in the Halden test reactor qualifies the overall thermal response of the code, and in particular the fuel cladding heat transfer model and gap size computation; it also serves to illustrate the extent of thermal feed-back due to fission gas release. The input data have been taken from the OECD database [12].

Figure 4 highlights the thermal feedback effect. Since the initial helium pressure is only 1 bar, xenon pollution is immediate. Opposite the thermocouple, the gap remains open more or less throughout the irradiation. At the beginning of life, fragment relocation is the main phenomenon affecting the thermal behaviour.

Rim effect

A high burn-up base irradiated fuel rod illustrates the thermal effect induced by the build-up of porosity in the peripheral part of the pellet. The rod was irradiated in the BR3 reactor for four cycles and reached a maximum axial burn-up of 83 GWd/tU. The average power level does not exceed 250 W/cm and reaches 150 W/cm at the end of life. The gas release is rather low and does not affect the thermal response.

Figure 4. IFA432

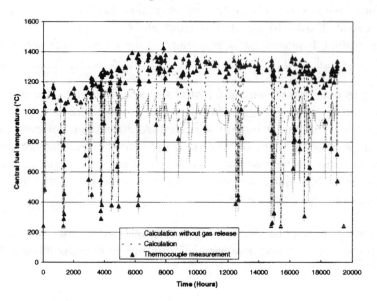

SEM examinations reveal a xenon depletion zone approximately 1 mm thick at the pellet periphery. The rim model [13] computes the porosity in the external zone of the pellet and the resulting fuel thermal conductivity degradation induces a fuel temperature increase. This increase is presented in Figure 5 where the radial temperature distribution is plotted at the maximum flux level, at the end of life.

Figure 5. Radial temperature distribution at 83 GWd/tm with and without the "rim" model

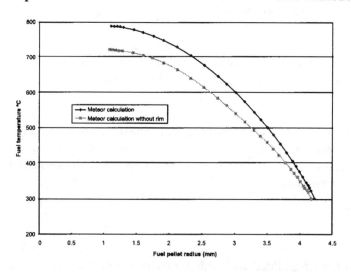

Mechanical modelling

Originally, the TRANURANUS mechanical model assumed visco-elastic hypotheses for the fuel. A new model [14] has been developed with a more comprehensive approach to fuel cracking; as the stress computation is improved it becomes possible to compute fuel creep.

The GDGRIFF experiment for a gadolinium doped fuel was carried out in the Siloe test reactor. This experiment shows how the mechanical model can affect the thermal response of the code. At high temperature, fuel creep induces gap and central hole closures; the effect on fuel centreline temperatures can be seen in Figure 6.

Figure 6. GDGRIF_02: Fuel central temperature

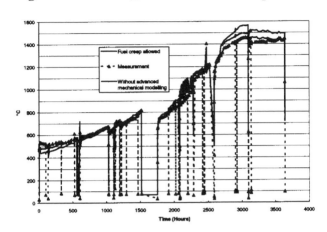

Conclusion

As regards fuel relocation, a three-dimensional model yields displacements that are close in magnitude to those associated with mono-dimensional empirical relocation models. This would not only tend to argue in favour of the standpoint whereby relocation models are palliatives for three-dimensional deformation mechanisms but also argue against the frequently expressed idea that relocation is stochastic in nature, hence unpredictable.

The corollary is that the way is paved for the development of a relocation model based on three-dimensional calculations. These calculations can be regarded as part of a more general methodology aimed at eliminating uncertainties related to fuel relocation when switching to an unfamiliar fuel rod design.

Although the models pertaining to the heat transfer across the gap and the fuel conductivity are of primary importance, other models are seen to significantly influence the thermal response of the code. This paper illustrates the extent to which the fission gas release model, the RIM porosity or fission gas porosity calculations and the mechanical behaviour model also affect this response.

Acknowledgements

The authors wish to express their gratitude for the technical support of F. Bentejac and N. Hourdequin (CEA/Saclay) who, through frequent and useful discussions helped elaborate the paper.

REFERENCES

[1]   Struzik, Moyne, Piron, "High Burn-up Modelling of UO$_2$ and MOX fuel with METEOR/TRANSURANUS Version 1.5", ANS Portland, March 1997.

[2]   I. Palmer *et al.*, "A Model for Predicting the Radial Power Profile in a Fuel Pin", AIEA Specialist Meeting on Water Reactor Fuel Element Performance Computer Modelling, Preston 1992.

[3]   Eric Fédérici, P. Blanpain, P. Permezel, "A Model for the Description of Pu Agglomerates in MOX fuels", IAEA Technical Committee Meeting, Bowness on Winderemere, UK, Sept. 1994.

[4]   Harding and Martin, "A Recommendation for the Thermal Conductivity of UO$_2$", JNM 166, 223-226 (1989).

[5]   Lucuta, Matzke, Verall, "Modelling of UO$_2$ Based SIMFUEL Thermal Conductivity: The Effect of Burn-up", JNM 217, 279-286 (1994).

[6]   Lucuta, Matzke, Hasting, "A Pragmatic Approach to Modelling Thermal Conductivity of UO$_2$ Fuel: Review and Recommendation", JNM 232, 166-180 (1996).

[7]   Bourreau S., this meeting, "Temperature Measurements in High Burn-up UO$_2$ Fuel: Extrafort Experiment".

[8]   K. Lassman, "The Revised Urgap Model to Describe the Gap Conductance Between Fuel and Cladding", *Nuclear Engineering and Design*, 103, 215-221 (1987).

[9]   G. Delette, this meeting, "Phenomena Affecting Mechanical and Thermal Behaviour of PWR Fuel Rods: Some Recent Experiment and Modelling Results".

[10]   Lanning, Mohr, Pannisko, Stewart, "Gapcon Thermal-3 Code Description," PNL 2434/NRC-1,3.

[11]   Brochard, Bentejac, Hourdequin, "Non-Linear Finite Element Studies of the Pellet Cladding Mechanical Interaction in a PWR Fuel", SMIRT 14, France, 17-22 August 1997.

[12]   Chantoin, Sartori, Turnbull, "The Compilation of a Public Domain Database on Nuclear Fuel Performance for the Purpose of Code Development and Validation", ANS Portland March 97.

[13]   B. Hermitte, D. Baron, J.P. Piron, "An Attempt to Simulate the Porosity Build-up in the Rim at High Burn-up," AIEA Technical Committee Meeting, Toranomo Pastral, Tokyo, Japan 26 October-1 November 1996.

[14]   Garcia, Moyne, "Modelling the Steady State and Transient Mechanical Behaviour of Fuel Rods," SMIRT 14, France ,17-22 August 1997.

PHENOMENA AFFECTING MECHANICAL AND THERMAL BEHAVIOUR OF PWR FUEL RODS AT BEGINNING OF LIFE: SOME RECENT EXPERIMENTS AND MODELLING RESULTS

Laurent Caillot, Gérard Delette
Commissariat à l'Énergie Atomique
DRN/SECC Grenoble – France

Abstract

The thermal and mechanical behaviour of fuel rods is affected by pellet fragmentation and relocation effects, especially at beginning of life due to a subsequent decrease in the gap size. These phenomena are quantified here through a comparative analysis of two experimental programmes conducted at the Commissariat à l'Énergie Atomique (CEA), with the co-operation of FRAMATOME Nuclear Fuels and Électricité de France (EDF). The discussion concerns the modelling of gap heat transfer and pellet clad mechanical interaction considering relocation effects.

Introduction

Fuel fragmentation and relocation of pellet fragments occur in a PWR fuel rod in normal operating conditions and have to be quantified for the purpose of modelling. Since fuel cracking is a result of thermal stresses, it depends on operating conditions. The number of pellet fragments that can be observed after irradiation on fuel macrographs increases with the maximum power level reached in operation (Figure 1). Furthermore the fragments are liable to take new positions within the fuel rod when they are no longer confined by the cladding. These displacements may be caused by mutual interactions between fragments occurring during thermal expansion and swelling stages. These causes may be coupled to cladding ovalisation or external effects such as vibration in the core or rod handling after shutdown. The greater apparent volume filled by the pellet due to cracks opening and fragment displacement compared to that of the solid pellet (no fragments) is globally described as relocation.

Two effects of relocation have to be considered:

- due to the decrease of the pellet clad gap the thermal transfer is enhanced;

- as a result of disorientations of fragments and mutual locking, the mechanical interaction with the clad can increase.

The quantification of relocation phenomena is necessary in order:

- to analyse with good accuracy the thermal behaviour of fuel rod elements at beginning of life (BOL);

- to describe the pellet clad contact especially in a context of reactor manoeuvrability studies.

A comparative analysis of two experimental programmes, GRIMOX 2 and RECOR, conducted at the CEA in co-operation with the EDF and FRAMATOME gives a basis for modelling the fuel relocation. Both programmes were presented separately in earlier ANS and SMIRT international meetings [1,2].

The GRIMOX 2 programme

The GRIMOX 2 programme aimed toward characterising the thermal behaviour and the fission gas release of MOX fuel at BOL. For that purpose it included the monitoring of the centreline temperature of two fresh UO_2 and MOX fuel rods during an irradiation up to 4.5 GWd/t_m. A set of complementary characterisations (power radial distribution, thermal diffusivity, density) was also performed with the aim of doing a fine thermal analysis.

GRIMOX 2 experimental conditions

A short fuel rod of 17×17 PWR diametrical geometry, including 16 MIMAS AUC MOX pellets (Pu/Metal = 7.8%) and five dry route UO_2 pellets, was irradiated in the SILOE reactor in Grenoble. The device used, GRIFFON, consists of a boiling capsule pressurised to 13 MPa placed at the edge of the SILOE reactor core. When nucleate boiling occurs (for powers greater than 230 W/cm) this capsule reproduces the standard PWR thermal and neutron conditions.

A final burn-up of 4.5 GWd/t_m was reached at a mean nominal power of 245 W/cm in the MOX and 230 W/cm in the UO_2. The centreline temperature of the two fuels was measured continuously at nominal power and at intermediate power levels ranging from 140 to 380 W/cm.

In order to analyse the thermal behaviour of both fuels, a very accurate determination of the irradiation power is essential. Due to the thermohydraulic characteristics of GRIFFON, this determination can not be achieved by a thermal balance. That is why nuclear methods are used:

- for experimental control, the power is determined on-line by computation, based on the thermal neutron flux measured by rhodium self-powered detectors;

- after irradiation, the value determined on-line is cross-checked by quantitative gamma spectrometry, specifically at the elevation of the thermocouples.

Thermal analysis

The GRIMOX 2 experiment gave evidence of a 6% higher centreline temperature in the MOX fuel compared to UO_2 for identical operating power levels (Figure 2). This global result must be analysed finely considering the different contributions to the thermal response. Most of them were directly characterised by out-of-pile measurements in order to minimise uncertainties associated with models.

Fuel thermal conductivity integral

Thermal conductivity of UO_2 and MOX fuels is given by thermal diffusivity measurement (laser flash method) performed on pellets coming from the same batch as those used in the experiment. For these fuels, the results are in good agreement with previous recommendations for UO_2 and $(U,Pu)O_2$. Porosity was quantified before and after irradiation by hydrostatic density measurement.

The radial power distribution which depends on fuel composition and neutron flux spectrum (thermalisation is higher in SILOE than in PWR) was measured by gamma spectrometry on a polished cross-section of each fuel. The non-mobile fission product ^{95}Zr was scanned with a step of 0.25 mm (Figure 3). The ratio $P_{centre}/P_{external}$ was 0.66 for MOX and 0.80 for UO_2. This examination give also some evidence of caesium migration (trough ^{137}Cs radial scanning) at the high temperature reached in the MOX fuel.

Gap heat transfer

The initial geometry of the cladding material and that of each pellet was perfectly characterised by fine metrology performed during the rod fabrication. It means that the initial gap size was determined with an accuracy better than 5 μm.

Elsewhere the in-pile redensification is well evaluated using density measurements performed before and after irradiation, and the kinetic law was deduced from the DENSIMOX programme [1]. In this programme, periodic measurements on neutronographs of the shortening of a MIMAS AUC MOX fuel stack operating at the same power as in GRIMOX 2 (225 W/cm) showed that the maximum redensification was reached at 2.5 GWd.t_m^{-1}.

This set of data was used to perform a complete calculation of the GRIMOX 2 irradiation with the METEOR/TRANSURANUS code developed in the CEA. This code allows to compute a 1-D thermo-mechanical and physical analysis of a fuel rod [3].

As shown on Figure 4a for UO_2 fuel and Figure 4b for MOX fuel, the code over-predicts the temperature by 80°C if relocation is not taken into account. This over-prediction is observed from the first rise in power, for UO_2 as well as for MOX fuel.

As the other main contributions to the thermal response were very well characterised in GRIMOX 2, this comparison between calculations and measurements gives a good indication of the amplitude of the relocation effects on the gap width reduction and the subsequent thermal conductance enhancement. The calculation performed with the relocation model developed by the CEA, describing an instantaneous contribution of relocation during the first rise in power and relocation accommodation, confirmed this conclusion, as shown in Figure 4, which displays good agreement with experimental points.

Beside these calculations, an indication of the relocation effect is given by considering the difference in the thermal response of the fuel:

- during the first rise in power, where fragmentation and relocation is assumed to begin;

- after the first rise in power, where the enhancement of the gap thermal transfer is obtained.

This point is illustrated by Figure 5, which shows that the relationship between $(T_c - T_{ext})$, the difference between the measured centreline fuel temperature and the external clad temperature, and the linear power is slightly changed just after the first rise in power.

The RECOR programme

The RECOR programme was designed to better characterise the impact of pellet fragmentation and relocation on the behaviour of a fuel rod submitted to grid regulation operation. The Pellet Clad Mechanical Interaction liable to occur in this type of operation depends on the power level and the power history:

- At intermediate power levels, if the pellet-cladding gap widens as a result of differential thermal expansion, the cladding contracts by creep under the external pressure exerted by the coolant on one hand, and, on the other hand, fuel fragments can move and occupy a greater volume (relocation). These two specific phenomena lead to a reduction in pellet-cladding gap and, consequently, a greater degree of interaction on returning to higher power operation.

- Conversely, after a power rise and as soon as the pellet-cladding gap closes, the cladding relieves the stresses induced by the deformation imposed by the fuel, and any relocated pellet fragment may move inward under the effect of the contact pressure; this process is called relocation accommodation.

RECOR experimental conditions

The RECOR experiment consisted in the irradiation of an experimental fuel rod in the DECOR device for *in situ* cladding diameter measurement [4]. This means that relocation effects were studied through their effects on cladding strain.

The DECOR device is used to make profile diameter recordings of a PWR fuel rodlet (total length = 360 mm) along a constant pair of diameter lines with a measurement step of 0.5 mm. DECOR is installed in the GRIFFON boiling capsule. The technology used for diameter measurement is based on strain gauges. In order to obtain a high accuracy with regard to the phenomena studied and to eliminate strain gauge drifts, the measurement system is calibrated for each profile recording. The measurement accuracy is ±1 μm relative (between two profile recordings) and ±3 μm absolute. In order to analyse and understand the fragmentation-related phenomena, the RECOR experiment was performed on a fresh fuel rod of perfectly known characteristics and dimensions.

The size of the pellet-cladding gap was chosen to simulate that of a fuel rod with a burn-up of 10 to 20 $GWd.t_u^{-1}$. This is the burn-up level at which fragmentation effects are greatest; the gap is then sufficiently wide to allow the relative displacement of fragments at low power. On the other hand, these gap dimensions promote pellet-cladding interaction during power rises. In addition, the power history was designed to accentuate the phenomena studied:

- periodic returns to low power levels of variable duration in order to open up the gap again and thus allow relocation of fragments;

- increasingly high power holding periods to quantify the low-power induced relocation when the gap closes, and to characterise the relocation accommodation kinetics of fragments during the high-power holding periods.

Complete profile recordings of the fuel rod were made at different power levels with a frequency suited to the kinetics of the phenomena studied. The main results of the RECOR experiment are presented in Figures 6 and 7 which show different profile recordings and the time-dependent changes in mean cladding diameter (calculated on each profile recording) at inter-pellet and mid-pellet positions.

Relocation analysis

Figure 6 illustrates the possible effects of fragmentation in low-power operating conditions. The profile recording *14* was obtained with a power of 40 W/cm while the fuel rod had not been subjected to a higher power level. Considerable deformation of the cladding is visible, indicative of pellet-cladding contact, whereas the gap at this power level is close to the initial gap; this illustrates relocation of the fragments. The next profile recording (*16*) obtained at 200 W/cm shows a cladding diameter less than that measured at 40 W/cm, typical of relocation accommodation. After a short plateau at 200 W/cm, the cladding diameter returns more or less to its original value, at 40 W/cm. These observations give rise to the following interpretation:

a) Fragmentation is initiated between 0 and 40 W/cm since the effects of relocation are observed. This value is consistent with the threshold calculated by Oguma [5]. Theoretically, at this power level, the pellet is axially divided into two identical fragments. In practice, it may be assumed that two to four fragments are created.

b) Given the size of the pellet-cladding gap at low power, the relocation produced with a few fragments is accompanied by relatively wider cracks, thereby allowing disorientation of the fragments with respect to their ideal positions. This configuration may induce mutual locking of fragments. Repetition of this process might explain the increasing deformation of the cladding on profile recordings *11* to *15*.

c) The number of fragments has a direct influence on relocation accommodation in as much that, during the first power rise to 200 W/cm, the thermal expansion alone would have led to a 30 µm increase in cladding strain, whereas on the contrary, a reduction is observed at the short plateau at 200 W/cm. This change can be linked to pellet fracturing into ten or so slight disoriented fragments (Figure 1) which, under the effect of the contact pressure (thermal expansion) are rapidly accommodated. This point is confirmed on returning to 40 W/cm, since no particular cladding strain is observed.

Relocation is therefore equivalent to, on one hand, an overall radial displacement of fragments and, on the other hand, to the occurrence of inter-fragment locking points, depending on the relative orientation of the fragments concerned. It is this latter factor which restricts the possibilities for relocation accommodation.

It should be pointed out that, while they are certainly of interest for understanding the phenomena, these observations are specific to the RECOR fuel rod. In a power reactor, a situation with simultaneous occurrence of a few fragments (beginning-of-life situation) and an intermediate size pellet-cladding gap (2nd PWR cycle situation) would never arise.

Relocation accommodation analysis

Figure 7 then shows the mean inter-pellet and mid-pellet diameter of RECOR fuel rods during two long plateaus (each lasting two days) at 250 and 300 W/cm interrupted by a return to low power level. At each high-power plateau, the cladding deformation is found to reduce gradually according to an exponential-type kinetic law. These observations can be interpreted by considering the following mechanisms of fragment relocation and subsequent accommodation:

a) The initial deformation at the beginning of the 250 W/cm plateau, which is greater than that induced by thermal expansion, results from the relocation that occurred during the preceding holding period at 40 W/cm. The clear indication of primary ridges (profile recording *19*) shows that there was a marked interaction between the fragmented pellets and the cladding.

b) At the start of the 300 W/cm plateau, no additional relocation would seem to have occurred compared to the end of the 250 W/cm plateau. The difference in diameter between the end of the 250 W/cm plateau and the start of the 300 W/cm plateau corresponds to thermal expansion of the fragmented pellet linked to the 50 W/cm rise in power. The return to low power conditions has not therefore induced any relocation despite reopening of the gap; this observation can be explained by the short duration of the holding period at 40 W/cm (~1 hour).

c) The reduction in diameter during the 250 W/cm plateau is assumed to be due to relocation accommodation of the fragments under the cladding pressure effect, given that it cannot be attributed to other phenomena such as UO_2 densification or end-dish filling by fuel creep. The kinetics gradually decreases with the increase in pellet compactness and with reduction in contact pressure.

d) At the start of the 300 W/cm plateau, the relocation level is the same as at the end of the 250 W/cm plateau (cf. b) although the kinetics is once again faster. This is due to the increase in cladding pressure on the fragments (differential expansion between UO_2 and Zy4).

As shown in Figure 7, obtained by averaging the results of over 19 pellets, despite the random nature of pellet fracturing and fragment displacement, pellet behaviour was found to be generally homogeneous, a feature that was not expected at the beginning of the RECOR experiment. Even so, some pellets exhibited singular behaviour on a local level which reflects the random component of the fragmentation processes.

Conclusions

As a synthesis of GRIMOX 2 and RECOR experimental results it can be stated that:

- relocation effects observed through fuel temperature and diameter measurements seem to occur at the very beginning of the rod life as a consequence of pellet cracking during the first rise in power, the threshold for the first crack is confirmed to be around 40 W/cm;

- relocation of fragments, which depends on the available volume, tends to increase the apparent volume of the pellet with significant consequences on the gap thermal transfer and the pellet clad interaction during power rises;

- in spite of a rather homogeneous average behaviour of pellets, random aspects such as disorientation and mutual locking of fragments can locally cause some scatter with potential incidence on the thermal response;

- the kinetics of rearrangement which depends on the external contact pressure exerted by the cladding on the fragments, is also a function of the cracked pellet compactness.

These results which were used to develop, in the frame of the CEA METEOR-TRANSURANUS, an empirical model describing the evolution of the cracked pellet code form an interesting base for the qualification of codes and more mechanistic models dealing with the thermal and mechanical behaviour of fuel rods at beginning of life.

REFERENCES

[1]   L. Caillot, G. Delette *et al.*, "Analytical Studies of the Behaviour of MOX Fuel", Proc. of the ANS Meeting on Light Water Reactor Performances, Portland (USA), 1997.

[2]   L. Caillot, G. Delette, B. Julien, J.C. Couty, "Impact of Fuel Pellet Fragmentation on Pellet-Cladding Interaction in a PWR Fuel Rod: Results of the RECOR Experimental Programme", Proc. of the 14th International Meeting on SMIRT, Lyon (France), 1997.

[3]   C. Struzik *et al.*, "High Burn-up Modelling of UO_2 and MOX Fuel with METEOR/TU V1.5", Proc. of the ANS International Topical Meeting on LWR Fuel Performance, Portland (USA), 1997.

[4]   L. Caillot, C. Lemaignan and J. Joseph, "In Situ Measurements of Cladding Strain During Power Transients Using the DECOR Device", Proc. ANS International Topical Meeting on LWR Fuel Performances, West Palm Beach (USA) 1994.

[5]   M. Oguma, "Cracking and Relocation Behaviour of Nuclear Fuel Pellets During Rise to Power", *Nuclear Engineering and Design*, 76: 43-45, 1983.

Figure 1. Number of fragments versus maximum linear heat generation rate

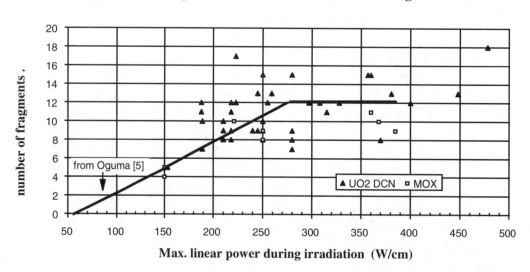

Figure 2. Centreline temperature of UO₂ and MOX fuel versus power in GRIMOX 2

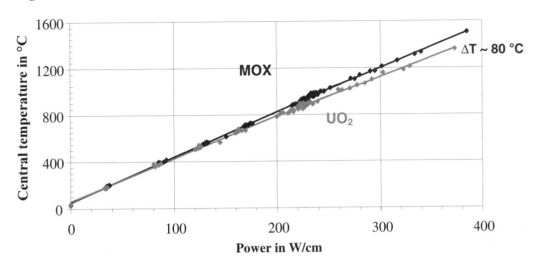

Figure 3. UO₂ and MOX radial power profiles in GRIMOX 2

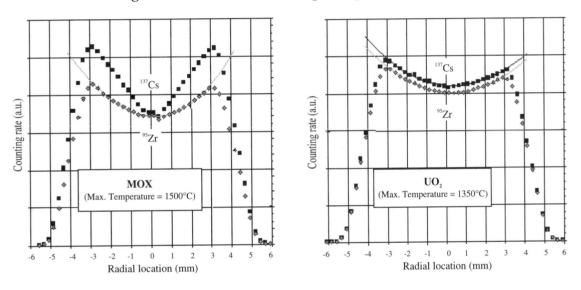

Figure 4a. Computed and measured UO₂ centreline temperature in GRIMOX 2 (up to cycle 3)

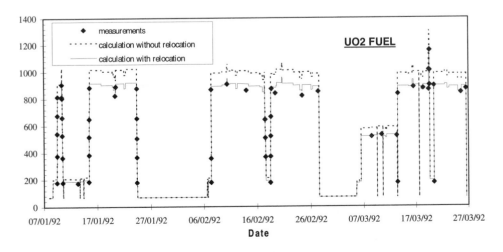

Figure 4b. Computed and measured MOX centreline temperature in GRIMOX 2 (up to cycle 3)

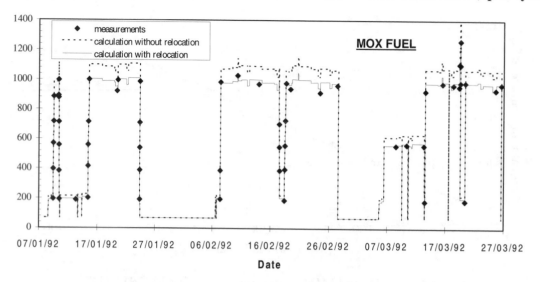

**Figure 5. Change in thermal behaviour of the
UO$_2$ fuel in GRIMOX 2 after the first rise in power**

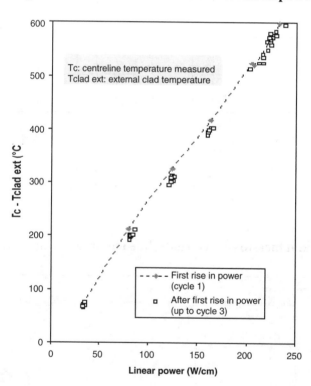

Figure 6. Relocation analysis in RECOR experiment

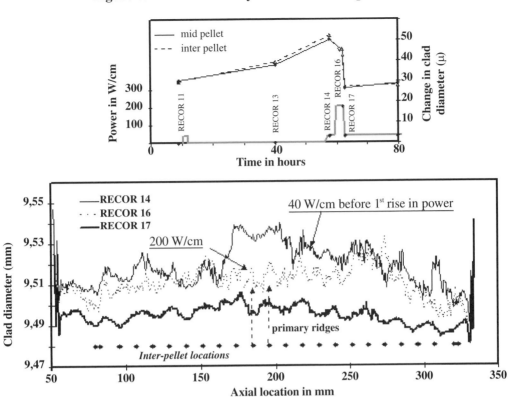

Figure 7. Relocation accommodation analysis in RECOR

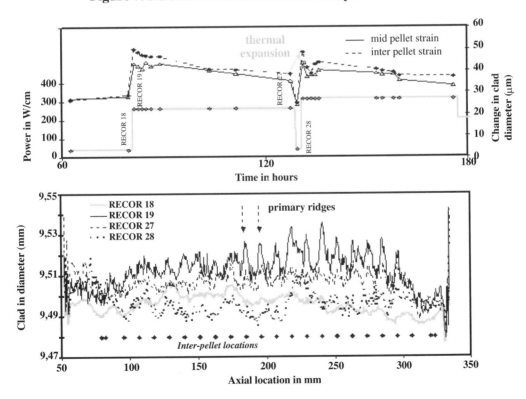

INTERNAL CORROSION LAYER IN PWR FUEL

Lionel Desgranges
Commissariat à l'Énergie Atomique
DRN/DEC/SDC/LEC, Cadarache, FRANCE

Abstract

An internal corrosion layer is formed by oxidation of the inner surface of cladding when contact occurs between pellet and cladding in PWR fuel rods. This paper gathers some experimental results obtained in CEA hot laboratories, to give a better understanding of this phenomenon. The internal corrosion layer is shown to form in two steps. In a first step stains 5 μm thick are formed. In a second step, the internal corrosion layer occupies the whole inner surface of cladding and gets monotonously thicker. Interpretation and consequences of these mechanisms are discussed. It is proved that the internal corrosion layer is important for thermal transfer from pellet to cladding at high burn-up, and that it would have a beneficial influence on fuel thermal conductivity by keeping fuel close to stoichiometry.

Introduction

The seeking of higher fuel performance in the nuclear industry has led to a great effort in the characterisation of high burn-up fuel. High burn-up fuel indeed exhibits specific features such as higher fission gas release or rim effect, as well as interaction between oxide and cladding which creates an internal corrosion layer. Up to now this layer has been little studied, because its thickness was believed to be too small to induce any problem. In higher burn-up fuel its thickness is not very much increased, but its interaction with the external rim of oxide pellets could induce some changes in the behaviour of fuel. Some recently published studies focused on the internal corrosion layer, providing a new characterisation for UO_2 BWR [1]and MOX PWR fuel [2].

Internal corrosion has been systematically observed in the frame of the French surveillance programmes on UO_2 and MOX fuels. The characterisation was often limited to some photos and systematic thickness measurements. Even if limited, these results are of interest because they provide information for a large range of burn-up. In this paper, some new outcomes gained from the CEA database will be presented. Some new results from EPMA and SEM experiments will also be presented. Special care will be taken to demonstrate the influence of the internal corrosion layer on the thermal properties of fuel.

This presentation will be divided in three parts:

- growth mechanisms;

- physical characterisation of internal corrosion layer;

- influence of internal corrosion layer on thermal behaviour.

Growth mechanisms

In different CEA laboratories, internal corrosion layer thickness has been measured over several years by optical microscopy on cross-sections at several azimuthal positions for different levels on MOX and UO_2 fuel rods.

Figure 1 shows the internal corrosion layer thickness as a function of burn-up. This figure clearly exhibits a discontinuity in thickness evolution. No internal corrosion layer has been measured with a thickness lower than 5 µm. Either small stains, 5 µm thick, appear, or no internal corrosion layer is observed Considering the number of experimental points, and the lack of correlation between the different measures, this can be considered as a true experimental fact.

Thus it is possible to describe thickness evolution as a two-step mechanism. In a first step, a 5 µm thick corrosion layer is formed so rapidly that it is not possible to observe intermediate thickness. In a second step thickness increases slowly with a great dispersion within the experimental points. Different irradiation histories can explain this dispersion, but it will be seen in Figure 4 that thickness of internal corrosion layer is not uniform at high burn-up, which makes difficult a precise determination of thickness.

The internal corrosion layer has been associated with the closure of the pellet-to-clad gap. The gap width has also been measured for a large number of PWR rods at the CEA.

**Figure 1. Thickness evolution of internal corrosion
layer according to burn-up evolution (MOX and UO₂)**

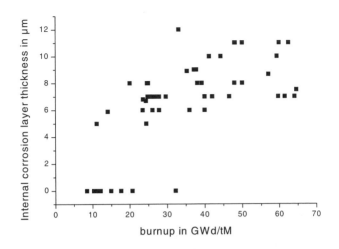

Figure 2 presents the evolution of gap width as a function of burn-up. On this figure a three-stage evolution can be observed. In a first stage, until approximately 20 GWd/tM, gap width decreases monotonously, which can be interpreted as creep down of the cladding towards fuel pellets. In a second stage gap, from 20-50 GWd/tM, width keeps constant between 10 and 20 µm. Considering that measurements are performed at room temperature, this residual gap can be viewed as a consequence of different thermal expansion of pellet and cladding, assuming that contact between pellet and cladding is achieved at irradiation temperature. In the last stage, over 50 GWd/tM, gap width is 0 meaning that the contact between pellet and cladding is kept even at room temperature.

Figure 2. Evolution of gap size according to burn-up evolution

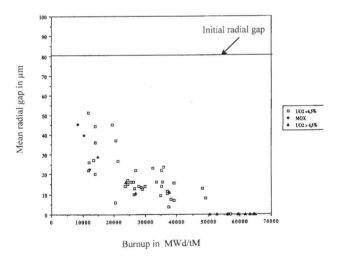

Comparison between gap closure and the internal corrosion layer formation in Figures 1 and 2, suggests that internal corrosion layer is formed once contact between pellet and cladding is achieved in operating conditions. A correlation with linear heat rate of fuel rod and burn-up was proposed to define the conditions of its apparition [3]. In order to illustrate this contact between cladding and pellet, two optical micrographies (Figures 3 and 4) are presented at two different stages of internal corrosion layer formation.

189

**Figure 3. Optical micrography of internal corrosion layer formed
in front of a Pu enriched agglomerate in a three-cycle MOX PWR fuel**

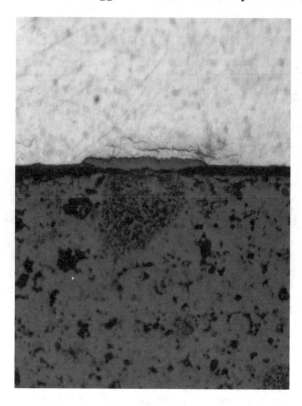

Figure 3 presents a cross-section in a three-cycle PWR MOX rod. The beginning of the internal corrosion layer formation is shown in front of an agglomerate enriched in plutonium. The agglomerate can be identified by its high content of fine porosity, its size being of the order of 80 µm in diameter. The internal corrosion layer is located in front of the agglomerate, nearly touching it, while elsewhere the gap is open. Its surface is rough on the zircaloy side and on the gap side. Zirconium hydride can be observed forming a line, parallel to the interface between internal corrosion layer and zircaloy.

Some authors [2] suggest that the formation of an internal corrosion layer in front of Pu-enriched agglomerates could be associated to higher oxygen content in agglomerate, because of more oxidising Pu fission. Making rough hypothesis on the composition and volume of the agglomerate and of the internal corrosion layer, we have calculated the oxygen content of the agglomerate and of the internal corrosion layer corresponding to Figure 5. This simple calculation shows that in the case of Figure 5, internal corrosion layer contains a third of the oxygen content of the agglomerate. It is a much greater quantity than the one that can be produced by fission. Fission indeed can only increase O/M by a few per cent. This calculation proves that oxygen diffusion in the whole pellet is needed to achieve the formation of an internal corrosion layer.

Formation of an internal corrosion layer in front of an agglomerate could be associated with mechanical contact between agglomerate and cladding. Pu-enriched agglomerates are known to swell under irradiation when emerging on a free surface. This swelling implies that mechanical contact between the internal corrosion layer and cladding occurs before contact between UO_2 matrix and cladding.

Once contact is achieved, the internal corrosion layer would be formed by solid-state diffusion of oxygen. However, the type of oxygen diffusion involved in formation of an internal corrosion layer must be specified: either thermal or athermal.

Oxygen thermal diffusion can be radially limited to the area damaged by recoil atom in cladding. Before any contact between oxide and cladding occurs, the inner cladding is already modified by fuel irradiation. The inner surface of the cladding receives neutron irradiation, and also recoil atoms created by fission in pellets. Neutron damage spreads throughout the whole cladding, while recoil atom damage is located in a thin layer at the inner surface. The stopping range of fission product recoil atoms is of the order 8.2-10.4 μm depending on the mass of atoms. They lead to a layer of about 10 μm depth bearing more than 2% of foreign atoms at standard burn-up [4]. Behaviour of inner cladding is of course modified by implantation of recoil atoms [4]. Oxygen thermal diffusion could be enhanced by these damages. This is why oxygen thermal diffusion should be restricted radially, but it should not be limited circonferencially. If oxygen thermal diffusion was involved, the internal corrosion layer would then also propagate quickly, during the first step of its formation, on the whole inner surface of cladding because thermal diffusion is isotropic.

Athermal diffusion can explain why the internal corrosion layer is restricted to an area close to contact between pellet and cladding. Fission recoil atoms leave pellet with a high energy which allows their implantation into cladding, but they also lose energy in pellet creating secondary recoil atom, among which oxygen. Athermal diffusion of oxygen should then be limited to an area corresponding to the path of recoil fission products through the contact zone.

Athermal diffusion requires the gap between pellet and cladding to be small. When it is filled with gas, collisions between gas and oxygen atoms will prevent the latter from reaching cladding. And after a certain amount of oxygen has reached internal cladding, the gap will close because of internal corrosion layer swelling. The volume occupied by the same number of zirconium atom increases by approximately 50% from zircaloy to zirconia due to oxygen incorporation. This swelling, together with the swelling of agglomerate, explains why the gap is smaller between the internal corrosion layer and pellet than between cladding and pellet in Figure 5. Swelling also induces stresses in cladding near the interface with internal corrosion layer. Zirconium hydride phases are known to precipitate preferentially along stress gradient. And evidence of hydride on Figure 5 could be associated with such stresses.

Thickness of internal corrosion layer increases with burn-up as shown on Figure 1, the corresponding swelling should also increase, which is demonstrated in Figure 4.

Figure 4 presents a cross-section in a UO_2 PWR fuel irradiated over five cycles. Zircaloy is located on the top of the figure. Just below, the internal corrosion layer, whose thickness is around 10 μm, is characterised by its wavy shape. Fuel pellet is characterised by its high content of fine porosity associated with rim effect. At the interface between internal corrosion layer and rim area, internal corrosion layer has a wavy shape with rolls. In Figure 6, interpenetration of rim area and internal corrosion layer is evidenced. This interpenetration can explain that on cooling no gap can open at the interface between rim area and internal corrosion layer as shown in Figure 2. A crack, clearly evidenced at the bottom of Figure 6, proves that it is easier to break rim area rather than to open the gap.

**Figure 4. Optical micrography of a five-cycle UO$_2$ PWR rod,
showing interpenetration of rim area and internal corrosion layer**

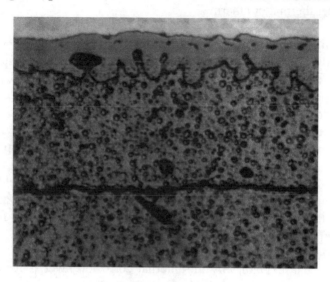

Physical characterisation of internal corrosion layer

In order to have a better description of internal corrosion layer, its behaviour will be characterised with regard to its composition, its electrical and mechanical behaviour in the following.

Once the internal corrosion layer is formed, its composition will reveal three features: original zircaloy, implanted fission products and oxygen coming from pellet which leads to oxidation of zircaloy. Figure 5 presents the results of an EPMA analysis on a line going from cladding (on the left side) to pellet (on the right side) through the internal corrosion layer for a five-cycle UO$_2$ PWR rod. On the upper part of the figure, the radial evolution of mass concentration is given for some selected fission products. In the lower part of the figure, uranium, zirconium and oxygen mass concentration is given as a function of radius.

The border between zirconium and the internal corrosion layer, symbolised by a dashed line on the left side of the figure, is evidenced a decrease in zirconium mass concentration and an increase of oxygen mass content. The border between the internal corrosion layer and uranium oxide, symbolised by a dashed line on the right side of the figure, is evidenced a zirconium mass concentration decrease and a uranium mass concentration increase.

Fission products profiles exhibit a monotonous increase from inside to the edge of pellet due to rim effect. Then it decreases from pellet to zirconium, in an area 10 µm depth inside pellet. Isotropic emission of recoil atom explains a decrease of FP concentration at pellet surface. The missing fission products are implanted in internal corrosion layer. Practically no fission products are detected outside of internal corrosion layer.

Some fluctuations of oxygen concentration are observed in internal corrosion layer. They can be associated to the coexistence of uranium and zirconium near pellet. Some authors interpret this coexistence as the evidence for the formation of a (U,Zr)O$_2$ phase [1]. This hypothesis will be modified with SEM observations performed in CEA hot laboratories.

**Figure 5. Evolution of mass concentration for
Ba, Mo, Nd, U, Zr and O between cladding and pellet**

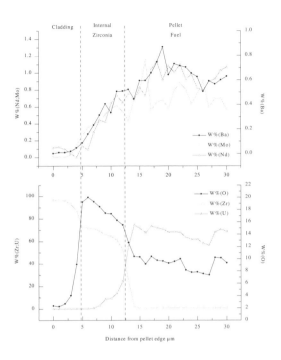

Figure 6 presents a SEM image showing the inside of an open fabrication pore in front of the cladding. The internal corrosion layer is easily visible with its bright aspect on the left side. This bright aspect can be interpreted as a charging effect under the assumption that the internal corrosion layer is an insulator. This hypothesis is confirmed by the black and bright strips at the interface between zirconium and internal corrosion layer. These strips can indeed be interpreted as potential contrast due to a conductor (zirconium)/insulator (internal corrosion layer) junction [5]. The same strips were also observed at the interface between internal corrosion layer and pellet.

Figure 6. SEM observation of internal corrosion layer in front of a pore

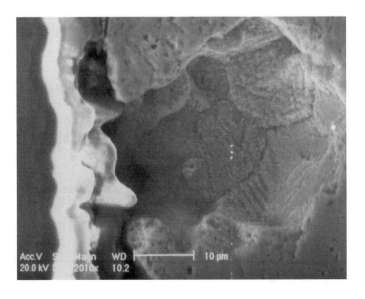

Uranium oxide and the internal corrosion layer seem to be well separated as seen in Figure 6, and there is no evidence on this figure that a $(U,Zr)O_2$ phase can be formed. EPMA measurements were performed on a polished cross were uranium oxide and internal corrosion layer are in contact. Considering that EPMA provides concentration values averaged over 1 μm^3, uranium and zirconium coexistence could also be interpreted as coexistence of uranium oxide and zirconium oxide in the same analysed volume because internal corrosion layer is not flat.

On its pellet side, internal corrosion layer exhibits rolls with smooth hills and valleys. This morphology is clearly seen in Figure 6 because of the absence of uranium oxide in front of the internal corrosion layer. This wavy aspect can also be observed at the interface between internal corrosion layer and pellet, on ceramographic photos taken on a polished cross-section of high burn-up fuel (see Figure 4). This seems to prove that the interface morphology is determined by internal corrosion layer morphology. Such a hypothesis would imply that the internal corrosion layer is harder than rim area in pellet, so that rim area would fit around the internal corrosion layer after the latter is formed.

The internal corrosion layer composition is close to that of zirconia; it is hard and insulating as zirconia. Its crystallographic structure can be expected to be similar to that of zirconia. X-ray diffraction experiments [6] were performed on the inner surface of cladding with internal corrosion layer on it. The obtained diffractogram revealed zirconium, and another quadratic crystallographic phase. This quadratic phase can be attributed to quadratic zirconia. Zirconia, usually monoclinic at room temperature can be stabilised by stresses or impurities. Evidence of impurities in internal corrosion layer is given by the EPMA (see Figure 5), and the existence of stresses was discussed with respect to Figure 2.

Influence of internal corrosion layer on fuel thermal behaviour

The results presented here underline two major outcomes that will be discussed independently in the following.

Internal corrosion layer determines pellet-to-cladding heat transfer at high burn-up

At high burn-up, when the internal corrosion layer occupies the whole inner surface of cladding, thermal transfer occurs via solid-solid contact, with no significant gas influence. While interface between internal corrosion layer and rim cladding is rather flat, interface between the internal corrosion layer and the rim pellet is affected by high roughness associated to interpenetration of these two materials. Rim area of the fuel pellet is characterised by its high content of fine porosity containing fission gazes. The remaining filling gas captured by gap closure can hardly be distinguished from rim porosity. As a consequence no gaseous gap can be said to exist between pellet and cladding. Because of interpenetration of internal corrosion layer and rim pellet no gaseous flux is expected to happen along the internal corrosion layer.

Composition, electrical and mechanical properties of internal corrosion layer are close those of zirconia, which explains why the internal corrosion layer is sometimes called "internal zirconia". As a matter of fact, its thermal properties should not be too far those of zirconia.

Internal corrosion layer as a stoichiometry regulator

Formation of the internal corrosion layer is associated with an oxygen flux from pellet to cladding. On a thermodynamical point of view, equilibrium is reached in a UO_2-Zr mixture when all Zr is oxidised. In fuel equilibrium is not reached because of the low irradiation temperature.

An athermal diffusion mechanism was proposed to explain oxygen migration from pellet to cladding. However this mechanism implies that oxygen diffusion should continue at the same speed until the whole area implanted with recoil atom is oxidised. Thickness evolution shows that it is not the case for the internal corrosion layer which suggests that the quantity of oxygen able to diffuse is limited.

Thickness evolution can be correlated to over-stoichiometric oxygen behaviour. With an athermal diffusion mechanism, diffusion can be limited by bounding energy of atoms. Interstitial over-stoichiometric oxygen are less bounded than oxygen belonging to fluorite structure. They could be implanted in cladding more easily. For a PWR fuel, an increase of 0.0013 in O/M by at% was estimated due to fission [7]. Making crude estimations, calculations prove that this rate is consistent with the observed evolution of internal corrosion layer thickness, with first a fast increase corresponding to migration of over-stoichiometric oxygen produced before contact is reached, and then a monotonous increase corresponding to the rate of over-stoichiometric oxygen production.

This assumption is justified by oxygen potential measurements made on high burn-up PWR fuel which showed that this fuel was close to stoichiometry [6], instead of being over-stoichiometric as expected if no internal corrosion layer is formed.

By keeping fuel near stoichiometry, the internal corrosion layer has a beneficial influence on fuel thermal conductivity, which is known to decrease with over-stoichiometry.

Conclusion

Considered as a negligible phenomenon for standard burn-up, the internal corrosion layer has proven to have more importance for higher burn-up. First, thermal transfer from pellet to cladding needs to be considered trough a suitable description of the internal corrosion layer. And second, it has a beneficial influence on fuel thermal conductivity by keeping fuel close to stoichiometry.

However, fast formation of the internal corrosion layer when contact between pellet and cladding occurs, still needs to be understood. Especially it can be wondered if such a phenomenon can occur during a transient. Experimental material is needed to answer this question.

Acknowledgements

The author thanks all people in CEA hot laboratories, in Cadarache, Grenoble and Saclay for experimental material which made this paper possible. I. Rochas and B. Pasquet, respectively, provided new SEM and EPMA results. N. Lozano, J. Noirot, D. Lespiaux and I. Canet have improved analysis by fruitful discussions.

REFERENCE

[1]   K. Nogita, K. Une and Y. Korei, "TEM Analysis of Pellet-Cladding Bonding Layer in High Burn-up BWR fuel", *Nuclear Instruments and Methods in Physical Research*, B 116, pp. 521-526 (1996).

[2]   C.T. Walker, W. Goll and T. Matsumura, "Further Observations on OCOM MOX Fuel: Microstructure in the Vicinity of the Pellet Rim and Fuel-Cladding Interaction", *Journal of Nuclear Materials*, 245, pp. 169-178 (1997).

[3]   J. Bazin, J. Jouan and N. Vignesoult, Trans. ANS 20, 235 (1975).

[4]   I. Schuster and C. Lemaignan, "Embrittlement, Induced by Fission Recoils, of the Inner Surface of PWR Fuel Cladding – A Simulation Using Heavy Ion", *Journal of Nuclear Materials*, 151, pp. 108-111 (1998).

[5]   Ch. Kittel, "Physique de l'état solide", 5th edition, Bordas Paris, p. 241 (1983).

[6]   C. Gibert, F. Couvreur, Private communication.

[7]   Hj. Matzke, "Oxygen Potential in the Rim Region of High Burn-up UO_2 Fuel" *Journal of Nuclear Materials*, 208, pp. 18-26 (1994).

SEPARATE EFFECT STUDIES AT THE HALDEN REACTOR PROJECT RELATED TO FUEL THERMAL PERFORMANCE AND MODELLING

W. Wiesenack
Institutt for Energiteknikk
OECD Halden Reactor Project, Norway

Abstract

The fuels and materials testing programmes carried out at the OECD Halden Reactor Project are aimed at providing data in support of a mechanistic understanding of phenomena, especially as related to high burn-up fuel. The investigations are focused on identifying long term property changes, and irradiation techniques and instrumentation have been developed over the years which enable to assess fuel behaviour and properties in-pile.

The paper reviews selected examples from separate effects studies with emphasis on the influence of the fuel-clad gap on thermal behaviour, pellet-clad mechanical interaction and aspects of gas communication and mixing. The data provide a basis for the understanding of fuel performance up to high burn-up and can be used for fuel behaviour model development and verification as well as in safety analyses.

Introduction

The fuels and materials testing programmes of the OECD Halden Reactor Project are aimed at providing data in support of a mechanistic understanding of phenomena associated with short and long term in-pile fuel performance and property changes. To this end, suitable irradiation techniques and instrumentation have been developed and applied for more than thirty years [1]. The data obtained can be used for fuel behaviour model development and verification as well as in safety analyses.

The knowledge of the temperature distribution in a fuel pellet as affected by fuel properties, operation conditions and irradiation induced changes of fuel and cladding are central to fuel behaviour modelling. A number of often interrelated phenomena need to be quantified to obtain adequate predictions, and separate effect studies provide efficient methods for assessing specific effects. Over the years, such studies have addressed the phenomena which are deemed most important for fuel behaviour modelling. Although the experimental focus has shifted in recent years to investigations of high burn-up effects, data based on fresh or low burn-up fuel are valuable as well, because they can often be transferred to high burn-up situations or are valid for worst case scenarios which may occur at low or medium rather than high burn-up. Since fuel behaviour modelling and safety analyses have to provide an adequate description for the entire in-core service, it seems important to maintain a complete picture of fuel behaviour from beginning to end-of-life.

In this paper, emphasis is put on data associated with the fuel-clad gap which in standard fuel designs is about 2% of the pellet diameter (150-250 mm) to accommodate thermal expansion and fuel swelling. A good knowledge of the gap size and its change with burn-up is essential for the prediction of both thermal and mechanical behaviour. For example, an underestimation of gap conductance at high burn-up will lead to overprediction of fuel temperatures (stored energy) and fission gas release and may thus severely impact safety assessments.

The selected examples from separate effects studies show the relation of the fuel-clad gap with thermal behaviour, pellet-clad mechanical interaction and aspects of gas communication and mixing. The eventual aim for the fuel modeller would be to provide a unified and seamless description of both the thermal and mechanical fuel behaviour.

Parameters influencing gap conductance

In modelling of the pellet-clad heat transfer, two main components are usually considered: conductance through the open gap and conductance through areas in intimate proximity under contact pressure. The dominant parameters in models for heat conductance through the open gap are obviously the gap size and the (fill) gas conductivity, while surface roughness and waviness become more important for small or tightly closed gaps. These influences have therefore been addressed over the years in a number of experiments carried out at the Halden Reactor.

Gap size and fill gas

The determination of gap conductance is affected with uncertainties which accumulate throughout irradiation. The problem can be alleviated by using – for experimental purposes – rod designs with small gaps (simulated gap closure due to fuel swelling and clad creep-down) and Xe fill gas (simulated fission gas release). A summary of the influence of gap size and fill gas type on fuel centre temperature determined in this way by a number of independent experiments is shown in Figure 1. The data represent the starting conditions where the gap size is not yet changed and uncertain due to fuel densification. As-fabricated diametral gaps ranged from 50 to 230 mm for Xe filled rods and 50 to 400 mm for He filled rods with about 10.5 mm pellet diameter, 95% of TD and 10% ^{235}U enrichment. In all cases, a centre hole of 1.8 mm is present to accommodate the thermocouple.

Figure 1. Typical fuel centre temperatures in He/Xe filled rods

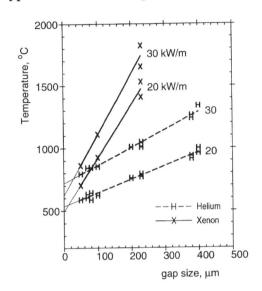

Pellet and clad surface roughness

It is of interest to note in Figure 1 above that the temperature difference between helium and xenon filled rods disappears at small initial gaps which are strongly closed at power. The pellet-clad surface roughness is an important parameter in this situation, both for modelling the contact conductance and the conductance through the residual open gap.

The influence of the mean pellet-clad surface roughness on gap conductance and hence fuel centre temperature has been studied in a separate effects experiment [2]. It contained pairs of rods equipped with thermocouples and "smooth" (<0.8 mm) and "rough" (~2 mm) pellet-clad surface combinations. Other characteristics were 10.59 mm pellet diameter and an as-fabricated cold diametral gap of 70 mm. Some of the rods were filled with xenon as fill gas in order to amplify any effect of the surface roughness on gap conductance.

The temperature data from this experiment did not show a systematic difference between smooth and rough pellets throughout the irradiation to about 10 MWd/kgU, as depicted in Figure 2 for the final power increase before discharge. At this point, the fuel-clad gaps were fully closed for average rod powers >25 kW/m as evidenced by simultaneous cladding elongation measurements. In this situation, as already mentioned before, the influence of fill gas type was found to be insignificant.

The results imply that gap conductance is dominated by solid-solid contact conductance when the gap is tightly closed. It seems therefore that gap conductance models should not retain a minimum separation between pellet and cladding due to roughness and waviness in the case of very hard contact. As will be shown later, also results obtained with high burn-up fuel confirm this point.

Figure 2. Temperature versus rating curves for rod pairs with rough and smooth pellet-clad interface surfaces; the gap is closed for powers >25 kW/m

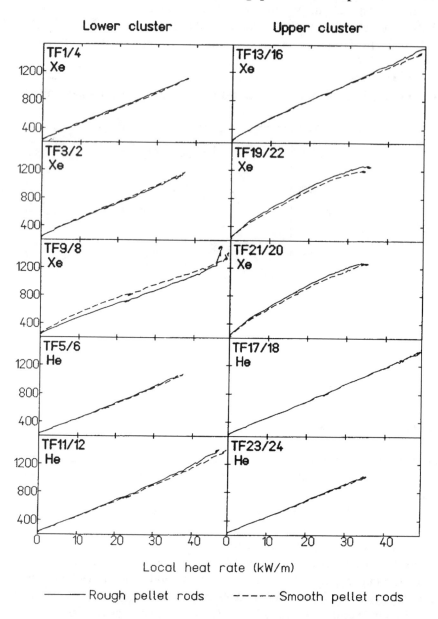

Eccentric position of fuel pellets within the fuel rod

Fuel modelling codes usually employ a geometry where the pellets are concentric with the cladding tube. While this leads to a convenient 1-D axi-symmetric description of, e.g. the temperature distribution, it is also clear that some observations cannot be explained with such an approach which actually assumes the most unlikely configuration. The effect of the location of a fuel pellet on the centre temperature has been the subject of experiments where some pellets were fixed in a concentric or eccentric position [3]. The fuel properties and dimensions are otherwise as given for Figure 1 (gap size 230 mm). The start-up temperature data are shown in Figure 3; they give a clear indication that eccentric pellets are running at lower temperatures compared to the concentric ones.

Figure 3. Fuel centre temperature in pellets which have been fixed to concentric and eccentric positions within the cladding tube; the rods are filled with xenon

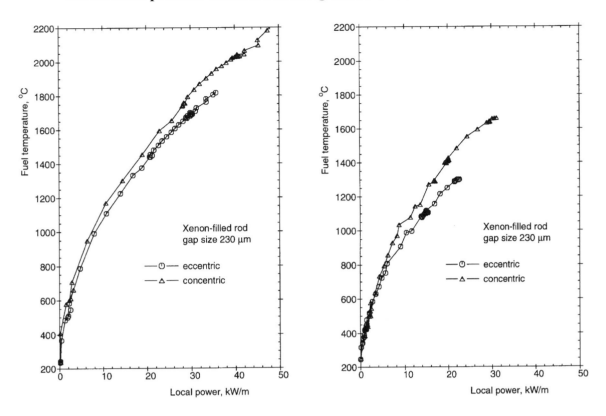

These results do not immediately follow from a 1-D approach since the average gap is the same in both cases. Considerations such as improved heat transfer in the contact area (although the contact pressure is low compared to when the gap is tightly closed) and a greater fuel fragment relocation into the larger gap space opposite of the point of contact have been used as an explanation. An off-centre location of the maximum temperature must also be taken into account.

Today, these results may appear more as a curiosity from a time when a basic understanding of factors influencing gap conductance was sought. However, they remain nevertheless a part of the complex picture that must be taken into account in the modelling of gap conductance.

Change of gap with power

It is well known that gap closure calculated with pellet thermal expansion alone would result in fuel temperatures which are higher than actually measured. Therefore, additional mechanisms are invoked to improve the fuel-clad heat transfer, e.g. the assumption that parts of the pellet surface are in contact with the cladding (contact area approach), or fuel fragment relocation. For the latter several assumptions are possible – one being that the gap is reduced by a certain fraction of the cold gap.

The change of gap size with power has been addressed in two ways in the experimental programme at the Halden Reactor: by so-called hydraulic diameter measurements and with a gap measurement rig where the cladding was squeezed onto the fuel while recording force and deflection. An example of hydraulic diameter data obtained for high burn-up fuel (75 MWd/kgU) and decreasing power is shown in Figure 4.

Figure 4. Change of hydraulic diameter in response to decreasing power

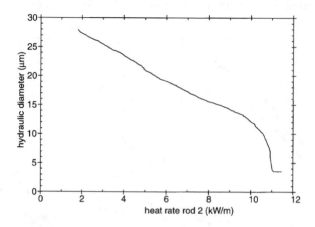

The hydraulic diameter is obtained from the gas flow through the fuel stack as a function of the pressure difference between the ends. As such, this quantity contains both the pellet-clad gap and cracks in the fuel, and the change of hydraulic diameter with power as shown in Figure 4 is nearly identical with thermal expansion. The abrupt transition from a closed to an open "gap" at the highest power should be noted and compared to the change of PCMI discussed later.

The squeezing technique can differentiate between the "open gap" and the component of cracks in the fuel which cannot be closed. The in-pile gap measurements [4] using the squeezing technique show a picture similar to the hydraulic diameter data (Figure 5). The start-up data (fresh fuel, 150 mm gap) show a basically linear decrease of gap size which is clearly more than gap closure calculated with thermal expansion alone. After return to zero power, most of the difference between thermal expansion and the gap measured at maximum power remains as permanent gap closure (relocation). It is also evident that relocation is a gradual process occurring during the first rise to power. The assumption that (initial) relocation is a certain fraction of the cold gap may, however, be quite adequate for all following power changes and steady state operation.

Figure 5. Change of gap size during first power increase and following decrease

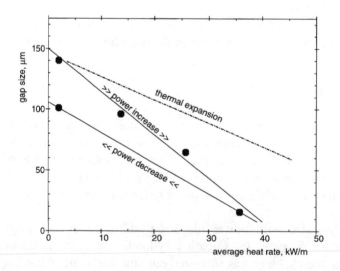

Change of gap with burn-up

In the following discussion, data from two sources are used: measurements of axial cladding elongation and, again, hydraulic diameter measurements. The latter contains the fuel rod of which measurements have been shown in Figure 4.

Figure 6. Development of the hydraulic diameter at zero power as a function of burn-up

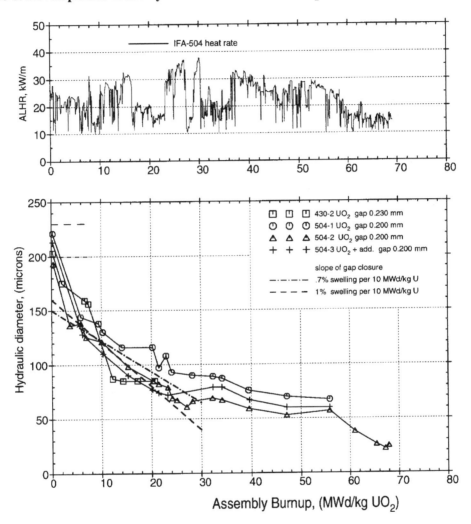

Figure 6 above shows a collection of zero power hydraulic diameter data obtained for several rods of similar design (gap 200-230 mm) with burn-up ranging from 0 to 68 MWd/kgUO$_2$. Several observations can be made:

- the gap decreases initially with an amount (40-80 mm) which, relative to the gap size, is comparable to what was previously shown in Figure 5;

- the gap closure for burn-up <30 MWd/kgUO$_2$ follows approximately the fuel swelling;

- between 30 and 55 MWd/kgUO$_2$, much less decrease is observed;

- and finally, for the last four measurements a further decrease can be noted.

It is of interest to correlate the development for burn-up >35 MWd/kgUO$_2$ with the power history which is shown in the upper part of Figure 6. The gap at zero power corresponds approximately to the pellet thermal expansion at operation power. Especially for the highest burn-up (>60 MWd/kgUO$_2$), this power is lower than during most of the previous history and the measured hydraulic diameter decreases accordingly. It seems that fuel and cladding accommodate to each other when the gap is sufficiently closed by fuel swelling.

Gap closure is also manifested in pellet-clad interaction which can be measured as clad elongation. The development of axial interaction with burn-up of four similar rods is shown in Figure 7 for five down-ramps. It is common that axial interaction during the first power cycle is strong and decreases during the following cycles. This can be attributed to random eccentric stacking of the pellets which are pushed to more central positions in contact with the cladding. It should be noted that thermal expansion is not sufficient to close the gap at start-up (not even with fuel swelling at end of life, 34 MWd/kgUO$_2$), thus conventional models based on concentric geometry would not calculate any interaction, obviously at variance with the experimental evidence.

**Figure 7. Cladding elongation of BWR rods; as power gradually decreases
due to burn-up, the point of contact stays close to the highest power**

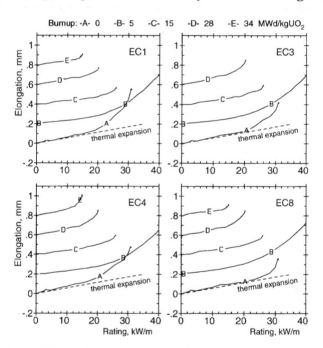

The release of contact between fuel and cladding is close to the highest power for the curves marked C, D and E in Figure 7. This corresponds to the development discussed with the hydraulic diameter measurements, where pellet and cladding seemed to accommodate to each other by reaching a balance between fuel swelling and creep caused by contact forces.

The situation when a period of operation is followed by one with higher power can be studied using an example from FUMEX-1 [5]. As shown in Figure 8, the onset of interaction moves to higher power for successive start-ups. PCMI starts at about the highest power reached during the preceding ramp for start-up 2 and 3. It seems that the pellet fragments can be relocated inwards up to a certain limit.

Figure 8. Cladding elongation response to power changes; the point of PCMI onset power moves to higher power for successive start-ups

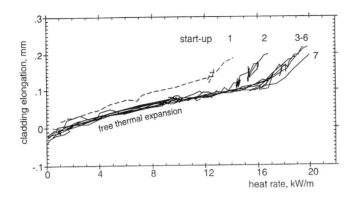

A consequence that follows from these observations is that power transients (RIA) will lead to (additional) cladding loading when the power level (or the thermal expansion) of the preceding operation period is exceeded.

Gas communication and mixing

The pellet-clad gap is not only a resistance for the heat transfer to the coolant, but also a path for released fission gas to the plenum. It is usually assumed in modelling of fuel rod behaviour that the released fission gas mixes completely (and instantaneously) with the helium fill gas. However, it has also been surmised that the mixing may be incomplete and that the fuel-clad gap may be filled preferentially with fission gas. This effect would produce an amplified thermal feedback of poor gap conductance on fuel temperatures and further fission gas release.

The effect has been investigated at the Halden Reactor Project both with a dedicated experimental set-up [6] and by provoking bursts of fission gas release in fuel rods [7]. Figure 9 shows the temperatures measured during an up-ramp on a time scale of minutes. It can be seen that a FGR burst occurred in Rod 1 (Curve 1) and Rod 4 (Curve 8). The overshoot in Rod 1 decayed in a few minutes, while FGR seemed to have been more continuous in Rod 4 where the temperatures decayed slowly over several days of operation at constant power (not shown). The analysis of this and other experiments has indicated that simple diffusion theory is sufficient for an adequate description of the dynamic process.

Figure 9. Measured temperatures during an up-ramp; the spikes are due to fission gas bursts

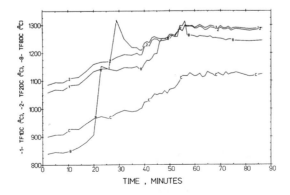

The strong change of gap conductance due to FGR requires that an appreciable gap be present at power. In the experiment presented above, the gap was about 70-80 mm (zero power, hot stand-by) as deduced from hydraulic diameter measurements.

Since the potential for fission gas release increases with burn-up, it is of interest to see the influence on the fuel temperature in a high burn-up fuel rod. Figure 10 shows release and temperature data which are related to a power cycle of high burn-up fuel (60 MWd/kgU) in the Halden reactor. During the down-ramp, considerable gas release was registered which lead to higher fuel centre temperatures compared to the preceding up-ramp. However, the difference is small, about 20 K around 15 kW/m. It is known from PIE that the gap in this fuel rod is essentially closed and that a bonding layer between fuel and cladding has formed. The temperature response to fission gas release is therefore much less pronounced than in the example shown in Figure 9. The results are more in line with the findings from experiments with small gap Xe filled rods discussed in the section entitled *Parameters influencing gap conductance*.

Figure 10. Temperatures for up- and down-ramp with fission gas transport through the gap to the plenum during the down-ramp

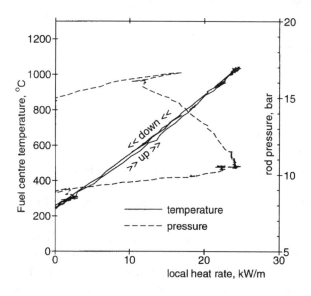

The observation that fission gas reaches the free volume (plenum) only during a power decrease is quite common for high burn-up fuel. The phenomenon contributes to alleviate the concern of an amplified temperature/gas release feedback since the contamination of the gap occurs when temperatures are lower due to lower power.

Other separate effects studies

The material presented above is only a limited subset from the overall experimental activities at the Halden Reactor Project. It was selected with special emphasis on the influence of the pellet-clad gap on fuel performance. The current three-year programme from 1997 to 1999 addresses, among others, the following subjects with separate effects studies:

- fuel conductivity degradation at ultra high burn-up;

- thermal behaviour, densification, swelling and fission gas release of gadolinia fuel;

- investigations of MOX fuel performance;

- effect of microstructure on fission gas release and PCMI;

- rod overpressure/clad lift-off.

They contribute to a better understanding of the behaviour of fuel variants with emphasis on high burn-up effects and provide the necessary data for modelling and safety analyses.

REFERENCES

[1] O. Aarrestad, "Fuel Rod Instrumentation", IAEA Meeting on In-Core Instrumentation and In-Situ Measurements in Connection with Fuel Behaviour, Petten, NL, 1992.

[2] M.R. Smith, "Final Report on the In-Pile Results from the Pellet Roughness Test in Rig IFA-562", HWR-245, November 1989.

[3] M.E. Cunningham, R.E. Williford, C.R. Hann, "Effect of Fill Gas Composition and Pellet Eccentricity: Comparison Between Instrumented Fuel Assemblies IFA-431 and IFA-432", NUREG/CR-0331, PNL-2729, April 1979.

[4] T. Johnsen, R.J.P. Cribb, K.O. Vilpponen, "Measuring Fuel/Cladding Gap Size In-Pile", HWR-9.

[5] W. Wiesenack, "Data for FUMEX: Results from Fuel Behaviour Studies at the OECD Halden Reactor Project for Model Validation and Development", IAEA Technical Committee Meeting on Water Reactor Fuel Element Modelling at High Burn-up and its Experimental Support, Windermere, UK, September 1994.

[6] C. Vitanza, T. Johnsen, J.M. Aasgard, "Interdiffusion of Noble Gases – Results of an Experiment to Simulate Dilution of Fission Gases in a Fuel Rod", HRP-295.

[7] F. Sontheimer, P. Wood, "Experimental and Modelling Aspects of Gas Mixing Dynamics in High Burn-up Rods of IFA-504/505 During Burst and Continuous Fission Gas Release, HWR-301, 1991.

SESSION IV

Experimental Databases

Chairs: C. Vitanza, J.P. Piron

TEMPERATURE MEASUREMENTS IN HIGH BURN-UP UO$_2$ FUEL: EXTRAFORT EXPERIMENT

S. Bourreau, B. Kapusta, P. Couffin, G.M. Decroix [CEA]
J.C. Couty [EDF]
E. Van Schel [FRAMATOME]

Abstract

The thermal conductivity of UO$_2$ is an important property considering that it controls the fuel operating temperature and therefore directly influences fuel performance and behaviour, such as fission-gas release and pellet-cladding mechanical interaction. This conductivity decreases with the burn-up due to the changes that take place in the fuel pellet during irradiation: solid fission products build-up, fission-gas bubbles formation, radiation damage, swelling. The evaluation of fuel thermal conductivity and its degradation with burn-up is an essential requirement for any code dedicated to the prediction of fuel rod behaviour. As yet only a few data are available concerning the degradation of the thermal conductivity of highly irradiated commercial UO$_2$, especially above 40 GWd/t.

In collaboration with FRAMATOME and the EDF, in-pile fuel centreline temperature measurements were performed at CEA Saclay in the experimental reactor OSIRIS, in order to estimate the thermal performance of fuel rods at end-of-life in pressurised water reactors, and to validate thermal conductivity models of high burn-up UO$_2$ fuel. In this experiment (EXTRAFORT), a FRAGEMA UO$_2$ fuel rod irradiated up to 62 GWd/t in a French PWR was re-fabricated and instrumented with a central thermocouple, using the drilling technique created in RISØ (Denmark). This technique was validated on non-irradiated fuel, in the CALIF experiment; it was shown that the RISØ drilling technique does not induce perturbations in fuel centreline temperatures, compared with temperatures measured in hollow pellets. In the ISABELLE 1 loop of OSIRIS, the EXTRAFORT fuel rod underwent different power levels, ranging from 100 to 305 W/cm at the thermocouple level, and power cycling. On-line average power was precisely determined using the thermal balance method. The local power at the axial location of the thermocouple is deduced from post-irradiation gamma scanning.

In the first part of this paper, the EXTRAFORT and CALIF experiments are described: fuel rods fabrication, experimental device and irradiation conditions are detailed. Fuel centreline temperature measurements during these irradiations are presented and discussed afterwards.

The second part of the paper presents post-calculations of the EXTRAFORT experiment, using different models for UO$_2$ thermal conductivity at high burn-up. These calculations have been carried out with fuel modelling codes considering a radial thermal conduction. In order to evaluate the influence of axial thermal conduction on centreline temperatures measured during the EXTRAFORT experiment, computations with a 2-D code using the finite elements method were performed. The temperatures measured turn out to fit the stabilised temperatures in the hollow pellet with a precision better than 20°C. Calculation results, using different expressions of the thermal conductivity for irradiated UO$_2$ are then presented and discussed. The analysis points out the fact that taking into account the burn-up effect, mainly fission product build-up which increases the number of lattice defects and consequently reduces the thermal conductivity of the fuel, especially in the low temperature region, leads to a better prediction of the centreline temperature.

Introduction

The knowledge of irradiation effects on UO_2 thermal conductivity at high burn-up is of practical importance for a better prediction of the fuel rod behaviour which strongly depends on the temperature profile in the fuel. As yet only a few data are available concerning the degradation of the thermal conductivity of highly irradiated commercial UO_2, especially above 40 GWd/t.

This paper presents the EXTRAFORT experiment which provides the first measurements of fuel centreline temperatures performed during irradiation on a French commercial PWR fuel rod irradiated up to 62.5 GWd/t. This experiment, realised at CEA Saclay in collaboration with FRAMATOME and EDF, aims to estimate the thermal performance of PWR fuel rods at high burn-up. The power history imposed on the EXTRAFORT rod consisted in power cycling at several power levels in order to point out a possible recovery or degradation of the thermal conductivity. In order to qualify the drilling technique used in the high burn-up rod to introduce a central thermocouple, the preliminary experiment CALIF was performed with a fresh UO_2 fuel rod and also consists of in-pile centreline temperature measurements. Post-calculations of the experiment carried out with a 1-D½ code are then presented; the ability of different expressions available to represent the thermal conductivity of UO_2 irradiated at high burn-up are tested in order to reproduce the temperatures measured during the EXTRAFORT experiment. In order to validate the fuel temperature measured via the thermocouple, the effect of axial heat transfer on experimental temperature was evaluated with a 2-D code using the finite elements method.

EXTRAFORT experiment

Description of the PWR "mother rod"

The PWR rod, designed by FRAMATOME, contains UO_2 pellets initially enriched up to 4.5% in ^{235}U. It was irradiated during five cycles in a French 900 MWe PWR and reached a mean burn-up of 57.2 GWd/tU. The mean rod linear density power experienced during the fifth cycle was 16.2 kW/m.

Fuel rod refabrication and instrumentation

The EXTRAFORT fuel rod was refabricated from a segment of the UO_2 fuel rod described above, at CEA Saclay, using the FABRICE technique. The mean burn-up of the FABRICE rod was 62.5 GWd/tU, which also was the maximum local value. The fuel column of the FABRICE rod was drilled over a 3 pellets height using the RISØ drilling technique. This technique has previously been validated on fresh fuel in the CALIF experiment which is presented below. The centreline temperatures were measured using a WRe 5%/WRe 26% thermocouple with rhenium cladding and insulated with HfO_2 (to avoid local heat-up by power production in the thermocouple). The thermocouple has an external diameter of 1.8 mm and was directly inserted in the 2.1 mm diameter drilled hole of the EXTRAFORT refabricated rod. The extremity of the thermocouple reached the middle of the third pellet (Figure 1) and the distance between the hot junction and the hole bottom was 3.5 mm. The correct positioning of the thermocouple was controlled by neutron radiography before and after the experiment.

Figure 1. Neutron radiography of the thermocouple inserted in the EXTRAFORT rod before the experiment

Experimental device

The CALIF and EXTRAFORT rods were irradiated in the ISABELLE 1 loop of the experimental reactor OSIRIS at CEA Saclay. This device is designed for experimental irradiation of PWR and BWR fuel under thermohydraulic and chemical conditions representative of prototype and commercial power reactors. Its design allows it to be inserted and removed while the OSIRIS reactor is operating.

The loop includes three main components:

- an in-pool section providing the containment and cooling of the sample to be tested and its positioning in the desired neutron flux near the core;

- an out-pool section which provides circuit pressurisation and supplies the fluid to the in-pool section;

- the connections between in and out-pool sections are provided by flexible metal tubes which are routed under water.

The experimental conditions of power ramps in the ISABELLE 1 loop have been qualified further to an extensive programme conducted from 1992 to 1994 [1]. This qualification concerns in particular the determination of the power experienced by the tested rod.

Validation of the drilling technique: CALIF experiment

The CALIF experiment was conducted to test whether the cryogenic RISØ drilling technique introduces cracks or changes in the fuel microstructure that could result in a modification of the fuel thermal conductivity. For this purpose, the CALIF fuel rod was manufactured at CEA Saclay. It was composed of two half-rods joined by a screwed piece, each of them made of fresh UO_2 pellets and instrumented with a central thermocouple. In the lower half-rod, the three inferior pellets were hollowed by classical drilling of full UO_2 pellets to receive the thermocouple whereas, in the upper half-rod, the three superior pellets were drilled using the CO_2 freezing technique created at RISØ. The pellets were manufactured by FBFC. All pellets were annealed in hydrogen in order to limit densification during irradiation. The diametral fuel-clad gap was limited in order to ensure the gap closure during irradiation. This fact was confirmed by rod diameter measurements performed after irradiation. The centreline temperatures were measured in both half-rods using WRe 5%/WRe 26% thermocouples with rhenium cladding and insulated with HfO_2. Both thermocouples were directly inserted in the 2 mm diameter holes of the hollow and drilled pellets. The extremity of the thermocouples reached the third hollow and drilled pellets, leading to a 4 mm gap between the hole bottom and the thermocouple clad. The correct positioning of the thermocouples was controlled by neutron radiography before and after the experiment.

The power history of the CALIF rod was composed of three sequences of power cycling, between 25 kW/m and 35 kW/m on the thermocouple level. The maximal power reached was determined by heat balance and gamma spectrometry after irradiation. This spectrometry, which provides the axial power profile, was also used to determine the local power on the thermocouple level. During the CALIF experiment, the temperatures measured in the two half-rods were in quite good agreement (Figure 2). The fact that slightly higher temperatures were systematically measured in the rod drilled by the RISØ technique is related to the upper position of this half-rod which triggers higher external temperatures for the clad. So, the CALIF experiment showed that the drilling technique by freezing did not introduce perturbations in fuel centreline temperatures.

Figure 2. Centreline temperatures measured during the CALIF experiment (fresh fuel)

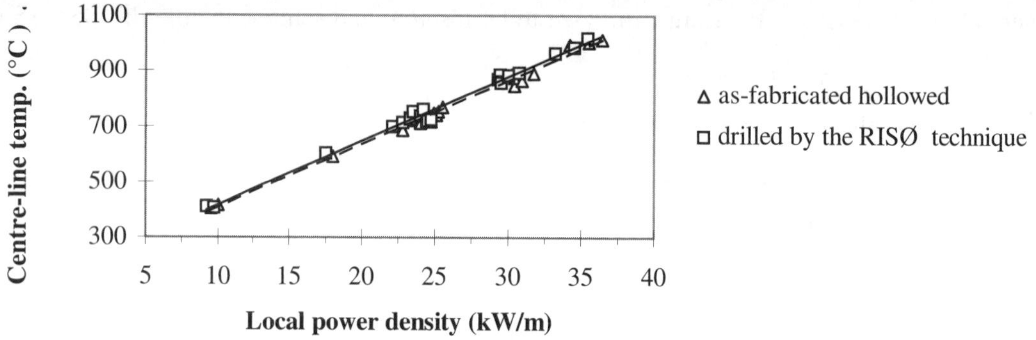

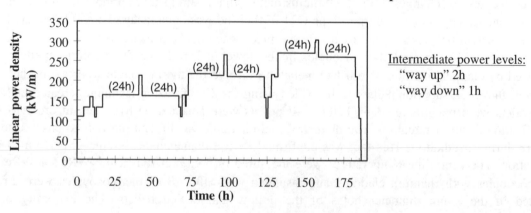

High burn-up temperature measurements: EXTRAFORT experiment

The power history (Figure 3) was composed of four sequences of power cycling between 12 kW/m and 35 kW/m on the maximum flux level and between 10.7 kW/m and 30.5 kW/m on the thermocouple level. Constant power levels were separated by slow power transients about 0.5 kW.m^{-1}.min^{-1}. The maximum power level was increasing at each cycle and maintained during two hours. For the last three sequences, before and after each maximum level, a constant power level of 24 hours was respected.

Figure 3. Power history of the EXTRAFORT experiment

During the EXTRAFORT experiment, temperature variations were directly related to power variations: the response of the thermocouple was not delayed. The plot of measured centreline temperatures as a function of linear density power on the thermocouple level is given for the EXTRAFORT experiment in Figure 4.

Temperature was stable during constant power levels, and no cycling effect was observed when coming back to a preceding power level. The EXTRAFORT experiment showed no recovery or degradation effect of the fuel thermal conductivity after an excursion at high power, even after a two-hour power step at 31 kW/m.

Figure 4. Temperatures measured during the EXTRAFORT experiment

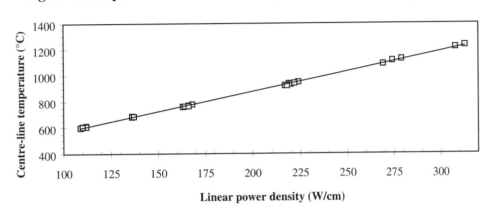

Post-calculations of the EXTRAFORT experiment

Post-calculations of the EXTRAFORT experiment have been carried out with the computer code METEOR/TRANSURANUS [2], developed by the CEA to calculate the thermal-mechanical behaviour of a fuel rod during nominal and incidental conditions. This 1-D½ fuel modelling code considers the thermal heat conduction as purely radial in fuel pellets and neglects the axial heat conduction in the fuel column although it is taken into account within the coolant.

Since the EXTRAFORT experiment focuses on the thermal aspect the main phenomena are those which directly influence the thermal calculation. Among them are to be found:

- The degradation of the fuel thermal conductivity with burn-up, which is taken into account in the code through various conductivity formulations.

- The rim effect which consists of an increase of the fuel porosity in the external pellet region leading to an overall fuel temperature increase.

- The fuel fragmentation and relocation. The effect of fuel cracks on the fuel diameter is taken into account by means of a supplementary fuel strain. The cracks are also taken into account to reduce the fuel thermal conductivity.

- The creation and behaviour of fission gas within the fuel rod. The influence of fission gas on the thermal calculation is taken into account in the model describing the pellet-clad gap conductance. In the case of the EXTRAFORT post-calculations, the URGAP model [3] was used.

- The corrosion of the external clad, which consists in the build-up of a zircon layer by oxidation of the zircaloy. The increase of the clad temperature due to the low conductivity of the oxide is therefore taken into account.

Fuel thermal conductivity

During the EXTRAFORT experiment many phenomena influence the thermal field of the refabricated rod. Nevertheless the radial jump in temperature inside the fuel pellets represents 90% of the total radial jump within the rod. For that reason our study focuses more specifically on fuel thermal conductivity modelling. Therefore five expressions of the fuel thermal conductivity were used to calculate the EXTRAFORT experiment. They are presented below. In these, unless specified, T represents the temperature in Kelvin, β the burn-up in at%, p the volume fraction of pores and bubbles, and σ a pore shape factor (with a value of 1.5 for spherical bubbles in the absence of other data).

Harding and Martin formulation with burn-up correction of Lucuta (1992)

This modelling of fuel thermal conductivity follows the recommendation of Harding and Martin for fully dense fuel [4]. The degradation of conductivity with burn-up has been adjusted on measurements performed by Lucuta [5] on SIMFUEL (simulated fuel with an equivalent burn-up of 3-8 at%). Porosity is also taken into account through the corrective term $(1-p)^{2.5}$.

$$\lambda = \left[\frac{1}{0.0375 + 2.165\ 10^{-4}\ T + 0.012\beta} + \frac{4.715\ 10^9}{T^2}\ \exp\left(-\frac{16361}{T}\right) \right](1-p)^{2.5} \tag{1}$$

Harding and Martin formulation with burn-up correction of Lucuta (1994)

This expression differs from the previous one in the choice of the burn-up correction applied. Here, the variation of fuel conductivity with burn-up stems from ulterior SIMFUEL experiments performed by Lucuta [6].

$$\lambda = \left[\frac{1}{0.0375 + (2.165\ 10^{-4} - 5\ 10^{-7}\beta\ )\ T + 0.0155\ \beta} + \frac{4.715\ 10^9}{T^2}\ \exp\left(-\frac{16361}{T}\right) \right](1-p)^{2.5} \tag{2}$$

CEA/LPCA formulation with burn-up correction of Lucuta (1994)

This expression applies the same burn-up correction than the previous one. But this time, the fuel thermal conductivity follows the recommendation of CEA/LPCA [7].

$$\lambda = \left[\frac{1}{0.062 + (2.068\ 10^{-4} - 5\ 10^{-7}\ \beta\ )\ T + 0.0155\ \beta} + \frac{6.157\ 10^9}{T^2}\ \exp\left(-\frac{16340}{T}\right) \right]\frac{(1-p)}{(1+2p)} \tag{3}$$

Halden formulation

This expression is based on a recommendation [8] at zero burn-up for 95% t.d. fuel identical with the MATPRO [9] formulation. The fuel thermal conductivity degradation with burn-up has been adjusted on in-pile measurements performed within the Halden Project [10]. T is expressed in °C and β in MWd/kgUO$_2$.

$$\lambda = \frac{1}{0.1148 + 2.475\ 10^{-4}(1 - 0.00333\ \beta)\,T + 0.0035\ \beta} + 0.0132\ \exp(0.00188\ T) \tag{4}$$

Lucuta formulation

To describe the thermal conductivity of irradiated UO$_2$ fuel, Lucuta [11] takes into consideration the effects of burn-up (dissolved and precipitated solid fission products), porosity and fission gas bubbles, deviation from stoichiometry and radiation damage. It results in an analytical expression including factors describing the above effects, which are applied to the expression for unirradiated UO$_2$ fuel thermal conductivity:

$$\lambda = K_{1d} \cdot K_{1p} \cdot K_{2p} \cdot K_{4r} \cdot \lambda_0 \tag{5}$$

$$\lambda_0 = \frac{1}{0.0375 + 2.165\ 10^{-4} \cdot T} + \frac{4.715\ 10^9}{T^2}\ \exp\left(-\frac{16361}{T}\right) \tag{6}$$

is Harding's expression for the thermal conductivity of unirradiated UO$_2$.

$$K_{1d} = \left(\frac{1.09}{\beta^{3.265}} + \frac{0.0643}{\sqrt{\beta}}\ \sqrt{T}\right) \arctan\left[\left(\frac{1.09}{\beta^{3.265}} + \frac{0.0643}{\sqrt{\beta}}\ \sqrt{T}\right)^{-1}\right] \tag{7}$$

quantifies the effect of the dissolved fission products.

$$K_{1p} = 1 + \frac{0.019\ \beta}{(3 - 0.019\ \beta)}\ \frac{1}{1 + \exp\left(-\frac{T - 1200}{100}\right)} \tag{8}$$

describes the effect of precipitated solid fission products.

$$K_{2p} = \frac{1 - p}{1 + (\sigma - 1)\,p} \tag{9}$$

is the modified Maxwell factor for the effect of the pore and fission gas bubbles.

$$K_{4r} = 1 - \frac{0.2}{1 + \exp\left(\frac{T - 900}{80}\right)} \tag{10}$$

characterises the effect of the radiation damage in the UO$_2$ matrix.

The thermal conductivity variations with temperature given by Eqs. (1-5) are plotted in Figure 5. The burn-up and porosity values chosen are representative of the EXTRAFORT experiment.

Figure 5. Fuel thermal conductivity given by Eqs. (1-10)
as a function of temperature for a burn-up of 6 at%

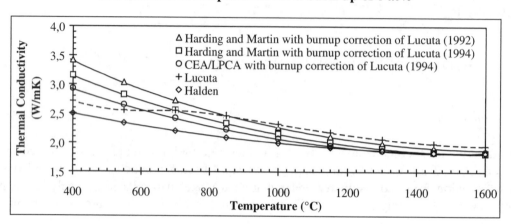

Calculation assumptions

The post-calculations of the EXTRAFORT experiment required the preliminary calculation of the "mother rod" irradiation up to 6 at% in a French PWR. Indeed the acquaintance of thermal stresses and deformations undergone by the "mother rod" up to the experiment is necessary to obtain a realistic representation of the FABRICE rod at the beginning of the experiment.

Pre-irradiation in PWR

This preliminary calculation has been carried out with a geometry including full and hollow pellets. Indeed the code does not allow to change full pellets into annular pellets during the calculation. Thus, the annular pellets of the FABRICE rod have to be introduced even during the calculation of the "mother rod" irradiation.

This introduction is realised with the conservation of the power density between the full and hollow pellets. The results provided by the computation of the PWR irradiation showed reasonable agreement compared with most of the "mother rod" examination results (fission gas release, fuel column lengthening, radial deformation of the clad, etc.). However, the non-conservation of the linear power density leads to an underestimation of the fuel-clad gap and of the clad oxidation layer thickness in the hollow pellets region. The corresponding underestimation generated on calculated centreline temperature does not exceed 15°C at the maximum power level of the EXTRAFORT experiment (30.5 kW.m⁻¹). Indeed, the pellet-clad gap is closed during this experiment and the radial jump in temperature due both to the gap and to the oxide layer represents under 5% of the total radial jump within the rod.

EXTRAFORT experiment

During the post-calculations of the EXTRAFORT experiment, the axial power profile determined by the post-irradiation gamma spectrometry is imposed (whereas the radial power distribution is calculated by the RADAR model [12]). The clad external temperature is imposed as well. Its value is calculated as a function of linear power density from temperature measurements in the coolant during the experiment and by considering a clad-coolant exchange coefficient responding to the Colburn relation [13-14].

Influence of axial heat transfer on measurements: 2-D code calculations

The METEOR/TRANSURANUS code does not take into account the axial heat transfer from full pellets towards hollow pellets. Yet during the EXTRAFORT experiment, temperature measurements were performed by a thermocouple which hot junction was close (3.5 mm) to the hole bottom. The interpretation of the experimental results consequently requires to study accurately the influence of the axial heat transfer on the temperature measured by the thermocouple.

Calculations with a 2-D code using the finite elements method CASTEM 2000 have been conducted to study the temperature field in the EXTRAFORT refabricated rod and then to evaluate the influence of axial heat transfer on the centreline temperature measured during the experimental irradiation. Two computation assumptions were considered: the first one, pessimistic, assumes a perfect heat transfer between fuel pellets and the thermocouple whereas the second one, more realistic, takes into account thermal conduction by helium and radiation. Calculations at several power levels show that the temperature decrease from the hole bottom is very sharp (cf. Figure 6).

Figure 6. Centreline temperature calculated with CASTEM 2000 in the full pellet and in the thermocouple for a linear power density of 32 kW/m in the hollow pellet

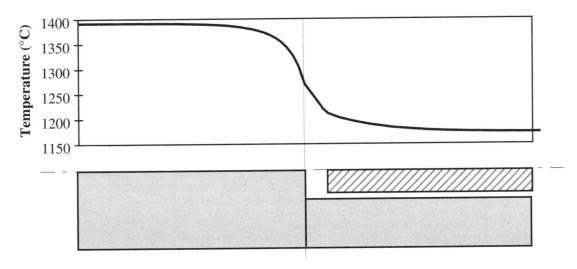

Nevertheless the temperature measured at the thermocouple hot junction still overestimates the stable temperature calculated in the hollow pellets far from the hole bottom. This overestimation linearly increases with the power level (cf. Figure 7). Even when considering a pessimistic assumption (perfect heat transfer), the temperature increase at the hot junction does not exceed 20°C at the maximum power level of the EXTRAFORT experiment (30.7 kW/m). Compared with the measured temperature of 1230°C, this value demonstrates that axial heat transfer weakly influences the temperature measurements.

Figure 7. Differences between the calculated temperature at the thermocouple hot junction and the stable temperature calculated in the hollow pellets far from the hole bottom given as a function of the linear power density on the thermocouple level

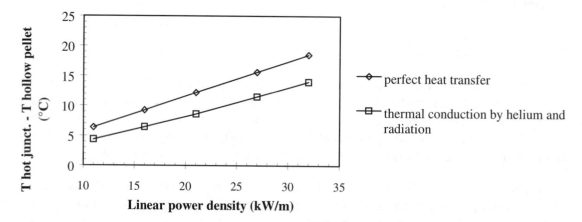

In the continuation of this paper, as the axial heat transfer is neglected in the 1-D½ computer code used, the thermocouple measurements will be corrected in order to be compared with the calculation results. This correction, which consists of a lowering of the measured temperature increasing with the power level, is around 10°C at 16 kW/m and 20°C at 32 kW/m.

Results and discussion

Post-calculations of the experiment have been carried out with the five fuel thermal conductivity formulations presented in Eqs. (1-5). The centreline temperatures obtained in the hollow pellets are compared to the EXTRAFORT measurements corrected of the effect of axial heat transfer. Differences in temperature between each calculation and the experimental results are plotted in Figure 8.

Figure 8. Differences in temperature between calculations with the fuel thermal conductivity given by Eqs. (1-5) and the EXTRAFORT measurements corrected of the effect of axial heat transfer

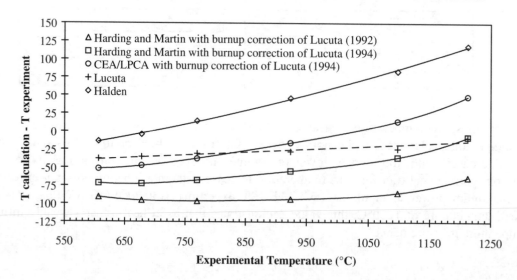

The calculations performed with the five conductivity formulations lead to centreline temperatures which differ from up to 180°C. The difference between calculations and measurements generally varies with temperature. At low temperature range calculations undervalue the experiment, whereas for high temperature the experiment is overestimated or at least less undervalued.

The Lucuta expression, contrary to the others, leads to an underestimation of the experimental results which varies very little with temperature. Indeed, Lucuta takes into account in his formulation the radiation damage through the K_{4r} factor given in Eq. (10), which induces a smoothing of fuel thermal conductivity dependence with temperature compared to the other formulations (cf. Figure 5). The results obtained with this expression seems to fit the temperature measurements with a maximal deviation of 40°C. Considering all the calculation assumptions, this result is quite remarkable. The remaining underestimation of the experiment could be partly attributed to the introduction of hollow pellets in the calculation of the "mother rod" irradiation, for a contribution around 13°C at the maximum power level of the experiment.

Conclusion

The centreline temperature measured in a commercial UO_2 fuel rod irradiated in PWR up to 62.5 GWd/tU, refabricated by the FABRICE technique and instrumented with a central thermocouple reaches around 875°C at 20 kW/m and 1200°C at 30 kW/m.

During the EXTRAFORT experiment, temperature was stable during constant power levels and no cycling effect was observed when coming back to a preceding power level. The experiment showed no recovery or degradation effect of the fuel thermal conductivity after an excursion at high power for a maximum of 24 hours, even after a power step at 31 kW/m on the thermocouple level.

Five expressions of the thermal conductivity of irradiated UO_2 were tested with a 1-D½ code in order to reproduce the temperature measurements of the EXTRAFORT experiment. The analysis points out the fact that taking into account the radiation damage in the thermal conductivity model, which reduces the temperature dependence of the thermal conductivity of the fuel, improves the centreline temperature predictions. Indeed, calculations carried out using the fuel thermal conductivity model proposed by Lucuta lead to a better prediction of centreline temperatures than the other expressions tested.

REFERENCES

[1]   A. Alberman, M. Roche, P. Couffin, S. Bendotti, D.J. Moulin and J.L. Boutfroy, *Nucl. Eng. Design*, 168 (1997) 293-303.

[2]   C. Struzik, J.C. Melis and E. Federici, Int. Topical Meeting on Light Water Reactor Fuel Performance, West Palm Beach, FL, 17-21 April 1994.

[3]   K. Lassmann and F. Hohleferd, *Nucl. Eng. Design*, 103 (1987) 215-221.

[4]   J.H. Harding and D.G. Martin, *J. Nucl. Mat.*, 166 (1989) 223-226.

[5]   P.G. Lucuta, Hj. Matzke, R.A. Verrall and HA Tasman, *J. Nucl. Mat.*, 188 (1992) 198.

[6]   P.G. Lucuta, Hj. Matzke and R.A. Verrall, *J. Nucl. Mat.*, 217 (1994) 279-286.

[7]   Private communication from the CEA/DRN/DEC/SPU/LPCA.

[8]   J.A. Christensen, Halden Int. Report, HIR-84.

[9]   D.L. Hagrman, G.A. Reymann (eds.), "MATPRO Version 11 – A Handbook of Materials Properties for Use in the Analysis of Light Water Reactor Fuel Rod Behaviour", TREE - NUREG - 1280, February 1979.

[10]   W. Wiesenack, "Assessment of UO_2 Conductivity Degradation Based on In-Pile Temperature Data", Proceedings of the ANS meeting, Portland, USA, 3-7 March 1997.

[11]   P.G. Lucuta, Hj. Matzke and I.J. Hastings, *J. Nucl. Mat.*, 232 (1996) 166-180.

[12]   I.M. Palmer, K.W. Hesketh and P.A. Jackson, IAEA Specialists' Meeting on Water Reactor Fuel Element Performance Computer Modelling, Preston, March 1992.

[13]   H.Y. Wong, Longman (eds.), "Handbook of Essential Formulae and Data on Heat Transfer for Engineers".

[14]   F. Kreith, Masson (eds.), "Transmission de la chaleur et thermodynamique".

THE COMPILATION OF A PUBLIC DOMAIN DATABASE ON NUCLEAR FUEL PERFORMANCE FOR THE PURPOSE OF CODE DEVELOPMENT AND VALIDATION

P.M. Chantoin
DRN/DEC/SDC/LEC
CEA Cadarache
Bat. 316
F-13108 St. Paul-lez-Durance
Cedex   France
+33 4 42 25 70 00

E. Sartori
OECD Nuclear Energy Agency
Le Seine St. Germain
12, Boulevard des Iles
92130 Issy-les-Moulineaux
France
+33 1 45 24 10 72

J.A. Turnbull
Independent Consultant
Cherry-Lyn
The Green
Tockington
Bristol BS32 4NJ
UK
+44 1454 620887

Abstract

This paper describes the compilation of a database on nuclear fuel performance for the purpose of code development and validation. The paper begins with the background behind its formation and the progress made to date in assembling 291 datasets from various sources encompassing PWR, BWR and WWER reactor systems. Agreement has been reached for the inclusion of further cases which will be processed into a common format for storage and retrieval in the OECD/NEA Data Bank.

Introduction

With increasing emphasis on economic as well as safe reactor operation, the approach to fuel performance assessment is now very much more detailed and closer to "best estimate" evaluation than in the past. Not only are the safety requirements more extensive, but the economics of nuclear power production are constantly under scrutiny to ensure minimum unit cost. This is particularly the case in countries where nuclear stations must compete with conventional ones. For these reasons it is no longer possible to support operation with anything other than well qualified code calculations where all aspects of fuel performance are treated simultaneously and in a self-consistent manner. Also, there is a move in some countries to "shadow" reactor operation with an on-line evaluation of fuel performance and a continuous assessment of core conditions against fixed limits for a large fraction of fuel assemblies in the reactor.

The principal function of a fuel performance code is to describe the behaviour of reactor fuel in the most accurate way possible under whatever conditions – both normal and off-normal – that are required by the licensing authority. By aiming to be a best estimate calculation or one that is intentionally biased, the uncertainties in the conditions under which the code is applied are under the control of the user.

The need for calculations to be best estimate necessitates that the code be developed and validated against good quality data. The most obvious source of these is the power reactors for which the calculations are to be applied. However, this source is insufficient, as data are also needed for fuel experiencing transients and other off-normal operating conditions which cannot be reproduced under experimental conditions in power reactors. For this reason, code development and validation must have access to both types of information and therefore it is of importance to include data obtained from dedicated experiments in test reactors.

This requirement was identified by the OECD/NEA Nuclear Science Committee (NSC) Task Force who, in their report [1] recommended the compilation of a public domain database on fuel performance for the express purpose of fuel performance code development and validation.

The Task Force ran concurrent with the IAEA FUMEX programme where "blind" predictions were compared with a selection of in-pile data from Halden Project experiments [2]. During this exercise it became apparent that many countries did not possess an adequate code for predicting LWR fuel rod behaviour, particularly during off-normal conditions. The principal reason for this was the inadequacy of the data available to these countries on which to develop and validate models. This was convincingly demonstrated by the improvements made by the end of the programme when advantage had been taken of the experimental data issued to participants.

Following the recommendations of the NSC Task Force, compilation of the International Fuel Performance Experiments Database (IFPED) commenced, and this paper provides a brief description of its structure and the progress made. The list of experiments incorporated in the database to date is given in Table 1. The database is operated by the OECD/NEA in co-operation with the IAEA. Although compiled for general use without restrictions, its success relies on participation. Therefore, incumbent on all users is the requirement to contribute whatever well-qualified data they too can make available. Only in this way can the objective of a standard comprehensive database be realised on which all codes can be developed and against which they can be judged.

Choice of data

From the outset it was recognised that the database should apply to all commercially operated thermal reactor systems and that the data should be both prototypic, originating from power reactor irradiations with pre- and post-irradiation characterisation, and test reactor experiments with in-pile instrumentation and PIE exploring normal and off-normal behaviour. It is recognised that experiments have been performed where the data remain of commercial interest, for example, much of the details of modern MOX performance remains proprietary to the manufacturers, and it is not the intention to compromise such arrangements. However, Zircaloy clad UO_2 pellet fuel can be largely regarded as a "standard product" and as such, release of what was previously proprietary data can only benefit the nuclear community at large.

A particular aspect of the compilation is the inclusion of data generated within internationally sponsored research programmes whose confidentiality agreements have expired. Such data, although available in principle, have not been widely used. The inclusion of such data is of particular importance where the originating organisation has changed its terms of reference. For example, the Risø laboratories in Denmark no longer perform nuclear research and find difficulty in resourcing the supply of information for their three fission gas release projects. In such cases, there exists the danger of losing access to the data altogether.

Extent of parameters included

The database is restricted to thermal reactor fuel performance, principally with standard product Zircaloy clad UO_2 fuel, although the addition of advanced products with fuel and clad variants is not ruled out. The data encompass normal and off-normal behaviour but not accident condition entailing melting of fuel and clad, resulting in loss of geometry. Recently the behaviour of defect fuel has been considered and it is planned to include such cases in the near future.

Of particular interest to fuel modellers are data on: fuel temperatures, fission gas release (FGR), clad deformation (e.g. creep-down, ridging) and mechanical interactions. In addition to direct measurement of these properties, every effort is made to include PIE information on the distribution of: grain size, porosity, Electron Probe Micro Analysis (EPMA) and X-ray Fluorescence (XRF) measurements on caesium, xenon, other fission product and actinides.

Emphasis has been placed on including well-qualified data that illustrate specific aspects of fuel performance. For example, cases are included which specifically address the effect of gap size and release of fission gas on fuel-to-clad heat transfer. Also in the context of thermal performance, the effect of burn-up on UO_2 thermal conductivity has been addressed. This is illustrated by cases where fuel temperatures have been measured throughout prolonged irradiation and at high burn-up where sections of fuel have been refabricated with newly inserted thermocouples. Regarding fission gas release, data are included for normal operations and for cases of power ramping at different levels of burn-up for fuel supplied by several different fuel vendors. In the case of power ramps, the data include cases where in-pile pressure measurements show the kinetics of release and the effect of slow axial gas transport due to closed fuel-to-clad gaps.

Brief description of data

At present, the database holds some 291 cases comprising the following:

Halden irradiated IFA-432	5 rods
The Third Risø Fission Gas Release Project	16 rods
The Risø Transient Fission Gas Release Project	15 rods
The SOFIT WWER fuel Irradiation Programme	12 rods
The High Burn-up Effects Programme	81 rods
Rods from Kola-3 FA-198 and 16 rods from FA-222	32 rods
Rods from the TRIBULATION programme	19 rods
Halden irradiated IFA-429	7 rods
Halden irradiated IFA-562.1	12 rods
Halden irradiated IFA-533.2	1 rod
Halden irradiated IFA-535.5 &.6	4 rods
Studsvik INTER-RAMP BWR Project	20 rods
Studsvik OVER-RAMP PWR Project	39 rods
Studsvik SUPER-RAMP PWR Project	28 rods

These are described briefly below highlighting some of the more important aspects of the data.

The first experiment to be addressed was the Halden irradiated IFA-432 commissioned by the USNRC for irradiation between December 1975 and June 1984. The main objectives of IFA-432 were measurement of fuel temperature response, fission gas release and mechanical interaction on BWR-type fuel rods up to high burn-ups. The assembly featured several variations in rod design parameters, including fuel type, fuel/cladding gap size, fill gas composition (He and Xe) and fuel stability. It comprised six BWR-type fuel rods with fuel centre thermocouples at two horizontal planes. Rods were also equipped with pressure transducers and cladding extensometers.

Data from five rods have been included in the database providing in-pile and limited PIE data up to 46 MWd/kg UO_2. An illustration of the type of data available on thermal performance is given in Figure 1. Here, the measured fuel centreline temperature normalised to 20 kW/m is plotted as a function of burn-up for rod 1 (230 micron gap) and the small gap rod 3 (80 micron). At the start of the irradiation, the difference in temperature is due to the 150 micron difference in gap size. This difference persists to such an extent that fission gas release commences in rod 1 at around 5 MWd/kg UO_2 resulting in a "poisoning" of the gap conductance and increased fuel temperatures, reaching a maximum at around 11 MWd/kg UO_2. Meanwhile, the low temperatures sustained in rod 3 prevented significant gas release and the slow monotonic increase in temperatures is predominantly due to degradation of the UO_2 thermal conductivity with irradiation damage and accumulation of fission products. Because of the difference in gap size, greater axial interaction was observed in the small gap rod 3 compared to the other larger gap rods.

The Risø Transient Fission Gas Release Project (Risø II) was executed in the period 1982-86. Short lengths of irradiated fuel were fitted with in-pile pressure transducers and ramped in the Risø DR3 reactor. The fuel used came from either IFA-161 irradiated in the Halden reactor or from segments irradiated in the Millstone BWR. Using this refabrication technique, it was possible to back fill the test segment with a choice of gas and gas pressure and to measure the time dependence of fission gas release by continuous monitoring of the plenum pressure. The short length of the test segment was an advantage because, depending on where along the original rod the section was taken, burn-up could be a chosen variable, and during the test the fuel experienced a single power. Some segments were tested without refabrication. Here the fuel stack was longer than in the case of the refabricated tests and hence the segments experienced a range of powers during the ramp depending on axial position in the test reactor. These "un-opened" segments were used to confirm that refabrication did not affect the outcome of the tests. Extensive hot cell examination compared the fuel dimensions and microstructure before and after the tests.

Some 17 tests were performed and all but one (which failed) have been included in the database and provide valuable information on fission gas release during power transients at high burn-up and also clad diametral deformation and fuel swelling as a function of ramp power and hold time. Figure 2 shows the evolution of fission gas release as a function of time during the power ramp for one of the tests using fuel from IFA-161. At each step, the fractional gas release shows a square root dependence on time which is characteristic of release by a diffusion type process. The sudden increase in release at the end of the test on decreasing power shows that towards the end of the hold time, there was contact between the fuel and the cladding, thus causing a restriction to the axial communication to the plenum where the pressure transducer was situated. This was confirmed by a comparison of diameter traces before and after the test which showed no significant diameter increase but a significant increase in permanent ridge height. The database also includes diametral profiles of retained fission products measured by EPMA and XRF. The radial position for the onset of release is clearly evident in Figure 3. The difference between the two types of measurements on retained xenon can be taken as the gas residing on grain boundaries; i.e. the difference between the total gas content (XRF) and the gas residing only in the matrix (EPMA).

The third and final Risø Project which took place between 1986 and 1990, bump tested fuel re-instrumented with both pressure transducers and fuel centreline thermocouples. The innovative technique employed for re-fabrication involved freezing the fuel rod to hold the fuel fragments in position before cutting and drilling away the centre part of the solid pellets to accommodate the new thermocouple. The fuel used in the project was from: IFA-161 irradiated in the Halden BWR between 13 and 46 MWd/kg UO_2, GE BWR fuel irradiated in Quad Cities 1 and Millstone 1 between 20 and 40 MWd/kg UO_2 and ANF PWR fuel irradiated in Biblis A to 38 MWd/kg UO_2. The data from the project are particularly valuable because of the in-pile fuel temperature and pressures measurements as well as extensive PIE. The database includes seven cases with ANF PWR fuel, six cases with GE BWR fuel and two cases using fuel from IFA-161. Within the test matrix it was demonstrated that the refabrication did not interfere with the outcome of the tests and that there was a correspondence with the results of the previous project.

Figure 4 shows a comparison of two tests using ANF fuel of near identical pre-test characteristics and power history during the power ramps. Both were refabricated with thermocouples and pressure transducers but one section was filled with helium whilst the other was filled with xenon. Diameter measurements before and after the ramp showed that the fuel-to-clad gap was closed. There is only a small difference in temperatures between the two cases, much smaller than predicted by many models. This shows that contrary to laboratory tests which demonstrate an effect of surface roughness on heat transfer between two surfaces in contact, at high burn-up, the surface roughness of the fuel and cladding for closed gap situations has little effect on fuel temperatures.

The HBEP was an international, group-sponsored programme managed by Battelle North West Laboratories whose principal objective was to obtain well-characterised data on fission gas release for typical LWR fuel irradiated to high burn-up levels. The programme was organised into three tasks, the first of which was a review of existing data. Tasks 2 and 3 comprised the experimental work carried out by the HBEP. Under Task 2, 45 existing fuel rods, either at moderate burn-up levels or undergoing irradiation to higher burn-up levels, were identified, acquired and subjected to PIE. Some rods were also subjected to power-bumping irradiations. Under Task 3, a series of fuel rods were built for irradiation in BR3 to high burn-up levels. Four design variations and three variations in operational history were used to study the effect of design and operation parameters on high burn-up FGR. This programme provides a substantial amount of data on fission gas release and the variations observed with different manufacturing routes, different reactor systems and different characteristics of the fuel. In the latter case it is possible to quantify the difference release between: high and low internal pressure, hollow versus solid pellets, the effect of different grain sizes and the effect of adding gadolinia to the UO_2 pellets. A total of 45 rods were considered under Task 2 and 37 rods for Task 3 and all have been included in the database.

The SOFIT programme was a series of experiments on WWER fuel carried out in the MR pool type research reactor at the Russian Research Centre Kurchatov Institute (IRTM) in co-operation with the Finnish Utility, Imatran Voima Oy (IVO). The programme was divided into three distinct phases, each addressing specific objectives:

SOFIT 1	SOFIT 2	SOFIT 3
Parametric fuel rod irradiations with basic steady state power histories up to moderate levels of burn-up as dictated by instrumentation endurance.	Parametric studies based on irradiation of re-instrumented high burn-up rods.	Irradiation testing under transient conditions.

At the moment, the database contains detailed information on six rods instrumented with centreline thermocouples and fission gas release data for two uninstrumented rods from SOFIT 1.1, centreline temperature data from two rods and fuel and clad elongation data from another two rods of SOFIT 1.3. It is hoped that further data from this programme will be released in the near future.

WWER rods are clad in ZR-1% Nb and contain hollow pellets. The temperature data are very much as expected, with the effect of differing gap size clearly seen in the first ramp to power. From the rods fitted with extensometers, the early life behaviour shows negligible change in the clad length apart from thermal expansion, whilst there is evidence of significant fuel column shrinkage due to densification, Figure 5.

The database also includes data for two pre-characterised standard WWER-440 fuel assemblies: FA-198 and FA-222 manufactured by the Russian fuel vendor Electosal and irradiated in the Kola-3 reactor. These assemblies were the centre of a programme called Blind Calculations for the WWER-440 High Burn-up Fuel Cycles Validation, initiated in Spring 1994 with the objective of testing the predictive capabilities of several Russian codes.

The maximum linear heat generation rate (LHGR) of FA-198 was <31 kW/m at the beginning of life and decreased to about 14 kW/m by the beginning of the fourth cycle and 11 kW/m at the end of life. In FA-222, peak LHGR values of 21 to 26 kW/m were experienced at the beginning of the second cycle, followed by steady state operation at LHGR of 10 to 22 kW/m before gradually decreasing to around 8 kW/m at the end of life. During the whole of the irradiation, both assemblies were located remote from any control rods. Consequently, the irradiation conditions are considered representative of base load operation for WWER-440 reactors.

The database contains details of 16 rods from each assembly; these are the two corner rods 7 and 120 and rods along the diagonal, Figure 6. As well as comprehensive pre-characterisation, the data include detailed 10 zone irradiation histories and PIE observations of dimensional changes and fission gas release. The measured gas release varied from ~0.5% for the diagonal rods to ~1.2% for the FA-198 corner rods, and 1-1.6% for the diagonal rods to 2.3-3.7% for the corner rods of the higher burn-up FA-222.

The objectives of the TRIBULATION programme were twofold. It was primarily a demonstration programme aimed at assessing the fuel rod behaviour at high burn-up, when an earlier transient had occurred in the power plant. The second objective was to investigate the behaviour of different fuel rod designs and manufacturers when subjected to a steady state irradiation history to high burn-up.

The first objective was met by irradiating fuel rods under steady state conditions in the BR3 reactor and under transient conditions in BR2. The effect of the transient was determined by comparing data from four identical rods tested as follows:

i)   BR3 irradiation followed by PIE;

ii) BR3 irradiation followed by BR2 transient then PIE;

iii) BR3 irradiation followed by BR2 transient and re-irradiated in BR3 before PIE;

iv) BR3 irradiation and continued BR3 irradiation to maximum burn-up before PIE.

The database contains data from 19 cases using rods fabricated by Belgo-Nucleaire (BN) and Brown Boveri Reactor GmbH (BBR). The matrix provides good data on clad creepdown and ovality as a function of exposure as well as the effect of different irradiation histories on fission gas release.

IFA-429 consisted of PWR type fuel rods assembled in three axially separated clusters of six rods and irradiated in the Halden Boiling Water Reactor. The eighteen-rod assembly had been designed to investigate gas absorption, fission gas release and thermal behaviour of UO_2 fuel during both steady state and a period of repeated rapid power transients. In order to achieve the objectives, the assembly was instrumented with nine vanadium neutron detectors, one cobalt detector. Two rods were instrumented with fuel centre line thermocouples and nine rods were instrumented with null-balance gas pressure transducers monitoring rod internal gas pressure.

This database contains power histories for seven rods as well as the measured temperature history for one of the middle cluster rods up to the burn-up level of 53 MWd/kg UO_2 and the measured internal pressure data for three upper and three lower cluster steady-state irradiated fuel rods subjected to two series of rapid power transients. Manually initiated gas pressure measurements were performed during the testing sequence allowing comparison of FGR data at three different fuel densities (91% TD, 93% TD and 95% TD), at two different grain sizes (6 and 17 microns) and at two different fuel-cladding gap sizes (200 and 360 microns) are possible. The power histories for the two series of power ramps are given schematically in Figure 7 whilst the measured rod internal pressure is shown as a function of cumulative time at high power for the first ramp series in Figure 8. The parabolic form of these curves is indicative of release by a diffusion controlled process.

IFA-562.1 contained twelve fuel rods assembled in two axially separated clusters irradiated to investigate the effect of pellet surface roughness on both fuel thermal and rod mechanical performances. As a second experimental objective the fuel grain growth caused by a power ramp at the end of the irradiation period was intended for investigation by PIE. To achieve the objectives fuel centre temperatures and clad elongation were measured in the fuel rods containing either rough or smooth pellets and filled with xenon or helium gas. To provide ideal conditions for the comparison, the rods were fabricated with small fuel-clad gap size and the fuel temperatures were kept below 1200°C during the base irradiation. The small initial gap size allowed gap closure, and the low fuel centre temperature prevented both fission gas release and grain growth. Nevertheless, no systematic differences were observed in either thermal or mechanical behaviour of the smooth and rough pellet rods. Consequently the effect of pellet surface roughness is insignificant in comparison with other influences such as the pellet-cladding gap size change, fuel densification and relocation.

After twelve years irradiation in the Halden Boiling Water Reactor two fuel rods (Rod 807 and Rod 808) were re-instrumented with fuel centre thermocouples and reloaded as IFA-533.2 into the reactor in order to investigate fuel thermal behaviour at high burn-up. The fuel rods were pre-irradiated with four other rods in the upper cluster of IFA-409. After base irradiation the four neighbouring rods were re-instrumented with pressure transducers and ramp tested in IFA-535.5 (slow) and IFA-535.6 (fast) providing useful data about FGR at two different ramp rates, Figure 9. As the irradiation history of IFA-533.2 in the first months was very similar to the history of the ramp tests, the fuel temperature and FGR data measured in the different IFAs complement each other, although the fuel-cladding gap sizes were slightly different and due to re-instrumentation the internal gas conditions were also dissimilar.

The database currently contains data from three Studsvik Ramping Projects, these are: INTER-RAMP, OVER-RAMP and SUPER-RAMP. Data are also available for DEMO-RAMP I & II and these will be included shortly. In all cases, PIE data are available after base irradiation and after the final ramp test, as well as whether or not clad failure occurred. The tests are therefore valuable for developing or validating code predictions of dimensional changes and FGR during power ramps as well as predictions of failure by PCI and SCC.

The objective of the INTER-RAMP project was to establish the PCI failure threshold of 20 standard un-pressurised BWR fuel rods on power ramping at burn-ups of 10 and 20 MWd/kgU. The fuel rods were supplied by ASEA-ATOM of Sweden. Nine rods were of identical reference design: 150 micron fuel-to-clad gap, 95% TD fuel pellets and re-crystallised cladding. The other 11 rods differed from these by way of gap size (80 and 250 microns), fuel density (one rod at 93% TD) and cladding heat treatment (8 rods stress relieved). The rods were base irradiated in boiling capsules in the Studsvik R2 reactor at powers ranging between 40 and 22 kW/m. The final base irradiation power was between 22 and 30 kW/m in all but two cases, where the final powers were ~38-39 kW/m. Power ramping took place in a pressurised loop simulating BWR conditions. Each test comprised a conditioning period of 24 hours at a power nominally identical to the final base irradiation power. This was followed by a fast ramp at a rate 4 kW/m/min to a terminal power in the range 41-65 kW/m, held for 24 hours or until failure was detected. In all, 11 of the 20 rods failed.

In the OVER-RAMP project, 39 rods from two sources were ramped under PWR conditions. Twenty-four rods were supplied by KWU/CE which had been irradiated at powers in the range 14-25 kW/m to burn-ups between 12 and 31 MWd/kgU in the Obrigheim reactor, Germany. The rods were in six groups of 4 rods covering different ranges of burn-up, diametral gap (137 to 171 microns) and different fuel grain sizes (4.5, 6 and 22 microns). A further 13 rods in three groups were supplied by Westinghouse after pre-irradiation in BR-3. The groups covered the burn-up ranges 18-24, 16-23 and 20-21 MWd/kgU with two different levels of fill pressure 13.8 and 24.8 bar. The power ramp was preceded by a conditioning phase lasting 72 hours where all rods were irradiated at 30 kW/m apart from 3 Westinghouse rods where 23 kW/m was chosen. The ramp rate was typically 10 kW/m/min but a small number of rods experienced slower rates. The ramp ended at a terminal power of typically ~42 kW/m but ranged between 37 and 53 kW/m. The maximum power was held for 24 hours or until failure was detected. Seven KWU/CE and 6 Westinghouse rods failed.

To date, only the PWR sub-programme of the SUPER-RAMP project has been included in the database; the BWR cases will follow soon. In the PWR sub-programme, 28 rods were divided into six groups of different designs:

- 2 groups of standard KWU/CE design PWR rods;

- KWU/CE standard design containing gadolinia doped fuel pellets;

- KWU/CE supplied large grain fuel;

- KWU/CE supplied annular pellet fuel;

- Westinghouse supplied standard design;

- Westinghouse supplied annular pellets.

The KWU/CE rods were pre-irradiated in the Obrigheim reactor and the Westinghouse rods were irradiated in BR-3. The time averaged power for all rods was 14-26 kW/m with final burn-ups in the range 33-45 MWd/kgU. The conditioning power was chosen to be 25 kW/m which was held for 24 hours. The ramp rate was 10 kW/m/min up to a ramp terminal powers of 39-50 kW/m which was held for 12 hours or until failure was detected. Two KWU/CE rods were tested at coolant temperatures lower than normal by 25 and 50°C and two other KWU/CE rods had conditioning times in excess of 24 hours. The results of the ramp tests were 2 out of 5 KWU/CE large grain fuel rods failed, 3 out of 5 Westinghouse standard rods and all KWU/CE annular rods failed.

Format of files

The criterion adopted for the file format was one primarily of simplicity. It was considered that users should be able to read the files independent of commercial software. It was recognised that the majority of codes were written in FORTRAN and therefore all files are in simple ASCII format for easy interrogation by file editors. Even text files are of this format despite the enforced limitations imposed by this approach. The adopted ASCII format does not preclude reformatting into a commercial database system sometime in the future if so desired. All databases have the following common elements:

- *Summary file*. This is a text file which describes the purpose of the experiment or test matrix and the scope of the data obtained.

- *Index file*. This is also a text file and lists all file titles and gives a brief summary of their contents.

- *Pre-characterisation*. This includes information on the fuel pellets and cladding used, their manufacturing route, dimensions and chemical composition including impurities. For the fuel, this is augmented by details of enrichments, porosity distribution, re-sintering test data and microstructure. For the cladding, additional information includes mechanical properties, corrosion characteristics and texture as and when available. Details of the fuel rod geometry include: relevant dimensions, fuel column length, weight, fill gas composition and pressure. Details of reactor irradiation conditions.

- *Irradiation histories*. All histories are in condensed form with care to ensure that all important features are preserved. Where there was a significant axial power profile the history is provided in up to 12 axial zones. For each time step, the data are provided as: time, time increment over which power was constant, clad waterside temperature and local heat rating for the prescribed number of axial zones. Information is provided to calculate the fast neutron flux and its spatial variation if the information is available.

- *In-pile data*. When applicable, separate files tabulate data from in-pile instrumentation as a function of variables such as: time, burn-up or power. For example, IFA-432 fuel rods were equipped with centreline thermocouples and cladding elongation detectors. Files were created tabulating temperatures at constant powers of 20 and 30 kW/m as a function of local burn-up throughout life. At several times during the irradiation, at approximately 5 MWd/kg UO_2 intervals, temperature is tabulated against local power during slow ramps. In this case, the data are reproduced directly from the original files where signals were logged every 15 minutes. A similar procedure was adopted for clad elongation measurements; during short periods of variable power, 200-5000 hours duration, elongation is tabulated against rod average power showing cases where there was pellet-clad mechanical interaction.

- *PIE data*. Where such examinations were made, data are recorded either in tabular or text form. Dimensional data include axial and diametral dimensions before and after irradiation, and post irradiation ridge heights where available. Data on fission gas release include rod averaged values obtained by puncturing and mass spectrometry, local values from whole pellet dissolution and across pellet spatial distributions as measured by gamma scanning, EPMA and XRF. Before and after porosity and grain size distributions are given as is the radial position for the onset of grain boundary porosity when such measurements have been made from metallographic examination.

All data files are held centrally on the OECD/NEA Data Bank computer system from which all files are dispatched. This single source for distribution is necessary for Quality Assurance purposes, particularly for the tracking and release of upgraded or corrected files. With the OECD's experience of data bank management, this arrangement ensures long term availability of the service.

Often PIE data are only available in graphical form and photomicrographs which are difficult to preserve in the current ASCII file format. For this reason, the possibility of scanning figures and photographs for storage and retrieval using the medium of the CD ROM was investigated. As a result, all reports available for compiling the datasets to date have been scanned and copied onto a single CD-ROM. The files are all in Tif-Group IV format and can easily be read and printed using a wide range of software.

First use of database

IAEA 4/012 programme

The database (at that time, 137 rods of diverse origins) was first proposed to countries for inclusion in the IAEA regional technical co-operation programme for Europe RER 4/012. The aim of this programme was to transfer to participants a nuclear fuel modelling computer code, and developing it for application to WWER reactor systems. Countries involved in this technical co-operation programme are Armenia, Bulgaria, Czech Rep., Hungary, Poland, Romania, Slovakia and Ukraine.

The choice of the European developing countries, which, as far as nuclear energy is concerned, means mainly the east European countries, was dictated by the urgency of the situation. In the period 1990-1992, following the political changes in eastern Europe, the situation in the area of nuclear safety was fairly precarious. Nuclear Regulatory Authorities were non-existent and nuclear fuel was loaded in the reactors without any licensing procedures. In addition, the quality assurance system was unknown.

With the Agency's help Regulatory Authorities were progressively set up in these countries. After completion of this task, the next step is to ensure that these national authorities function in a normal way and the Agency has therefore embarked on a series of programmes addressing different aspects of fuel behaviour modelling, with the objective of giving as much available information as possible on WWER fuel through literature or meetings or by providing the tools necessary to perform fuel and fuel computer codes licensing. Amongst these tools the transfer of a mature computer code (TRANSURANUS developed by TUI Karlsruhe) and of the fuel database were two corner stones of the project.

The transfer of the fuel database was made during an Agency training course, held at the Halden reactor project in September 1996, in which the data proposed to the modellers were analysed according to the type of utilisation to be made. On one hand, experiments were reviewed one by one. On the other hand they were reviewed according to their utilisation in models addressing the following parameters:

- thermal performance;

- fission gas release;

- pellet clad interaction.

In co-operation with TUI Karlsruhe and the help of the database, the exercise of developing the TRANSURANUS code for WWER applications is now underway in several countries.

For the future

The creation of the database has met with universal approval and consequently there has been no difficulty experienced in obtaining data for inclusion. Current work focuses on completion of the Studsvik ramping tests of which there are 16 rods in the SUPER-RAMP BWR sub-programme, 5 rods in the DEMO-RAMP I and 8 rods from DEMO-RAMP II BWR programmes. Agreement has been reached with Framatome, EDF and the CEA for the release of 3 Fessenheim rods of 3, 4 and 5 cycles irradiation, 2 experimental gas flow rods: CONTACT 1 & 2 and 10 experimental defect fuel rods irradiated in the Siloe research reactor. This will bring the number of datasets within the database to 335. In addition, discussions are in hand for the inclusion of WWER fuel ramped in the MR reactor at the Kurchatov Institute, Moscow, and CANDU data from both Romania and AECL Canada. Interest in the database has been shown from China, Argentina and India with promises of yet further data to add to the list of cases.

Finally, it may be concluded that the initiative shown for the creation of the publicly available database has been given world-wide acceptance from countries whose nuclear industries and fuel performance assessment capabilities are in various of stages of maturity. It is hoped therefore that wide dissemination of the database will contribute to a wholesale improvement in code capabilities and that these will work their way through to a safer and more economic use of nuclear fuel in all reactor systems.

Acknowledgement

The authors wish to acknowledge the co-operation of the following participating organisations for the supply of their data and assistance in preparing the database: Risø National Laboratory, Denmark; Halden Project, Norway; Imatran Voima Oy, Finland; The Kurchatov Institute, Russian Federation; Battelle North West, USA; Belgo-Nucleaire, Belgium and Studsvik Laboratories, Sweden. It should be noted that several organisations are preparing data for further inclusion. These are: CIAE, Beijing China; CEA, France; AECL, Canada; and INR, Pitesti Romania.

REFERENCES

[1]   "Scientific Issues in Fuel Behaviour," a report by an NEA Nuclear Science Committee Task Force, *OECD Documents*, January 1995. ISBN 92-64-14420.

[2]   Chantoin P., Turnbull J.A., Wiesenack W., "Summary of the Findings of the FUMEX Programme," Paper presented at the IAEA Technical Committee Meeting on Water Reactor Fuel Element Modelling at High Burn-up and its Experimental Support, Windermere, UK, September 1994, IAEA-TECDOC-957, pg. 19-37.

Table 1. International Fuel Performance Experiments Database (IFPE)

Operated by the OECD Nuclear Energy Agency in co-operation with the
International Atomic Energy Agency (IAEA)

List of compiled data as of May 1997

Origin of data	Data-set-name	Package-Id
IFPE/IFA, Halden Reactor Project Fuel Performance Experiments Data	IFPE/IFA-432	NEA 1488/02
	IFPE/IFA-429	NEA 1488/03
	IFPE/IFA-562.1	NEA 1488/04
	IFPE/IFA-533.2	NEA 1488/05
	IFPE/IFA-535.5 & 6	NEA 1488/06
IFPE/RISØ II, Fuel Performance Data from Transient Fission Gas Release	IFPE/RISØII-A	NEA 1502/01
	IFPE/RISØII-B	NEA 1502/02
	IFPE/RISØII-E	NEA 1502/03
	IFPE/RISØII-GE-A	NEA 1502/08
	IFPE/RISØII-GE-B	NEA 1502/09
	IFPE/RISØII-GE-G	NEA 1502/10
	IFPE/RISØII-GE-H	NEA 1502/11
	IFPE/RISØII-GE-I	NEA 1502/12
	IFPE/RISØII-GE-K	NEA 1502/13
	IFPE/RISØII-GE-L	NEA 1502/14
	IFPE/RISØII-GE-M	NEA 1502/15
	IFPE/RISØII-GE-N	NEA 1502/16
	IFPE/RISØII-GE-O	NEA 1502/17
	IFPE/RISØII-H	NEA 1502/04
	IFPE/RISØII-I	NEA 1502/05
	IFPE/RISØII-IFA161B	NEA 1502/18
	IFPE/RISØII-K	NEA 1502/06
	IFPE/RISØII-L	NEA 1502/07
IFPE/RISØ-III, Fuel Performance Data from 3rd Risø Fission Gas Release	IFPE/RISØIII-AN1	NEA 1493/01
	IFPE/RISØIII-AN10	NEA 1493/06
	IFPE/RISØIII-AN11	NEA 1493/07
	IFPE/RISØIII-AN2	NEA 1493/02
	IFPE/RISØIII-AN3	NEA 1493/03
	IFPE/RISØIII-AN4	NEA 1493/04
	IFPE/RISØIII-AN8	NEA 1493/05
	IFPE/RISØIII-GE2	NEA 1493/08
	IFPE/RISØIII-GE4	NEA 1493/09
	IFPE/RISØIII-GE6	NEA 1493/10
	IFPE/RISØIII-GE7	NEA 1493/11
	IFPE/RISØIII-IFA161	NEA 1493/16
	IFPE/RISØIII-II1	NEA 1493/12
	IFPE/RISØIII-II2	NEA 1493/13
	IFPE/RISØIII-II3	NEA 1493/14
	IFPE/RISØIII-II5	NEA 1493/15
IFPE/HBEP Batelle's High Burn-up Effects Programme for Fuel Performance	IFPE/HBEP-2	NEA 1510/01
	IFPE/HBEP-3	NEA 1510/02
IFPE/SOFIT, WWER-400 & 1000 Fuel Performance Experiments Database	IFPE/SOFIT-1.1	NEA 1310/01
	IFPE/SOFIT-1.3	NEA 1310/02
IFPE/KOLA-3, WWER-440 Fuel Performance Data from Kola-3 NPP	IFPE/KOLA-3	NEA 1532/01
IFPE/TRIBULATION Data for 11 BN and 8 ABB Fuel Rods Tested in the TRIBULATION Programme	IFPE/TRIBULATION	NEA 1536/01
IFPE/INTER-RAMP (Studsvik) Data from 20 ASEA-ATOM BWR Fuel Rods	IFPE/INTER-RAMP	NEA 1555/01
IFPE/OVER-RAMP (Studsvik) Data from 24 KWU/CE PWR Rods	IFPE/OVER-RAMP	NEA 1556/01
IFPE/SUPER-RAMP (Studsvik) Data from 24 Rods of Different KWU/CE and Westinghouse PWR Design	IFPE/SUPER-RAMP	NEA 1557/01

Figure 1. Fuel centreline temperatures at constant power as a function of burn-up for rods 1 (larg gap) and 3 (small gap) of the Halden irradiated IFA-432 experiment

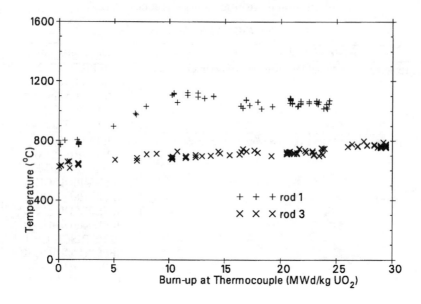

Figure 2. Fission gas release as determined from rod internal pressure for test Risø-b (Risø Transient Fission Gas Release Project) during the power ramp in the DR3 reactor

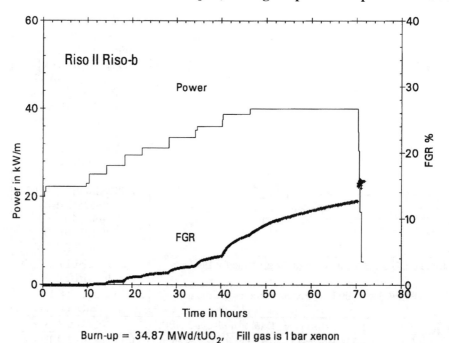

Burn-up = 34.87 MWd/tUO$_2$,   Fill gas is 1 bar xenon

Figure 3. Comparison of EPMA and XRF retained xenon profiles for test GEa of the Risø Transient Fission Gas Release Project

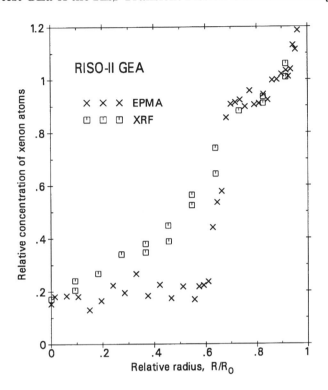

Figure 4. comparison of centreline temperatures of two nominally identical rods AN3 and AN4 filled with 15 bar He and 1 bar Ne respectively during the power ramp in DR3 for the Third Risø Fission Gas Project

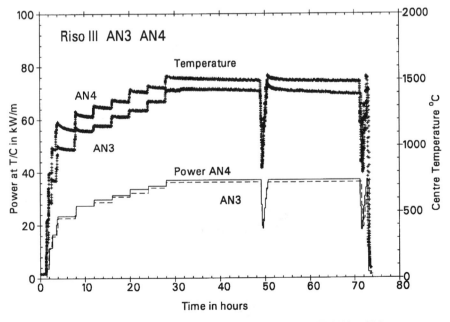

Figure 5. Axial extension of fuel cladding as a function of time for rod 3 of SOFIT 1.3

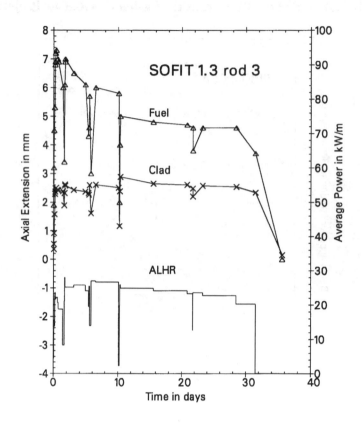

Figure 6. Arrangement and numbering of rods in the WWER-440 assemblies FA-198 and FA-222 irradiated in the Kola-3 reactor

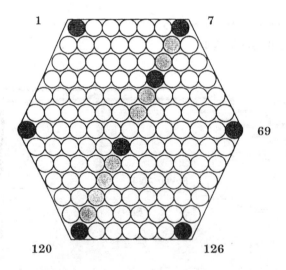

Figure 7. Schematic histories for the two series of power transients experienced by rods in IFA-429

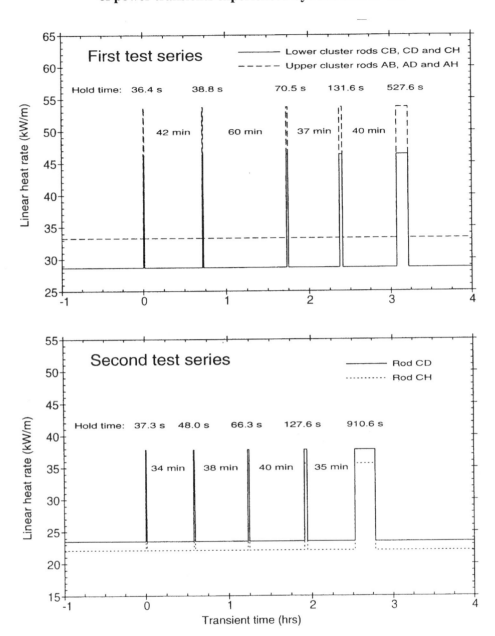

Figure 8. Measured rod internal pressure as a function of cumulative time at peak power for rods during the first series of power transients in IFA-429

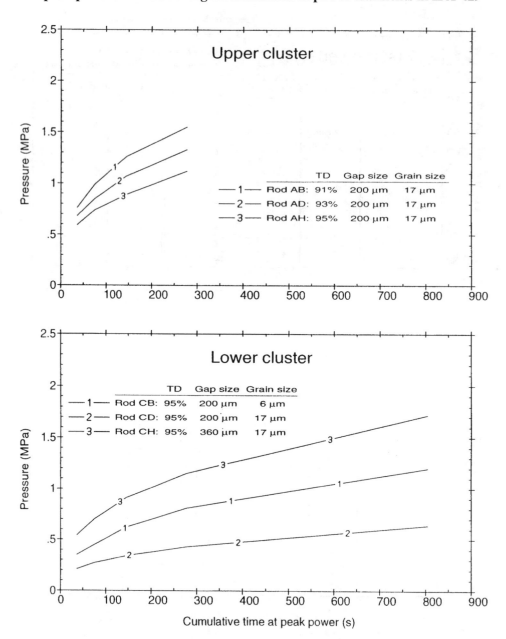

Figure 9. Fission gas released for the fast ramped IFA-535.5 and the slow ramped IFA-535.6 rod as a function of time

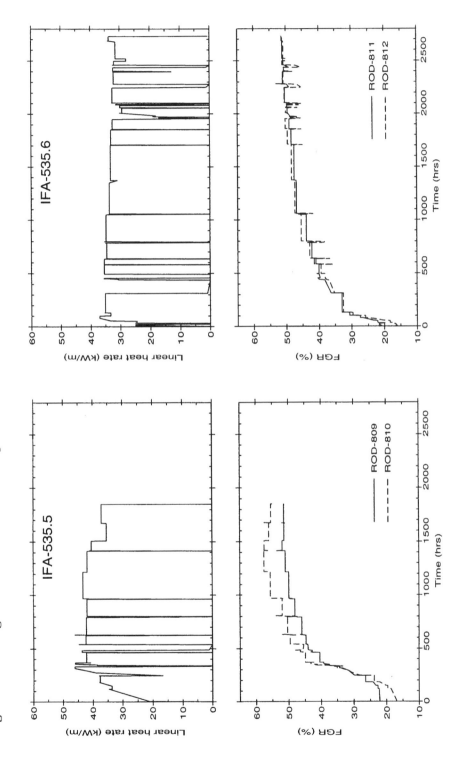

COMPARATIVE THERMAL BEHAVIOUR OF MOX AND UO₂ FUELS

Marc Lippens and the COMETHE Team
BELGONUCLEAIRE S.A. – Avenue Ariane, 4 – 1200 Brussels, Belgium

Abstract

The paper aims at providing a simulation of the in-reactor thermal behaviour of MOX fuel comparatively to UO_2. The MOX thermal conductivity is firstly reviewed. The introduction of Pu in UO_2 leads to a significant but small decrease of conductivity, of about 4% for 10% Pu. The introduction of Pu also modifies the radial power distribution. At beginning of life, the large absorption of thermal neutron flux on Pu isotopes increases the power depression, reducing the fuel central temperature. With burn-up increase, the preferential Pu consumption at pellet periphery leads to a relative increase of power at pellet centre, and thus an increase of fuel central temperature, all other parameters being kept constant.

Besides thermal conductivity and radial power profiles, the other variables controlling fuel thermal behaviour are not directly sensitive to Pu content. Modelling of these variables (gap closure, fuel restructuring, ...) is made as for UO_2.

The thermal conductivity of MOX fuel degrades with burn-up. Modelling of fuel central temperature with the COMETHE code, using the conductivity degradation set up for UO_2 provides a good agreement with fuel central temperature measured in the FIGARO International Programme.

Introduction

The irradiation of MOX fuel in LWRs is today an industrial reality, as demonstrated by the large and increasing number of MOX fuel assemblies unloaded after having reached burn-ups of about 40 GWd/tM [1]. The objective of high burn-up achievement for MOX is meaningful, owing to large amounts of available Pu and the high cost of fuel fabrication. However, to achieve this objective, optimised power histories are required not to excessively increase the fission gas release within rod, and accurate fuel thermal calculations are needed so as not to consume the margins facing design criteria and penalise the reactor operation.

The present paper first analyses the influence of Pu on fuel thermal conductivity. The influence of Pu on radial power distribution is also evaluated. After that, the global MOX in reactor thermal performance is analysed and the conductivity degradation with burn-up assessed as compared with UO_2. Finally, the predictions of fuel central temperature using the COMETHE code [2] are compared to experiments made in the frame of the FIGARO International Programme [3,4].

The thermal conductivity

Expectations

The MOX fuel fabricated for LWRs is very close to stoichiometry and has a Pu content less than about 10% (Pu/U+Pu). However, the UO_2 and PuO_2 crystals have nearly the same characteristics [except a marked reduction in melting temperature for PuO_2] (Table 1), and completely dissolve into each other when stoichiometric.

Under these conditions, the conductivities (k) of both materials are expected to be similar, with a slight reduction for k_{MOX} in the low temperature range (< 1500°C) due to the perturbations in the phonon vibrations introduced by the presence of Pu.

Open literature data

The thermal conductivity of $(U_{1-y} Pu_y) O_{2 \pm x}$ (y = 0 - 0.3; 1.0) has been extensively measured, analysed and reviewed in the past [5-7], and has also been the subject of more recent studies [8-11]. From these varied works, no univocal conclusion can be obtained on the extent of conductivity decrease with Pu content. The reasons for this are multiple:

- the expected decrease of k with Pu is small;

- the influence of porosity (content, shape, spatial distribution) is not correctly taken into account;

- insufficient consideration is given to the influence of O/M and its change during testing;

- sample microstructure changes during testing (microcracks, densification), especially at high temperatures;

- the measurements are inaccurate;

- the use of indirect methods (thermal diffusivity combined with specific heat) increases the errors.

Models from open literature

A theoretical approach of the change of phonon conductivity with introduction of foreign atoms in a pure lattice can assist in assessment of which parameters are modified in the classical form $1/k = A+BT$ when introducing Pu.

The thermal resistance (A) due to lattice point defects can be described by the formula [12]:

$$A = \left(4\pi^2 \, \overline{V_o} \, \theta_D \, / c^2 h \right) \sum_i \sigma_i n_i$$

with

$$\sigma_i n_i = \frac{y}{12}\left[\left(\frac{M_i - \overline{M}}{\overline{M}}\right)^2 + \eta\left(\frac{r_i - \overline{r}}{\overline{r}}\right)^2\right] \tag{1}$$

and $\eta = 32(1+1.6\gamma)^2$.

$\overline{V_o}$	– average atomic volume
θ_D	– Debye temperature
c	– average phonon velocity
h	– Planck's constant
$\overline{M}$	– mass of host lattice
M_i	– mass of foreign atom
$\overline{r}$	– ionic radius of host lattice
γ	– Gruneisen constant
η	– constant representing the lattice straining due to foreign atom
$\sigma_i n_i$	– scattering cross-section for element in concentration $n_i \, (y \, for \, Pu)$

This equation indicates that Pu increases the A term mainly through its difference of radius as compared to U.

Eq. (1) has been used by D. Olander [13] to model the scattering of phonons by Pu atoms. He obtains a simple expression for the change ΔA in the A term: $\Delta A = 5.4 \, y$ cm K/W. The coefficient 5.4 represents the combination of several physical constants in the general determination of the conductivity and has been inferred from observed conductivity reduction associated to stoichiometry change.

A simple expression also exists to describe the intrinsic lattice thermal resistance BT [14]:

$$BT = \gamma^2 T / \left(3.81(k/h)^3 \overline{M} V_o^{1/3} \theta_D^3\right) \tag{2}$$

with: T – absolute temperature

k – Boltzmann's constant.

The influence of Pu on BT term can be obtained using similarity laws applying when comparing UO_2 and PuO_2 crystals, and using the Lindemann transformation expressing θ_D versus the melting temperature T_m [15]. One has:

$$B_{MOX} = B_o (M/M_o)^{1/2} (a/a_o)^2 (T_{mo}/T_m)^{3/2} (\gamma/\gamma_o)^2 \qquad (3)$$

with index 0 referring to UO_2.

One concludes from Eq. (3) that the B term is increased by the Pu presence through a reduction of the melting temperature (Table 1).

From this analysis of A and BT terms, one concludes that both terms are susceptible to be increased in the presence of Pu.

Conductivities proposed in open literature

Changes in conductivity are noticed in the presence of Pu. The data of R.L. Gibby [5] indicate a decrease of conductivity of about 6% for 10% Pu at 100°C and 9% for 20% Pu. This decrease is enhanced with temperature.

The A.R.G. Washington compilation [6] reviewed by D.G. Martin concludes to a reduction between 0 and 10% for 20% Pu in the range 500-1800°C, with a final recommendation of a 5% decrease.

Y. Philipponneau [10,11] measured himself and summarised numerous data on conductivity of non-stoichiometric mixed oxide for several Pu contents. He concludes that the conductivity slightly decreases for small amounts of Pu, but the accuracy of measurements is not sufficient to formulate a quantitative proposal.

The CEA/FRAMATOME/EDF view [8,9] on conductivity change in the presence of Pu is to neglect the influence of this one.

The MATPRO correlation [16] recommends a linear decrease of k with Pu of about 4% for 10% Pu. This recommendation results from an analysis of thermal diffusivity data obtained by R.L. Gibby [5], L.A. Goldsmith and J.A. McDouglas [17] and H.E. Schmidt [18]. The dependence of k versus Pu content is introduced mainly through the B term which is enhanced. The final decrease of conductivity is however slightly less pronounced due to a small increase of the specific heat with Pu content.

COMETHE correlation

The dependence of thermal conductivity vs. Pu content was assessed in the 70s and updated in the 80s. The lastest study is mainly based on Gibby's data. His data were selected for several reasons:

- the samples were stoichiometric, of high density, with several Pu contents up to 30%;

- the fuels were fabricated by co-precipitation;

- the test conditions guaranteed a good control of O/M.

Appropriate consideration of other published data and analyses was made by verifying the absence of contradiction between these ones and the obtained correlation. This one includes a linear dependence of Pu in the A and B terms, by the addition of the following terms respectively: dA/dy = 5.0 cm K/W and dB/dy = 0.005 cm/W.

The conductivity decrease versus Pu content is shown in Figure 1. It decreases with a slope of about 4% for 10% Pu. The sensitivity versus Pu content is slightly reduced with increasing temperature. This correlation is quite compatible with the MATPRO proposal.

The conductivity law used in COMETHE code takes into account the most important variables modifying the reference value for dense UO_2: porosity, Pu content, Gd content, non-stoichiometry and degradation with burn-up. Influence of Pu content and burn-up on fuel central temperature is analysed in the sections entitled *Burn-up influence on conductivity* and *The FIGARO experiment*.

Radial power profile

The presence of Pu increases the thermal flux depression towards pellet centre, due to large absorption of thermal neutrons on fissile and fertile Pu isotopes.

The Pu build-up from breeding of ^{238}U is on the other hand expected to be similar to that of UO_2 fuel.

These effects are introduced in the COMETHE code owing to a neutronic module allowing to calculate radial power profile and its change with burn-up. The validation of the module has been made by comparison with LWR-WIMS calculations and experimental results obtained by EPMA, SIMS and radial radiochemistry. Figure 2 illustrates the calculated and measured radial distributions of the Pu isotopes for a MOX fuel irradiated to high burn-up.

The influence of radial power profile on fuel temperature has been determined for UO_2 (4.5% ^{235}U) and MOX (0.2% ^{235}U; 8.8% Pu) fuels irradiated up to 60 GWd/tM. Chosen enrichments are compatible with burn-up objective. For the MOX fuel, it corresponds to a central rod in the MOX assembly. The calculations are made with k = k (UO_2, no burn-up dependence) to evince also the influence of the power profile.

Results are illustrated in Figure 3. The power depression at BOL and at centre varies from 0.95 for UO_2 to 0.78 for MOX. At rim, the Pu build-up is noticed in both fuels with burn-up increase. The local power is however more pronounced in UO_2 fuel due to higher reactivity of fissile Pu compared to ^{235}U. At pellet centre, the depression increases with burn-up for UO_2 (0.84), but decreases for MOX (0.90). The increase for UO_2 results from the compensation of overpower at rim when averaging the pellet power, whereas for MOX the lower depression is associated to the high residual fissile Pu at centre. Figure 2 confirms the marked and preferential consumption at pellet periphery of the as-fabricated Pu.

The fuel pellet central temperature at BOL is slightly lower for MOX (35-40°C/200 W/cm depending on gap size) compared to UO_2. At high burn-up, the situation is reverse, the MOX central temperature is enhanced (+20°C/200 W/cm) compared to UO_2.

Burn-up influence on conductivity

The observations of in-reactor behaviour of MOX fuel (fission gas release, central temperature) indicate that the fuel thermal conductivity degrades with burn-up. The extent of degradation is essentially the same as in UO_2. Such a situation is compatible with our understanding of the origins of degradation, i.e. the increasing number of fission products acting as phonon scattering centres in the UO_2 lattice when burn-up increases.

The MOX fuel being composed for more than 90% by UO_2 and the fission yields for Pu fissile isotopes being similar to those of ^{235}U, it is therefore normal to observe a conductivity decrease with burn-up similar, if not identical, in MOX and UO_2 fuels.

The COMETHE conductivity law includes a dependency with burn-up, leading to a linear increase of A term. The burn-up coefficient in the A term has been determined in the past on the base, amongst others, of the work made by R.C. Daniel and I. Cohen on UO_2 [19]. It has recently been increased in view of the new experimental data obtained in the frame of the Halden Reactor Project, indicating a degradation more important than the one of Daniel and Cohen.

Besides the parameters mentioned before, i.e. small influence of Pu on conductivity and the radial power profile, no other parameter is retained as being Pu-dependent for the evaluation of MOX thermal performance. Gap closures and conductance, as well as fuel porosity development are processed as for UO_2.

Table 2 illustrates some of the features mentioned above and having an impact on the fuel central temperature. The table shows the fuel average (i.e. average between MOX and UO_2) central temperatures, and compares the difference of central temperature between MOX and UO_2 for different types of conductivity at BOL and EOL for different operating conditions.

At BOL, one finds a lower temperature for MOX compared to UO_2, in agreement with the analysis in the section entitled *Radial power profile*. At high burn-up, the situation is the reverse, and is amplified when burn-up degradation of conductivity is taken into account.

One observes that the reduction of conductivity with Pu does not markedly influence the fuel thermal performance. This one is dominated by the linear power and the conductivity decrease with burn-up, as in UO_2.

The FIGARO experiment

Fuel central temperatures and kinetics of fission gas release were measured on highly burned MOX fuel in the frame of the FIGARO International Programme [4].

The programme consists mainly of base irradiation in Beznau 1 PWR (Switzerland) of MOX fuel fabricated by BELGONUCLEAIRE following the MIMAS process. After having reached a peak pellet burn-up of 48 GWd/tM, two fuel rods were extracted from a mother assembly, and four fuel segments cut and sent to Kjeller hot cells (Norway) for refabrication with pressure transducer and fuel central thermocouple. Then followed short irradiations in the Halden HBWR according to power history designed for observation of fission gas release (Figure 4). Temperature and pressure were recorded with power changes. An extensive programme of post-irradiation examination has been made after base irradiation and power transient. Figure 5 illustrates the dependency of temperature versus power measured for one of the segments. This dependency is quite usual. One notices with time a slight power decrease when returning to temperature close to 800°C (plateau in temperature around 200-220 W/cm).

The fuel central temperatures calculated with the COMETHE code are in good agreement with observations. The plateau in temperature is also reproduced by the code. It has for origin a small degradation of the gap conductance when large amounts of fission gas were released during regular excursions to high power.

Rod power history in LWRs

At similar power, the MOX and UO_2 fuels operate under very similar thermal conditions, as concluded from previous sections.

In commercial LWRs, both fuels will however sustain different temperature histories due to basic differences in the neutronic properties of isotopes of U and Pu.

The rod power history of a fuel rod in commercial LWR depends on, amongst other variables, the fuel reactivity drop with burn-up. In MOX fuels, the initial presence of fertile Pu isotopes (indirectly ^{239}Pu and directly ^{240}Pu) leads to an in-reactor production of fissile Pu isotopes, with for consequence a reduction of reactivity drop with burn-up compared to UO_2 (Figure 6).

The relative higher reactivity of MOX fuel with burn-up increase potentially allows to reach higher power than in UO_2 during second part of life. The reduced possibility to operate at low power for MOX fuel during part of its reactor stay is penalising for the achievement of high burn-up.

To minimise the consequences of that situation, it is necessary to optimise the assembly neutronic design (limitation of maximum Pu content), to have an appropriate core management policy, and to perform accurate power history calculations.

Conclusions

The review of the MOX thermal properties concludes that they are very close to those of UO_2. Significant but small changes in fuel central temperatures for MOX fuel result from:

- a decrease of thermal conductivity with Pu content (4% for 10% Pu corresponding to +20°C/200 W/cm at BOL);

- a different radial power profile, more depressed and less depressed at BOL (- 20°C/200 W/cm) and EOL (<+50°C/200 W/cm) respectively.

A difference of temperature history between MOX and UO_2 could arise in commercial LWRs due to difference of power histories associated to changes of reactivity drop with burn-up. This drop is less pronounced for MOX fuel. This imposes appropriate consideration for high burn-up achievement due to potentially higher power, and thus fission gas release, during the second part of life.

REFERENCES

[1] D. Haas and M. Lippens, "MOX Fabrication and In-Reactor Performance", GLOBAL'97, Yokohama, Japan, 5-10 October 1997.

[2] N. Hoppe *et al.*, "COMETHE, Version 4-D release 022(4.4-022), Volume 1 "General Description", BN9409844/220-A, April 1995.

[3] J. Basselier *et al.*, "FIGARO International Programme, Technical Proposal", FIG 93/01, BN Ref. 220/83173, rev. April 1997.

[4] L. Mertens, M. Lippens and J. Alvis, "The FIGARO Programme: The Behaviour of Irradiated MOX Fuel Tested in IFA-606 Experiment, Description of Results and Comparison with COMETHE Calculation", Enlarged HPG Meeting on High Burn-up Fuel Performance, Lillehammer, Norway, 15-20 March 1998.

[5] R.L. Gibby, "The Effect of Plutonium Content on the Thermal Conductivity of $(U,Pu)O_2$ Solid Solutions", *J. Nucl. Mat.*, <u>38</u> (1971) 163-177.

[6] A.R.G. Washington, UKAEA Report TRG Report 2236 (D) (1973).

[7] D.G. Martin, "A Re-appraisal of the Thermal Conductivity of UO_2 and Mixed (U,Pu) Oxide Fuels", *J. Nucl. Mat.*, <u>110</u> (1982) 73-94.

[8] P. Blanpain, M. Trotabas, P. Menut and X. Thibault, "Plutonium Recycling in French Power Reactors: MOX Fuel Irradiation Experience and Behaviour", IAEA TCM on Recycling of Plutonium and Uranium in Water Reactor Fuel, Windermere, UK, 3-7 July 1995.

[9] D. Baron and J.C. Couty, "A Proposal for a Unified Fuel Thermal Conductivity Model Available for UO_2, $(U,Pu)O_2$ and $(U,Gd)O_2$ PWR Fuel", IAEA TCM on Water Reactor Fuel Element Modelling at High Burn-up and its Experimental Support, Windermere, UK, 19-23 September 1996.

[10] Y. Philipponneau, "Conductivity Measurement on $(U,Pu)O_2$ Unirradiated Fuel", Minutes of the AG71 Meeting, Harwell, UK, September 1987.

[11] Y. Philipponneau, "Thermal Conductivity of $(U,Pu)O_{2-x}$ Mixed Oxide Fuel", EMRS 1991 Fall Meeting, Strasbourg, France, 4-8 November 1991.

[12] V. Ambegaokar, *Phys. Rev.*, <u>114</u> (1959) 488-489.

[13] D.R. Olander, "Fundamental Aspects of Nuclear Reactor Fuel Elements", TID-26711-P1 (1976).

[14] G. Leibfried and E. Schlömann, Akad. Wiss. Gottingen, Math.-Physik, KI II-AC (1954) 71.

[15] F.A. LINDEMANN, *Phys. Z.*, 11 (1910) 609.

[16] MATPRO, NUREG/CR-5273, EGG-2555, Vol. 4 (1990), Section 2.3.

[17] L.A. Goldsmith and J.A. McDouglas, "The Thermal Conductivity of Plutonium Uranium Dioxide at Temperatures Up to 1273K", *J. Nucl. Mat.*, 43 (1972) 225-233.

[18] H.E. Schmidt, "Die Warmeleitfähigkeit von Uran and Uran-Plutonium Dioxyd bei hohen Temperaturen", Forschung, Ingenieur-Wesen, 38 (1972) 149-151.

[19] R.C. Daniel and I. Cohen, "In-Pile Effective Thermal Conductivity of Oxide Fuel Elements to High Fission Depletions", WAPD, April 1964.

Table 1. Comparison of UO_2 and PuO_2 properties having an influence on thermal conductivity

Material	UO_2	PuO_2
Crystallographic structure	Fluorite SC: O^{2-} ; FCC: M^{4+}	
Lattice parameter (Å)	5.470	5.396
Melting temperature (°C)	2840	2290
Cation atomic radius (Å)	0.97	0.93
Density (g/cm^3)	10.96	11.45
Molecular mass	~270	~271

Table 2. Comparison between MOX and UO_2 fuel central temperatures for different types of MOX thermal conductivities[1]

k type	k(UO_2, Bu=0)	k*(UO_2, Bu)	k*(MOX, Bu)
$\delta(T_c-T_s)(°C)$[2]			
BOL			
150 W/cm	-25		-16
250 W/cm	-51		-33
EOL			
150 W/cm	11	17	27
250 W/cm	24	65	85
$T_c(°C)$[3] EOL			
150 W/cm	570	840	850
250 W/cm	770	1250	1260

[1] UO_2 (4.5% U5) and MOX (0.2% U5; 8.8% Pu) up to 60 GWd/tm

[2] $\delta(T_c-T_s)= (T_c-T_s)MOX-(T_c-T_s)UO_2$

   T_s ~ 500°C BOL – open initial gap

      350°C closed gap (400°C at EOL/250W/cm for k=k*)

[3] $T_c=(T_c(MOX)+T_c(UO_2))/2$

Figure 1. Reduction of fuel thermal conductivity with Pu content calculated following COMETHE code correlation

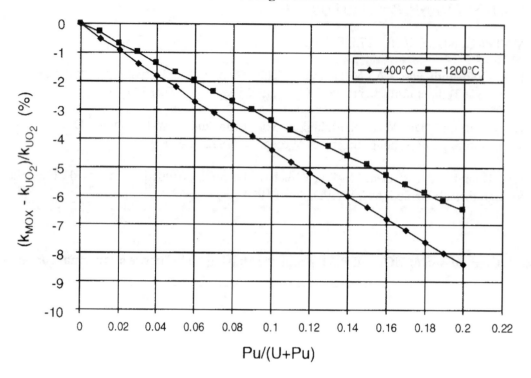

Figure 2. Comparison between experimental and calculated (COMETHE) radial distribution of Pu isotopic composition in MOX fuel irradiated to high burn-up

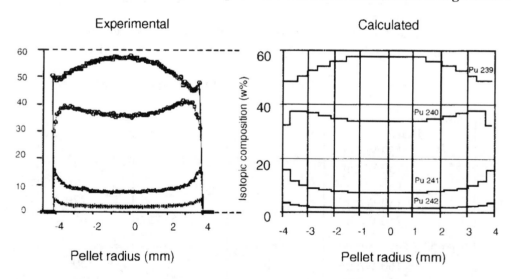

**Figure 3. Radial distribution in (a) UO$_2$ (4.5% U5) and in
(b) MOX (0.2% U5, 8.8% Pu) fuel pellet irradiated in a PWR up to 60 GWd/tM**

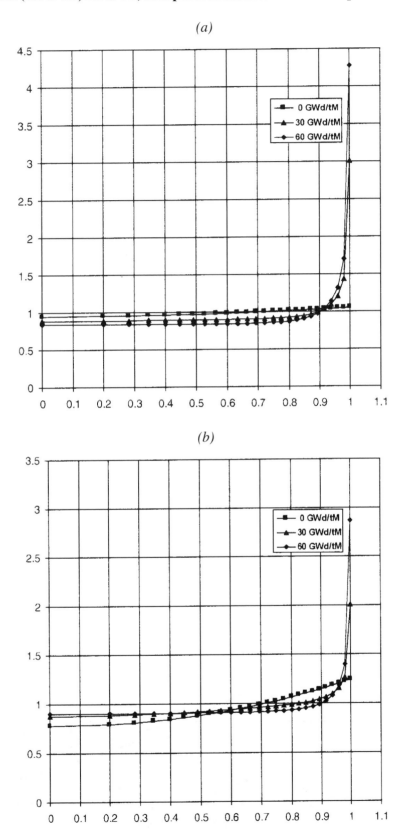

Figure 4. FIGARO experiment (IFA-606); power history used for Phase II of Halden irradiation

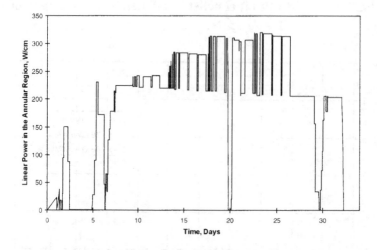

Figure 5. FIGARO experiment (IFA-606); temperature in the annular region during Phase II of Halden irradiation

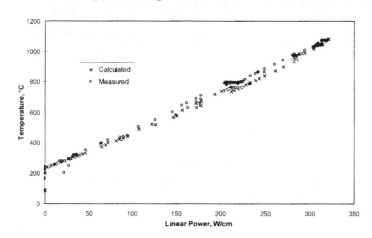

Figure 6. Drop of reactivity as a function of burn-up for MOX and UO₂ fuel

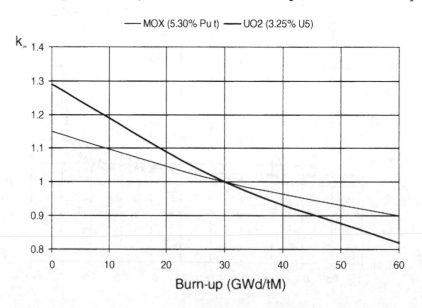

SESSION V

Advances in
Code Development
on Thermal Aspects

Chairs: M. Billaux, Ph. Garcia

QUALIFICATION OF THE NRC'S SINGLE-ROD FUEL PERFORMANCE CODES FRAPCON-3 AND FRAPTRAN

C.E. Beyer, M.E. Cunningham, D.D. Lanning
Pacific Northwest National Laboratory

Abstract

The GAPCON, FRAPCON and FRAP-T codes, developed in the 1970s and early 1980s, were used by the US Nuclear Regulatory Commission (NRC) to predict fuel performance during steady-state and transient power conditions, respectively. The newest versions of the codes are called FRAPCON-3 and FRAPTRAN and are intended to provide best-estimate predictions under steady-state and fast transient power conditions up to extended fuel burn-ups (> 65 GWd/MTU). The improvements to FRAPCON-3 are complete. An assessment has been made against a database independent of the database used for code benchmarking, a peer review has been performed, and code documentation has been published. Estimates of the FRAPCON-3 predictive uncertainties have been made based on comparisons to benchmark and independent databases. The FRAPTRAN code development and preliminary assessment will be completed in 1998. The peer review process for FRAPTRAN will begin in 1998 and final code assessment against both benchmark and independent databases will be complete in 1999.

FRAPCON-3

The FRAPCON-3 code is an updated version of FRAPCON-2 that will be used by the NRC to audit vendor fuel performance codes with an emphasis on thermal, fission gas release, and rod internal pressure analyses. A code assessment of FRAPCON-3 has recently been concluded along with a peer review process that concentrated on those areas where the code will be applied for assessing licensing analyses, i.e. thermal and fission gas release. The code benchmarking database includes thermal, fission gas release, internal rod void volumes, and cladding corrosion data that have previously been presented by Lanning, Beyer, and Painter [1] and in the code integral assessment document [2]. The code has also been assessed against an independent thermal and fission gas release database that is provided in this paper. The differences in FRAPCON-3 and FRAPCON-2 predictions are illustrated by comparison to fission gas release data from experimental light water reactor fuel rods at moderate burn-up levels.

The independent thermal data are divided into beginning-of-life (BOL) data (Table 1) and data as a function of nominal to high burn-up levels (Table 2). The independent fission gas release data are summarised in Table 3. Presenting predicted temperatures minus measured temperatures as a function of linear heat generation rate (LHGR) at BOL for both the benchmarking and independent databases (Figure 1) demonstrates that there is no bias in predictions. There is a slightly larger scatter in the comparison to independent data than to the benchmark data but the scatter is relatively small with a standard deviation of 24.5°C for the helium-filled rods increasing to 31.4°C when xenon-filled rods are included.

Table 1. Independent data for BOL fuel temperatures
(all BWR-size rods in Halden Reactor, Ref. [3])

Diametral gap size, μm (gap-to-diameter ratio, %)	Initial fill gas type and room temperature pressure, psia (MPa)	Maximum rod-average LHGR, kW/ft (kW/m)
50 (0.47)	He 14.7 (0.10)	9 (30)
100 (0.94)	He 14.7 (0.10)	9 (30)
200 (1.9)	He 14.7 (0.10)	9 (30)
50 (0.47)	Xe 14.7 (0.10)	9 (30)
100 (0.94)	Xe 14.7 (0.10)	9 (30)
200 (1.9)	Xe 14.7 (0.10)	9 (30)

Table 2. Independent data for fuel temperatures at nominal-to-high burn-up

Reactor and type (and reactor for ramping), reference	Rod-average burn-up, GWd/MTU	Rod identification	Diametral gap size, μm (gap-to-diameter ratio, %)	Initial fill gas type and room temperature pressure, psia (MPa)	Maximum rod-average LHGR kW/ft (kW/m)
Ringhals BWR (Halden), [a]	67	Halden, Rod 2	265 (2.5)	He 73 (0.50)	7.6 (25)
Quad Cities BWR (DR-2), [4]	42, 24	GE-2, GE-4 from RISØ-III	225 (2.1)	He 73 (0.5)	12.6 (41) 13.2 (43)
Halden (Halden) [5]	39	FUMEX-4A	220 (2.1)	He 44 (0.30)	15.7 (52)

[a] Halden Reactor Project, 1997. Personal communication with US NRC.

Table 3. Independent data for FGR at nominal to high burn-up

Reactor and type (and reactor for ramping), reference	Rod-average burn-up, GWd/MTU	Rod ID	Diametral gap size, μm (gap-to-diameter ratio, %)	Initial fill gas type and room temp-erature pressure, psia (MPa)	Maximum (bump terminal) LHGR, kW/ft (kW/m)	Hold time, hours	Measured FGR, %
Quad Cities BWR (DR-2) [4]	43	GE-2 from RISØ-III	225 (2.1)	He, 97 (0.66)	12.6 (40.7)	41	24.6
Quad Cities BWR (DR-2) [4]	22	GE-4 from RISØ-III	225 (2.1)	He, 97 (0.66)	13.2 (43.3)	34	27.0
Quad Cities BWR (DR-2) [4]	42	GE-6 from RISØ-III	225 (2.1)	He, 97 (0.66)	11.6 (37.9)	140	26.0
Halden (Halden) [5]	48.8	FUMEX Case 6s	260 (2.5)	He, 370 (2.5)	~15 (50)	83 days	50
Halden (Halden) [5]	48.8	FUMEX Case 6f	260 (2.5)	He, 370 (2.5)	~12 (40)	150 days	45
ANO-1 PWR (Studsvik), [6]	62.3	R1	188 (2.0)	He, 400 (2.7)	12.0 (39.5)	12	9.3
ANO-1 PWR (Studsvik), [6]	62.3	R3	188 (2.0)	He, 400 (2.7)	12.9 (44)	12	11.2

Figure 1. FRAPCON-3 predicted-minus-measured centreline fuel temperature at BOL as a function of LHGR for both independent and benchmark data sets

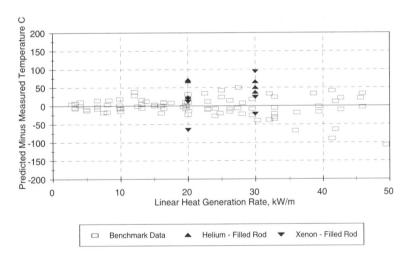

Predicted temperatures minus measured temperatures at the fuel centreline as a function of burn-up are provided in Figure 2 for both the benchmark and independent data. Because there are large variations in the LHGRs of the fuel rod data associated with this figure, this makes it difficult to assess the relative degree of under/overprediction. Therefore, the ratio of the difference between predicted and measured temperatures to the difference between measured centreline and coolant temperatures as a function of rod-average burn-up is provided in Figure 3. As demonstrated by Figures 2 and 3, FRAPCON-3 significantly underpredicts (17 to 25%) the two RISØ-III rods at rod-average burn-ups of 22 and 42 GWd/MTU. The reason for the underprediction is unknown and does not agree with the relatively good prediction of the FUMEX rod from the independent data at 39 GWd/MTU and the benchmark data at burn-ups below 45 GWd/MTU. The independent data from the Halden ramped rod at a rod-average burn-up of 67 GWd/MTU is underpredicted by 7.5 to 17.5% which is consistent with the one rod from the benchmark data (Halden Ultra-High Burn-up rod) at burn-ups >45 GWd/MTU.

**Figure 2. Predicted-minus-measured fuel centre temperatures
as a function of burn-up for benchmark data and independent data sets**

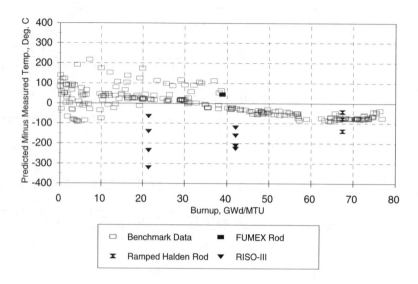

**Figure 3. Ratio of FRAPCON-3 predicted-minus-measured fuel centre
temperatures divided by measured fuel centreline-minus-coolant temperatures
as a function of fuel burn-up for benchmark data and independent data sets**

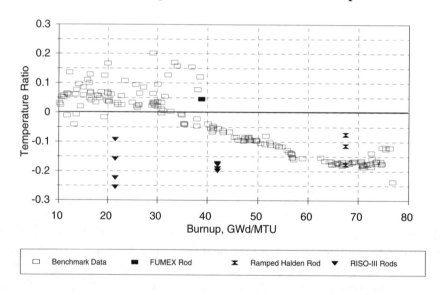

The code fission gas release comparisons to both the benchmark and independent fission gas release data are provided in Figures 4 and 5. The independent data are predicted to within 5% release (absolute) of the measured values, except the RISØ rod at 22 GWd/MTU burn-up. The relatively good prediction of the RISØ rod at 42 GWd/MTU contrasts with the fact that centreline temperatures for this rod were significantly underpredicted (>200°C) at the ramp terminal powers. The standard deviation on fission gas release is 5.4% for both benchmark and independent data, if rods with unstable (densification prone, >2.5% TD) fuel and BWR commercial rods (with large uncertainty in rod powers) are eliminated from the benchmark data.

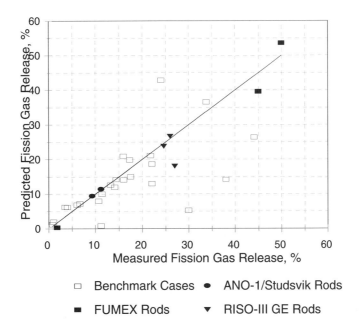

Figure 4. Predicted versus measured fission gas release for benchmark steady-state/power-ramp data and independent data sets

☐ Benchmark Cases ● ANO-1/Studsvik Rods

■ FUMEX Rods ▼ RISO-III GE Rods

Figure 5. Predicted-minus-measured FGR as a function of burn-up for benchmark steady-state/power-ramp data and independent data sets

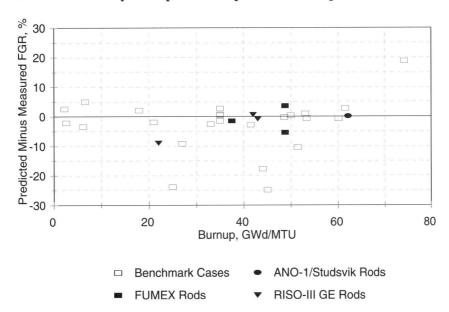

☐ Benchmark Cases ● ANO-1/Studsvik Rods

■ FUMEX Rods ▼ RISO-III GE Rods

A comparison is also provided to demonstrate the predictive differences between the new updated FRAPCON-3 code and the FRAPCON-2 code. The difference between FRAPCON-3 and FRAPCON-2 predicted fission gas release is illustrated in Table 4 where predicted values for each code are compared to actual measured values of fission gas release from four rods with moderate burn-up levels between 30 to 49 GWd/MTU. The latter two rods in Table 4 (IIIi5 and F14-6) are from the benchmark data while the first two rods (M2-2C and PA 29-4) are from Ref. [7]. This table shows that FRAPCON-2 significantly underpredicts fission gas release while FRAPCON-3 predicts these

261

rods relatively well. It is noted, though not shown here, that the FRAPCON-3 code predicts significantly higher fuel temperatures at high burn-ups (>45 GWd/MTU) than FRAPCON-2 because of conductivity degradation.

Table 4. Comparison of fission gas release predictions from FRAPCON-2 and FRAPCON-3 with FRACAS-1 and "MASSIH"

Rod number	Measured FGR, %	FRAPCON-2 predicted FGR^1,%	FRAPCON-3 predicted FGR^2, %	Burn-up, GWd/MTU	FRAPCON-2 predicted minus measured FGR, %	FRAPCON-3 predicted minus measured FGR, %
M2-2C	35.6	23.5	36.5	43	-12.1	0.9
PA29-4	48.1	28.5	43.6	40.9	-19.6	-4.5
111i5	14.4	3.1	14.2	48.6	-11.3	-0.2
F14-6	22.1	0.16	12.7	30	-21.94	-9.4

[1] Using PARAGRASS fission gas release and FRACAS-2 models
[2] Using Massih fission gas release and FRACAS-1 models

In summary, the FRAPCON-3 code provides a relatively good prediction of all the thermal data below a burn-up of 45 GWd/MTU except for the RISØ-III rods and begins to underpredict the two rods from the benchmark and independent data above 45 GWd/MTU. The code provides a very good prediction of fission gas release for fuel rods with stable fuel (low fuel densification, <1.5% TD) and accurate estimates of rod power. FRAPCON-3 provides a much better prediction of fission gas release at moderate to high burn-up levels than FRAPCON-2 and better thermal predictions at high burn-up due to the inclusion of the effects of fuel thermal conductivity degradation. The code documentation including the model description document, code manual with input instructions, and code integral assessment [1,2,8] have been published and are available on the NRC's web site. Information on code availability can also be found on the NRC web site. A FRAPCON-3 users group will be formed by PNNL to offer support and to identify and correct problems with the code.

FRAPTRAN

FRAPTRAN is an updated version of the FRAP-T (Fuel Rod Analysis Program-Transient) code series developed by the Idaho National Engineering Laboratory (INEL; the INEL is now known as the Idaho National Engineering and Environmental Laboratory – INEEL) for the NRC to calculate the transient fuel performance of single fuel rods. As with FRAPCON-3, the objective of the work is to implement improved burn-up-dependent thermal and mechanical models. FRAPTRAN is being developed from FRAP-T6, the sixth version in the FRAP-T code series, that was released in May 1981 [9] with an update in 1983 [10]. Since that time there have been few modifications, with the result that FRAP-T6 has not incorporated changes to accommodate high burn-up fuel behaviour which have been incorporated in other codes.

Work on FRAPTRAN began by PNNL in FY-1997. The updates are to be parallel and consistent with the changes that have been done to FRAPCON-3 [1]. As already noted, the emphasis of the work is on improving the high burn-up predictive capability of the code. Other requirements placed on the FRAPTRAN development work include: retaining the capability to model both pressurised-water and boiling-water reactor conditions; retaining the capability to model the thermal effects of mixed oxide fuel; being capable of modelling a wide range of transients such as reactivity initiated accidents, loss of coolant accidents, and anticipated transient without scram; and making the code as independent of computer platform as possible.

Among the issues and drivers for developing FRAPTRAN are that general updates were needed to account for the effect of high burn-up on properties, models, and fuel rod behaviour during transients. Previous assessments of the FRAP-T6 code had observed that the code: overpredicted cladding hoop strain and ballooning strains; overpredicted fuel temperatures when a rod was filled with fission gas; overpredicted transient fission gas release using the PARAGRASS model; and that the FRACAS-II mechanical subcode needed work.

Modifications to the code are being grouped into three general areas. First, general coding improvements to address known errors, ensure consistency across the code, and to delete undesirable or no longer needed coding/models. Second, model updates to existing models to account for data, knowledge, etc., gained since FRAP-T6 was released (e.g. radial power distribution, contact conductance, and burn-up dependent material properties like fuel thermal conductivity). And third, new model additions to extend the applicability of FRAPTRAN (e.g. modelling to account for fission gas release during fast transients).

The FRAPTRAN modifications began with FRAP-T6, Version 21. This code (plus Version 22) was transferred to PNNL and installed in September 1996. The FRAPTRAN modifications accomplished during FY-1997 have included the following:

- An initialisation link has been established between FRAPCON-3 and FRAPTRAN. This link consists of a file written by FRAPCON-3 that contains selected burn-up dependent values such as radial power and burn-up profiles for each axial node, gas composition and pressure, permanent cladding and fuel strains, and other variables based on steady-state power operation. This file is read by FRAPTRAN to initialise the burn-up dependent variables. To be consistent with the ability to read this data from the FRAPCON-3 file, gadolinia concentration for the fuel and axially varying radial power and burn-up profiles can now also be entered through the input data file.

- A revised fuel thermal conductivity model has been implemented using the same model used in FRAPCON-3 [1]. This model has both gadolinia and local burn-up dependencies.

- The change in contact gap conductance implemented in FRAPCON-3 [1] has been implemented in FRAPTRAN. The balance of the gap conductance modelling was reviewed to assure consistency.

- The MATPRO-11 [11] versions of gas thermal conductivity and gas viscosity have been implemented to replace the MATPRO-9 versions used in FRAP-T6. This maintains the compatibility with FRAPCON-3.

- Selected coding options have been deleted from FRAPTRAN and general code clean-up. These include deleting the uncertainty sensitivity analysis, the failure mode analysis (FRAIL subcode package), and the licensing assistance evaluation model package and maintaining consistency of units and constants.

Work still needing to be done, and planned for FY-1998, includes modifying properties and models that are specific to fuel and cladding behaviour during transients such as: incorporating strain rate dependency on mechanical properties and models; reviewing and revising as necessary the high temperature cladding oxidation models; developing and implementing a transient fission gas release model to replace PARAGRASS; and other general improvements to the coding. Preparing draft documentation and a draft code assessment is also planned for FY-1998.

Test cases and experimental data are being collected for the future verification and validation of FRAPTRAN. Cases that are being reviewed and proposed for this effort include: cases used in the assessment of FRAP-T6 [12], the large-break LOCA tests run in the NRU reactor (MT-1, MT-3, MT-4, and/or MT-6A), selected tests conducted in the Halden Boiling Water Reactor (e.g. IFA-508), selected tests conducted in the Power Burst Facility (e.g. RIA tests, LB-LOCA tests, and operational transient tests), pellet-cladding interaction ramp tests conducted by the Fuel Performance Improvement Program, recent RIA experimental tests such as those conducted at the Cabri and NSRR facilities [13], and others yet to be identified. Measured parameters that will be of interest include measured fuel and cladding temperatures, cladding strains, fill gas pressure, and fuel rod rupture times.

Although formal assessment of FRAPTRAN has not yet begun, some preliminary comparisons of FRAPTRAN against data have been conducted. One example is the initial power ascension of Rod 1 from IFA-432 irradiated in the Halden Boiling Water Reactor. Rod 1 was a simulation of a standard boiling water reactor rod. Presented in Figure 6 is a comparison of measured fuel centreline temperature as a function of linear heat generation rate against temperatures calculated by FRAPCON-3, FRAP-T6, and FRAPTRAN. It may be seen that the FRAPTRAN calculated temperature is in fair agreement with the measured temperature and the temperature calculated by FRAPCON-3. A second example is the effect of initialising FRAPTRAN from a FRAPCON-3 run prior to calculating a power ascension. Presented in Figure 7 are fuel radial temperature profiles at 20 kW/m based on different initialisation burn-up levels as calculated by FRAPCON-3. The effects of decreased thermal conductivity and highly peaked radial power profile are well demonstrated by comparing the zero and high burn-up radial temperature profiles.

**Figure 6. Comparison of measured and predicted fuel
centreline temperature for initial power ascension of Rod 1, IFA-432**

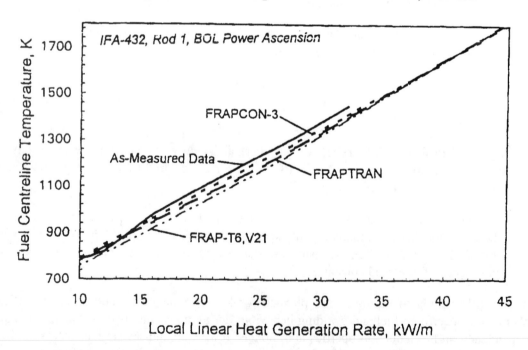

Figure 7. Comparison of beginning-of-life and end-of-life radial temperature profiles

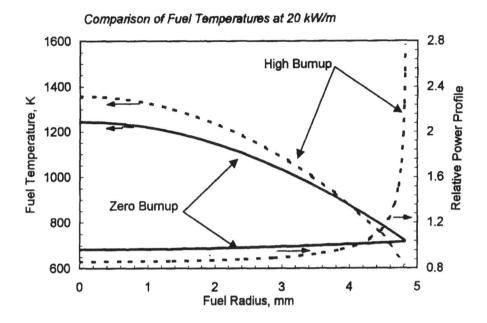

Comparison of Fuel Temperatures at 20 kW/m

REFERENCES

[1] Lanning D.D., C.E. Beyer, and C.L. Painter, "FRAPCON-3: Modifications to Fuel Rod Material Properties and Performance Models for High Burn-up Applications", NUREG/CR-6534, Volume 1 (PNNL-11513, Volume 1), Prepared for the US Nuclear Regulatory Commission by Pacific Northwest National Laboratory, Richland, Washington, 1997.

[2] Lanning, D.D., C.E. Beyer, and G.A. Berna, "FRAPCON-3: Integral Assessment", NUREG/CR-6534, Volume 3 (PNNL-11513, Volume 3), Prepared for the US Nuclear Regulatory Commission by Pacific Northwest National Laboratory, Richland, Washington, 1997.

[3] Wiesenack, W., "Review of Halden Reactor Project High Burn-up Fuel Data that can be Used in Safety Analyses", presented at the NRC-Research 23rd Water Reactor Safety Meeting, NUREG/CR-0145, Vol. p. 127, 1996.

[4] Knudsen, P., C. Bagger, M. Mogensen, and H. Toftegaard, "Fission Gas Release and Fuel Temperature During Power Transients in Water Reactor Fuel at Extended Burn-up", Proceedings of a Technical Committee Meeting held in Pembroke, 28 April-1 May 1992, Ontario, Canada, IAEA-TECDOC-697, International Atomic Energy Agency, Vienna, Austria, 1993.

[5]   Chantoin, P.M., J.A. Turnbull, W. Wiesenack, "How Good is Fuel Modeling at Extended Burn-up? – The IAEA's FUMEX Programme Provides Some Answers", *Nuclear Engineering International*, September 1997.

[6]   Wesley, D.A., K. Mori, and S. Inoue, "Mark BEB Ramp Testing Program", presented in the Proceedings of the 1994 ANS/ENS International Topical Meeting on Light Water Reactor Fuel Performance, West Palm Beach, Florida, American Nuclear Society, p. 343, 1994.

[7]   Bagger, C., H. Carlson, and P. Knudsen, "Details of Design, Irradiation and Fission Gas Release for the Danish UO$_2$-Zr Irradiation Test 022," RISØ-M-2152, RISØ National Laboratory, Denmark, December 1978.

[8]   Berna, G.A., C.E. Beyer, K.L. Davis, and D.D. Lanning, "FRAPCON-3: A Computer Code for the Calculation of Steady-State, Thermal-Mechanical Behavior of Oxide Fuel Rods for High Burn-up", NUREG/CR-6534, Volume 2 (PNNL-11513, Volume 2), Prepared for the US Nuclear Regulatory Commission by Pacific Northwest National Laboratory, Richland, Washington, 1997.

[9]   Siefken, L.J. *et al.*, "FRAP-T6: A Computer Code for the Transient Analysis of Oxide Fuel Rods", NUREG/CR-2148 (EGG-2104), Idaho National Engineering Laboratory, Idaho Falls, Idaho, 1981.

[10]   Siefken, L.J. *et al.*, "FRAP-T6: A Computer Code for the Transient Analysis of Oxide Fuel Rods", NUREG/CR-2148, Addendum (EGG-2104), Idaho National Engineering Laboratory, Idaho Falls, Idaho, 1983.

[11]   Hagrman, D.L., G.A. Reymann, and R.E. Mason, "MATPRO-Version 11 (Revision 2). A Handbook of Materials Properties for use in the Analysis of Light Water Reactor Fuel Rod Behavior", NUREG/CR-0479 (TREE-1280, Rev. 2), Prepared for the US Nuclear Regulatory Commission by EG&G Idaho, Inc., Idaho Falls, Idaho, 1981.

[12]   Chambers, R. *et al.*, "Independent Assessment of the Transient Fuel Rod Analysis Code FRAP-T6", EGG-CAAD-5532, Idaho National Engineering Laboratory, Idaho Falls, Idaho, 1981.

[13]   American Nuclear Society, "Proceedings of the 1997 International Topical Meeting on Light Water Reactor Fuel Performance", Portland, Oregon, 2-6 March 1997.

OVERVIEW OF EPRI FUEL PERFORMANCE CODE FALCON

Suresh K. Yagnik
Nuclear Power Group, EPRI, Palo Alto, CA 94303 (USA)

Abstract

The status of FALCON, which has steady state and transient analytical capabilities in a single finite element-based numerical platform, is highlighted. FALCON provides the ability to shift between steady and transient modes at any time during burn-up history, in a manner transparent to the user. This has been accomplished by reducing the two-dimensional set of equations into two sets of equations each having a one-dimensional structure (one for the radial solution and the other for the axial solution). This greatly reduces the steady state solution time (base condition of the fuel rod) without adversely impacting the two-dimensional spatial details that are necessary for transient analysis. The requisite thermo-mechanical coupling, especially at high burn-ups, has also been incorporated in FALCON. In addition, several high burn-up models based on relatively recent R&D results are being implemented in FALCON, including the evolution of the rim structure with burn-up, the fuel thermal conductivity degradation, and the cladding mechanical properties and failure limits.

Introduction

The need for a fuel performance code for high burn-up applications with a first-principle modelling approach and robust numerics has long been recognised. Such a code has many attributes; while it is expected to accurately model the steady state and the transient behaviours at high burn-ups, an ability to analyse fuel design specific features, that are being rapidly introduced in many LWR fuel rods, is also highly desirable. Further, since thermal and mechanical responses of the fuel rod are highly coupled at high burn-ups, the code must have them adequately coupled. In addition, the practice of "analysis by correlation", commonplace in many fuel performance and design codes, must be avoided. Finally, any extrapolation far beyond correlation database and the use of the code beyond validation range must be avoided.

With this perspective on mind, EPRI has undertaken the development of FALCON (Fuel Analysis and Licensing COde – New). The specific objectives behind development of FALCON are to: (a) incorporate the latest high burn-up data and material models; (b) expand validation range to ~65 MWd/KgU rod average or higher; and (c) migrate to an efficient finite element based numerics.

FALCON capabilities and features

FALCON builds on two existing EPRI codes: the ESCORE code for steady state analysis, and the FREY code for transient analysis. The new combined code is not a mere merging of the ESCORE and the FREY, but rather an innovative construct of robust numerics and relevant material models. ESCORE is a steady-state stacked 1-D code with finite difference numerics [1]. It was reviewed and approved by the US NRC in 1990 and has been maintained under Appendix B Quality Assurance plan. ESCORE predictions are excellent up to 40 MWd/KgU burn-up. On the other hand, FREY is a "best estimate" finite element code [2] for analysing normal and off-normal operations. It has a 2-D (r-θ or r-z) transient capability with single rod/single channel approach. FREY can comprehensively model complex thermal and mechanical responses in fuel pellet and cladding. It has undergone detailed design review and been maintained under Appendix B Quality Assurance plan. Both these existing codes are current EPRI products and continue to serve the industry well. However, recent EPRI experience in analysing RIA tests in the CABRI reactor [3] and BWR fuel degradation [4] encountered certain difficulties in applying the steady-state ESCORE results up to a certain base irradiation condition of the fuel rod to initialise the FREY for the transient analysis. This clearly underscored the need to seamlessly merge ESCORE and FREY capabilities. Thus automatic transition from steady-state to transient analysis and vice-versa – at any time during burn-up history in a manner transparent to the user – was considered essential in FALCON development.

Thermal (temperature) and mechanical (displacements) solutions are inherently coupled, especially at high burn-ups. Such coupling manifests itself in many ways: radial and axial heat conduction and displacements; cladding deformation behaviour and creepdown; pellet swelling, cracking, plasticity and creep; and, finally, pellet cladding mechanical interactions (PCMI). Experience has shown that commonly used fuel rod performance parameters depend directly and significantly on such coupling. For example, large differences in parameters such as fuel centreline and cladding temperatures, rod internal pressure, rod axial growth, and thermal margins (via thermal hydraulics) were observed in coupled and de-coupled cases during FALCON formulations.

The following high burn-up effects are being specifically implemented in FALCON:

- burn-up dependent radial power;

- burn-up dependent pellet conductivity;

- FGR and formation of porous rim;

- cladding waterside corrosion and waterside oxide conductivity;

- cladding hydriding and mechanical properties changes;

- PCMI and fuel-clad bonding.

Models for fuel design specific optional features such as IFBA, high wt.% gadolinia, newer cladding materials other than Zircaloys (e.g. Zirlo, Duplex, etc.) will also be implemented in FALCON as data become available from on-going programmes on such fuels in collaboration with fuel users and suppliers.

The thermal and mechanical properties as well as fuel behaviour models in FALCON are listed in Table 1.

FALCON's numerical structure

The spatial numerical structure selected for FALCON is the finite element method for both the thermal and the mechanical solutions, as well as for the steady state and transient modes of analysis. The power of the finite element method in providing the spatial details required for high burn-up effects has proven itself indispensable through the FREY code's experience [2]. However, as is well known, the two-dimensional finite element approach, which is the requirement for transient is computationally time consuming Thus, it would seem counterproductive that the transient analysis requirements should penalise the steady state capability to a level that hinders the use of the code.

On the other hand, while the number of transient analyses that are needed are close to a dozen, steady state analysis requires several dozens of computations to support fuel performance evaluation. Therefore, in order to preserve compatible spatial structure between the steady state and the transient analysis modes, the finite element structure was maintained for both. But an innovative approach was developed to reduce the two-dimensional set of equations in the steady state mode to two sets of equations each having a one-dimensional structure (one for the radial solution and the other for the axial solution). In this manner, the solution time was greatly reduced without adversely impacting the two-dimensional spatial details. This also allowed an easy transition between the two modes by preserving the number and spatial position of all state variables and eliminated the need for cumbersome procedures (and associated uncertainties) for the interpolation and transfer of variables.

Figures 1 and 2 show, respectively, full nine node finite element model for transient analysis and de-coupled radial and axial models for steady state analysis. In the latter case, FALCON ignores axial conduction but this has been shown to have only minor impact on the accuracy. The single element geometry, representing common nodes (•) and integration points (x) for the two cases is compared in Figure 3. Nodal points are those where temperature (t) and displacements (u and w) are computed and integration points are those where state variables (quantities requiring volume) are computed.

This approach facilitated faster convergence of steady-state solution without any loss of accuracy. As an example, Figure 4 shows fuel centreline temperature calculations for two linear heat rates using the FREY and FALCON numerics and compares them with analytical solutions.

Conclusions

Two key requirements of a high burn-up fuel performance code are adequately coupled thermal and mechanical solutions and incorporation of the latest properties data for critical high burn-up phenomena. EPRI's FALCON capitalises on ESCORE and FREY experiences while providing for seamless transfer between steady-state and transient conditions with an efficient numerics scheme. The code is designed to address current and future industry needs.

Acknowledgements

FALCON is being developed for EPRI at Anatech, San Diego, California. The author wishes to acknowledge many fruitful discussions with Dr. Y.R. Rashid and Dr. R. Montgomery.

REFERENCES

[1] I.B. Fiero *et. al.*, "ESCORE – the EPRI Steady-State COre Reload Evaluator Code: General Description", NP-5100-L-A, April 1991.

[2] Y.R. Rashid, R.O. Montgomery, A.J. Zangari, "FREY-01 Fuel Rod Evaluation System", EPRI NP-3277, August 1994.

[3] R.O. Montgomery, Y.R. Rashid, O. Ozer, "Evaluation of Irradiated Fuel Rods during RIA Simulation Experiments", EPRI TR-106387, August 1996.

[4] S.K. Yagnik *et. al.*, "Assessment of BWR Fuel Degradation By Post-Irradiation Examinations and Modeling in the DEFECT Code", Proceedings of Intl. Topical Meeting on LWR Fuel Performance, Portland, Oregon (USA), 2-6 March 1997, pp. 329-336.

Table 1. Sources of thermal and mechanical properties and behavioural models for FALCON

Thermal properties	Source
Fuel and cladding specific heat	MATPRO
Fuel and cladding thermal expansion	MATPRO
Fuel thermal conductivity	NFIR
Cladding thermal conductivity	MATPRO
ZrO_2 thermal conductivity	NFIR
Cladding emissivity	MATPRO
Gas thermal conductivity (Individual Constituents)	MATPRO
Melting temperature	MATPRO
Mechanical properties	**Source**
Cladding irradiation growth	Open literature
Cladding meyer hardness	MATPRO
Cladding elastic modulus	NFIR
Cladding plastic behavior	NFIR
Cladding yield stress and fracture strain	NFIR
Cladding and fuel creep	Open literature
Fuel elastic modulus	MATPRO
Fuel cracking and relocation	FREY
Behaviour models	**Source**
Fuel swelling and densification	Halden
Fission gas release	Halden
Fuel grain grown	Open literature
Cladding corrosion and hydriding	PFCC model
Radial power distribution	RADAR-G/TUBRNP model
Gap conductance	FREY
Rim-zone evolution	Open literature

Figure 1: Full 2-D nine-node finite element model

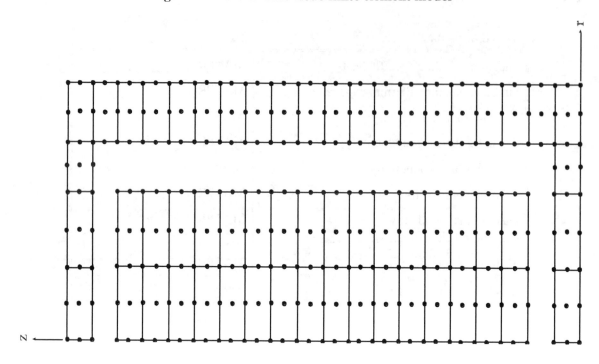

Figure 2. De-coupled 2-D finite element model

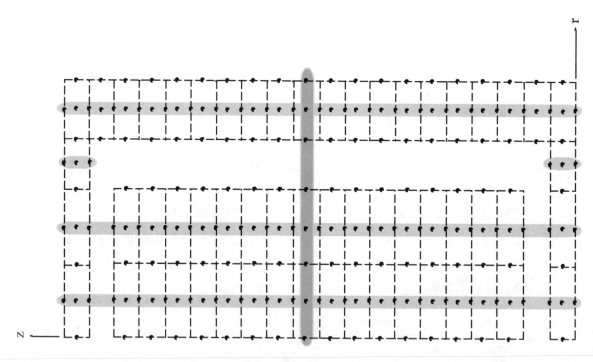

Figure 3. Representation of nodal points and integration points in full 2-D and de-coupled 2-D element geometry

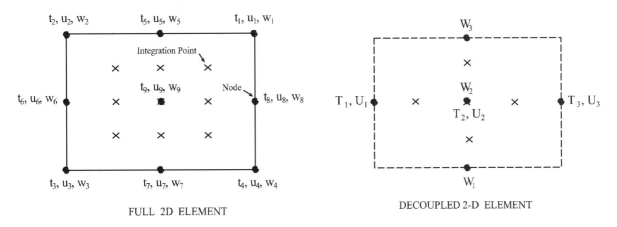

FULL 2D ELEMENT

DECOUPLED 2-D ELEMENT

Figure 4. Comparisons of numerical approaches

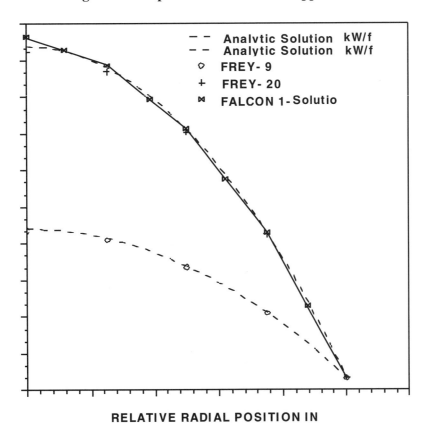

RELATIVE RADIAL POSITION IN

THERMAL PERFORMANCE MODELLING
IN CANDU-TYPE FUEL CODES AND ITS VALIDATION

V.I. Arimescu, A.F. Williams
Atomic Energy of Canada Ltd., Chalk River Laboratories, Canada

Abstract

Temperature is a critical parameter in fuel modelling because most of the physical processes that occur in fuel elements during irradiation are thermally activated. The focus of this paper is the temperature distribution calculation used in the computer code ELESIM, developed at AECL to model the steady-state behaviour of CANDU fuel. A validation procedure for fuel codes is described and applied to ELESIM's thermal calculation. The effects of uncertainties in model parameters, like UO_2 thermal conductivity, and input variables, such as fuel element linear power, are accounted for through an uncertainty analysis using Response Surface and Monte Carlo techniques.

Introduction

Temperature is a critical parameter in fuel modelling because most of the physical processes that occur in fuel elements during irradiation are thermally activated. The focus of this paper is the temperature distribution calculation used in the computer code ELESIM, developed at AECL to model the steady-state behaviour of CANDU fuel. A validation procedure for fuel codes is described and applied to ELESIM's thermal calculation. The effects of uncertainties in model parameters, like UO_2 thermal conductivity, and input variables, such as fuel element linear power, are accounted for through an uncertainty analysis using Response Surface and Monte Carlo techniques.

The magnitude of the fuel-to-sheath gap has a significant impact on fuel temperatures through its effect on the fuel-to-sheath heat transfer coefficient. Previous versions of ELESIM assumed a closed gap for the heat transfer coefficient calculation. While this is appropriate for most conditions typical for CANDU fuel and nominal pellet-to-sheath gap sizes, inclusion of a finite gap is necessary for low power situations or for gap sizes at the upper bound of the tolerance interval. For light water reactor (LWR) fuel, "fragment relocation" is important in determining fuel-to-sheath heat transfer. The physical process proposed for fragment relocation is incomplete re-fitting of fuel pellet fragments caused by cracking during a thermal cycle, and it is typically modelled in LWR fuel codes with semi-empirical correlations. It has been suggested that the collapsible cladding for CANDU fuel would assure an almost permanent restraint on the pellet and prevent major fragment relocation. One of the goals of this study was to assess the possible need of a fragment relocation model for CANDU fuel.

Pellet-to-sheath gap model

The latest developmental version of ELESIM includes a pellet eccentricity model for the fuel-to-sheath gap. When the gap is open the extent that the pellet is off-set from the sheath is considered a model parameter. By definition, full pellet off-set occurs when the pellet and the sheath are touching at one point for free-standing cladding, or at two diametrically opposed points for collapsible cladding. The typical condition for CANDU is collapsible, but large internal gas pressures can change the condition to free-standing.

To evaluate the pellet eccentricity model, ELESIM was used to simulate experiment FIO-142 with different options from a closed gap to a fully-eccentric gap. Experiment FIO-142 was a well-instrumented single element irradiation with measurements of both centreline temperature and neutron flux at two locations along the element.

Experiment FIO-142 consisted of a single fuel element irradiated in the NRU reactor in pressurised water coolant at a linear power of up to 55 kW/m for one operating cycle of 17 days. The fuel element was instrumented with thermocouples to measure sheath, end cap and upper and lower centreline fuel temperatures. The centreline fuel temperatures were measured with Type C thermocouples, while the remaining thermocouples used in the experiment were Type K, both having a precision of ±1% of the measured reading. Two flux detectors were also mounted on the fuel stringer.

Following irradiation, the fuel element was destructively examined with two sections of the element being analysed for chemical burn-up. Chemical burn-up for the two sections was reported as 18.7 MWh/kgU and 20.6 MWh/kgU, with a measurement uncertainty of ±5% (2σ).

The procedure for estimating the linear power of the fuel element is based on the assumption that the power is directly proportional to the flux detector readings throughout the irradiation. In reality, the proportionality constant, alpha, will change during the irradiation due to factors such as fission product build-up and plutonium disposition. Nevertheless, in the simulation of FIO-142, the proportionality is assumed to remain constant as the fuel element was irradiated to only a low burn-up.

Since the chemical burn-up is known at two positions in the fuel element, the corresponding values of alpha can be determined. The chemical burn-up was determined by assuming an energy per fission for ^{235}U of 200.8 MeV. The reported energy per fission for ^{235}U in the NRX experimental reactor is about 185.0 MeV. While this value will vary among reactors, it is used in calculations for the NRU experimental reactor. As a result, the chemical burn-up must be multiplied by the ratio (185/200.8) to correctly account for the burn-up in the NRU reactor.

The power history was reduced from more than 30,000 points to a more manageable number by averaging over eight hour intervals. Intervals were made smaller if the change in power was in excess of 2 kW/m from one step to the next. The measured centreline temperatures corresponding to the averaged power in each interval were also recorded. This resulted in a power history of just 136 steps, and a data set of 136 corresponding fuel centreline temperature measurements. The power histories resulting from this "condensing" process were used to complete an ELESIM input file for FIO-142. The geometry data used in the input file came from the fabrication report for this test element.

Simulation results

ELESIM test cases were created to account for different fuel-to-sheath gap conditions. All power histories were simulated for (a) an open, concentric gap; (b) an open, eccentric gap; and (c) a closed gap. In terms of fragment relocation models used for LWR fuel, Case (a) corresponds to no relocated fuel fragments, Case (b) corresponds to partial relocation, while Case (c) corresponds to 100% relocated fuel fragments.

An initial comparison revealed that the measured and calculated results agreed reasonably well except during times when the flux (and hence power) was changing rapidly. It was concluded that this discrepancy was due to the response time of the flux detectors (the rhodium flux detectors have a response time of about 5 min), rather than the performance of the code and the decision was made to reject data points that occurred during periods when the flux was changing rapidly i.e. during power ramps and shut downs.

For the open concentric fuel-to-sheath gap (Case (a)), the calculated centreline temperatures follow the measured temperature trends throughout the irradiation for both power histories. For the first power history (Figure 1), the calculated temperatures are higher than measured with an average difference between calculated and measured temperatures of 47 K. The second power history (Figure 2) also results in temperatures that are higher than measured with an average difference of 67 K. For both power histories, the fuel-to-sheath gap is calculated to be open for most of the irradiation, resulting in a low fuel-to-sheath heat transfer and higher fuel temperatures. In addition, the difference between measured and calculated temperatures increased with time at power for both histories.

The calculated temperatures for Case (b), the open eccentric gap (Figures 3 and 4), show better agreement with the measured temperatures for both power histories. The average difference between calculated and measured temperatures is 27 K for the first power history. The calculated temperatures are slightly higher than measured for the second power history with an average temperature difference of 37 K. As for the open concentric gap case, the fuel-to-sheath gap is calculated to be open for most of the irradiation history. However, the gap is smaller resulting in better agreement between calculated and measured temperatures relative to Case (a). For the second power history, Figure 4 shows the difference between calculated and measured temperatures to be increasing with time at power.

For the closed gap case, the calculated centreline temperatures are lower than the measured values for most of the irradiation (Figures 5 and 6). The average temperature difference for both power histories is about 47 K. For the closed gap option, ELESIM calculates the fuel-to-sheath heat transfer coefficient for a closed fuel-to-sheath gap, even if the gap may be open. This results in a higher heat transfer coefficient and lower fuel temperatures relative to the other gap cases. As seen in the other simulations, the difference between calculated and measured temperatures increases with time for both power histories.

The increasing difference between calculated and measured temperatures over time likely indicates that the scaling factor, alpha, used in the power estimation may not be constant throughout the irradiation as originally assumed. To account for this, the FIO-142 power histories should be generated by a reactor physics code to include the effects of fission product build-up and plutonium disposition in the fuel.

The uncertainty on the measured chemical burn-up was estimated to be ±5%, which translates directly to a ±5% uncertainty in the estimated power. To investigate the effect on calculations, several of the ELESIM simulations were rerun using power histories that were modified by ±5%. Note that only the power history corresponding to the lower burn-up was used for these simulations.

Figure 7 compares the calculated centreline temperatures for the original and the ±5% power histories with the measured temperatures for the open eccentric gap case. Changing the power history by ±5% results in calculated temperatures that bracket the measured temperature.

For the closed gap, the calculated temperatures were slightly lower than measured throughout the irradiation. Therefore, the simulation was rerun using a power history that was increased by 5%. This resulted in temperatures that are slightly higher than measured in the first half of the irradiation, and in good agreement with those measured thereafter (Figure 8).

The validation procedure

These initial temperature comparisons did not take into account other uncertainties associated with the models used to calculate the fuel temperatures. If these uncertainties can be properly included, our confidence in the agreement between the measured and calculated temperatures for the estimated power histories improves. To account for uncertainties in measured values of both input and output parameters, we are proposing a validation methodology that investigates the differences between a set of experimental data and the corresponding code calculations. If both the experiment and code were perfect, the differences would be zero. In fact, there are uncertainties in the experimental results which are stochastic in nature, and we expect the differences between calculated and measured values to be non-zero, but distributed around the zero value. That is, we expect the mean of the differences to be zero. In addition, the models implemented in the code rely on approximations and idealisations, and can contain mistakes that induce systematic errors. If the code has a tendency to over- or under-predict, the mean of the differences will not be zero. Therefore, one requirement for validation is that the mean of the relative differences must be less than a specified value.

It is possible that the code both over- and under-predicts values for the set of validation data. In this case, the mean of the differences between the calculated values and measured values may be close to zero. Therefore, it is not sufficient for the mean to be close to zero to consider the code validated. The differences must be further analysed in terms of the distribution around the mean. Our proposed technique assesses the response of the code to uncertainties in the experimental and input data, and this information is used to predict the variance in the differences between measured and calculated values. If the code is accurately simulating the behaviour of the relevant physical phenomena, the predicted variance between the measured and calculated values should match the observed variance. If the predicted variance is less than the observed variance, there are differences between the calculated and measured results that can not be explained by uncertainties in the experimental conditions and input values.

The complete validation procedure is outlined in this section as a series of steps.

1) The code simulations are compared to the validation data using the dimensionless parameter, ε. The ε values are calculated according to:

$$\varepsilon_i = \frac{y_i^c - y_i^m}{y_i^m}$$

where, y_i^c, is a calculated value and y_i^m, is the corresponding measured value.

2) The frequency distribution of the ε values is plotted and the mean value of ε is examined. If this is greater than a specified maximum value, the code results and the experimental results disagree, i.e. the code fails the validation exercise.

3) The response of the code [1] is sampled around each data point, with a variance on each input parameter using a Plackett and Burman [2] two-level experimental design. A linear response surface is fitted to the response values.

4) Backwards elimination is used to determine which of the input parameters contribute to the response of the code. To do this some measure is needed of the contribution to the fit of each input parameter, and such a measure is the t ratio:

$$t = \left| \frac{\hat{\beta}_i}{s_{\hat{\beta}_i}} \right|$$

where $s_{\hat{\beta}_i}$ is the standard deviation of all the possible β_i values (the coefficient of the term related to the i-th parameter in the response function). The code is sampled in greater detail with variances only on the reduced parameter set determined previously. A combination of forward selection and backwards elimination is then used to fit a second order response surface to the new sampled code results.

5) The fit of the response surface is checked by re-sampling the code, and the response surface is modified if necessary.

6) The frequency distribution calculated from the code response and the distribution associated with the uncertainty in the experimental measurements of the output parameters are combined by Monte Carlo sampling to produce a theoretical distribution for ε. The theoretical variance in ε is compared with the observed variance in ε. If the theoretical value is greater than the observed value, any discrepancy between the code and the experimental data can be explained by uncertainties in the input parameters and the code can be said to be valid.

The measured temperatures and corresponding calculated values were used to calculate the dimensionless parameter ε and the resulting values are plotted in Figure 9. For this exercise only ten input parameters of the ELESIM (Table 1) code were chosen for study, based on expert judgement and previous knowledge of fuel behaviour. The ten input parameters include 1) parameters that have an influence on the centreline temperature; and 2) parameters that were thought to have no (or very little) effect on the fuel temperature. The latter were included to illustrate the ability of the regression fitting process to identify which parameters are necessary for the analysis and which may be rejected from the study. Details of intermediate stages of the response surface, backward regression and final uncertainty ranges for the remaining input parameters are listed in Tables 3, 4 and 5. As can be seen from Table 4, the thermal conductivity and linear power have the most significance, whereas parameters like grain size and sheath thickness have been dropped as having negligible significance (remaining parameters are listed in Table 3).

Having completed the calculation of a theoretical set of ε values, the theoretical frequency distribution of ε was plotted. Figure 10 shows a graph of both the theoretical and measured distributions. It is readily apparent from the graph that the theoretical distribution straddles the measured distribution indicating that the differences between the code calculated and measured values can be explained as uncertainty in the input parameters. This is further supported by the values for the variances of the distributions:

$$\sigma_{th} = 0.05 \quad \text{and} \quad \sigma_{cm} = 0.03$$

The validation criterion, $\sigma_{th}^2 > \sigma_{cm}^2$, is fulfilled. Therefore, the fuel centreline temperature calculations of ELESIM have been successfully validated against the experimental data. The minimum uncertainty that can be quoted for the fuel centreline temperature is a variance of 5% (based on $\sigma_{th} = .05$).

Conclusions

The thermal module used in estimating the thermal performance of CANDU-type fuel has been validated by using a detailed uncertainty analysis to account for uncertainties in both measured input variables and measured temperature.

Different models of pellet-to-sheath gap conductance with respect to pellet eccentricity and relocation have been tested and the conclusion was that a pellet eccentricity model provides the best agreement with data.

REFERENCES

[1]   Box G.E.P, and Draper N.R., "Empirical Model-Building and Response Surfaces", J. Wiley & Sons, ISBN 0-471-81033-9, 1987.

[2]   Plackett R.L., and Burman J.P., "The Design of Optimum Multifactorial Experiments", Biometrika, 33, 305-325 and 328-332, 1946.

Table 1. Ranges of the ten initial input parameters

Number	Parameter	+	-
1	Power	+5%	-5%
2	Pellet/sheath gap (mm)	0.15	0.05
3	Thermal conductivity of UO_2	+15%	-15%
4	Fill Gs volume	+40%	-40%
5	Roughness of pellet (µm)	3.0	0.5
6	Roughness of sheath (µm)	1.4	0.8
7	Grain size (µm)	30	7
8	Dish depth (mm)	0.25	0.18
9	Density of pellet (Mg/m^3)	10.8	10.4
10	Sheath thickness (mm)	0.45	0.38

Table 2. The Plackett and Burman sampling matrix

Sample	Input									
	1	2	3	4	5	6	7	8	9	10
1	+	-	-	-	+	-	-	+	+	-
2	+	+	-	-	-	+	-	-	+	+
3	+	+	+	-	-	-	+	-	-	+
4	+	+	+	+	-	-	-	+	-	-
5	-	+	+	+	+	-	-	-	+	-
6	+	-	+	+	+	+	-	-	-	+
7	-	+	-	+	+	+	+	-	-	-
8	+	-	+	-	+	+	+	+	-	-
9	+	+	-	+	-	+	+	+	+	-
10	-	+	+	-	+	-	+	+	+	+
11	-	-	+	+	-	+	-	+	+	+
12	+	-	-	+	+	-	+	-	+	+
13	-	+	-	-	+	+	-	+	-	+
14	-	-	+	-	-	+	+	-	+	-
15	-	-	-	+	-	-	+	+	-	+
16	-	-	-	-	-	-	-	-	-	-

Table 3. The coefficients of the remaining five parameters at the end of the fitting of the linear response surface

Parameter	t ratio	beta	R squared
0	7.967022	2475.528	0.986737
1	8.006105	9.375833	
2	7.7487	907.439	
3	24.37161	-9.51375	
5	3.29679	15.44329	
9	3.871562	-113.348	

Table 4. The final twelve term response surface

Parameter	t ratio	beta	R squared
1	12.08267	1087.722	0.986736
2	7.165223	721.2384	
3	30.25039	-908.9	
5	3.161279	10.18684	
9	10.94431	-219.42	
x_2^2	7.00198	-1061.38	
x_3^2	8.683666	2107.001	
$x_1 x_2$	3.739828	-2017.3	
$x_3 x_9$	3.183308	428.9158	
$x_2 x_5$	2.27098	146.9969	
$x_1 x_9$	1.737267	-702.145	
$x_2 x_3$	1.53563	828.3301	

Table 5. Distributions assigned to the five parameters for Monte Carlo sampling

	Parameter	Type	Mean	Sigma	Range
1	Power	Normal	1	2.5%	–
2	Radial gap	Square	0.102	–	0.082-0.122
3	UO_2 thermal conductivity	Normal	1	7.5%	–
5	Pellet roughness	Square	0.81	–	0.65-0.97
9	Pellet density	Square	10.7	–	10.6-10.8
Output	Fuel centreline temperature	Normal	y_i^c	3.3%	–

Figure 1. Centreline temperature vs. time for first power history, open concentric gap

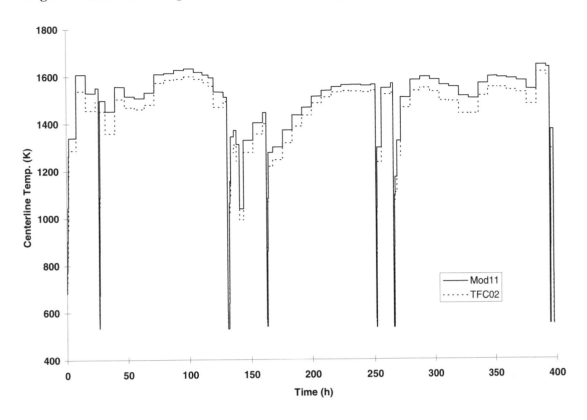

Figure 2. Centreline temperature vs. time for second power history, open concentric gap

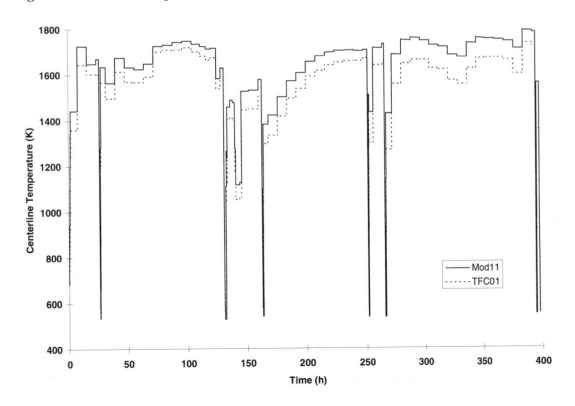

Figure 3. Centreline temperature vs. time for first power history, open eccentric gap

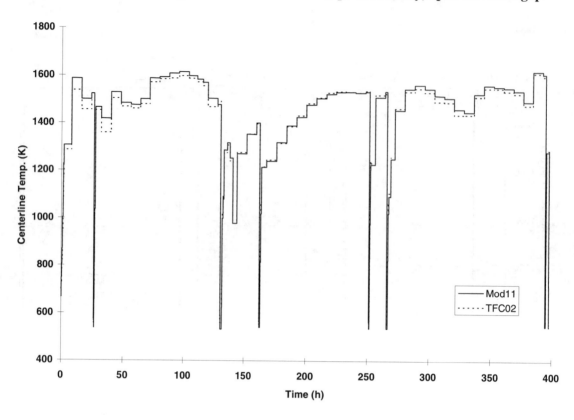

Figure 4. Centreline temperature vs. time for second power history, open eccentric gap

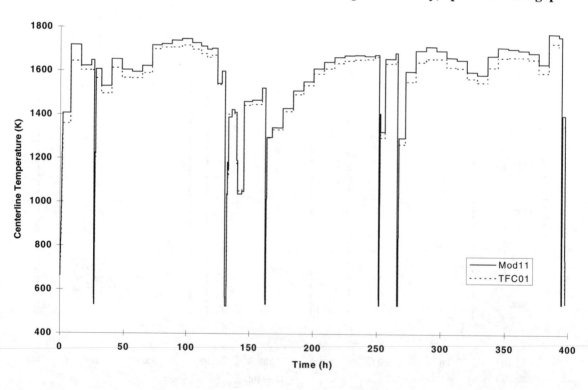

Figure 5. Centreline temperature vs. time for first power history, closed gap

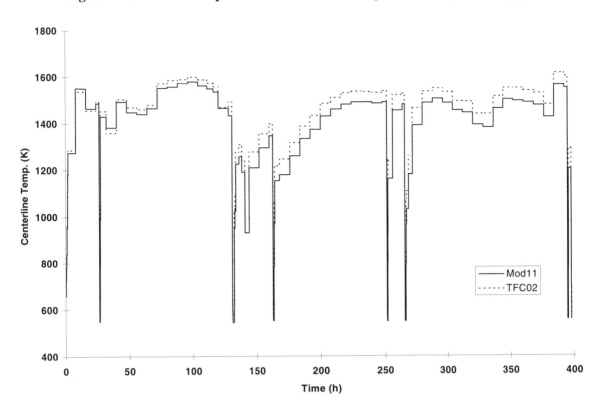

Figure 6. Centreline temperature vs. time for second power history, closed gap

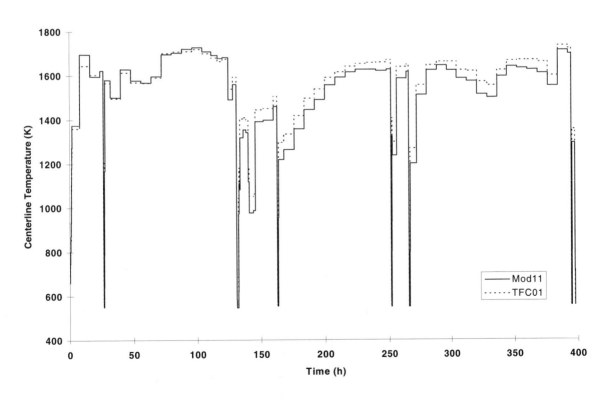

Figure 7. Centreline temperature vs. time for second power history, power varied by ±5%, open eccentric gap

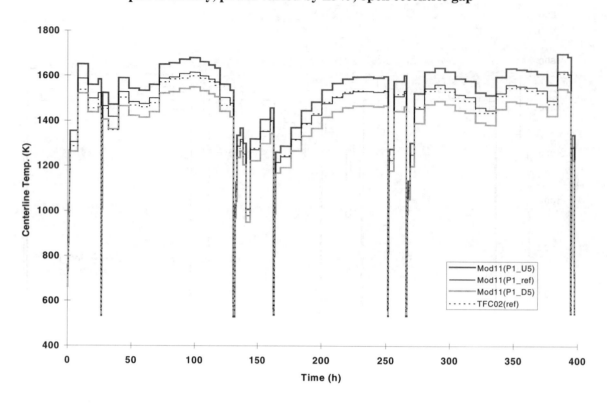

Figure 8. Centreline temperature vs. time for second power history, power increased 5%, closed gap

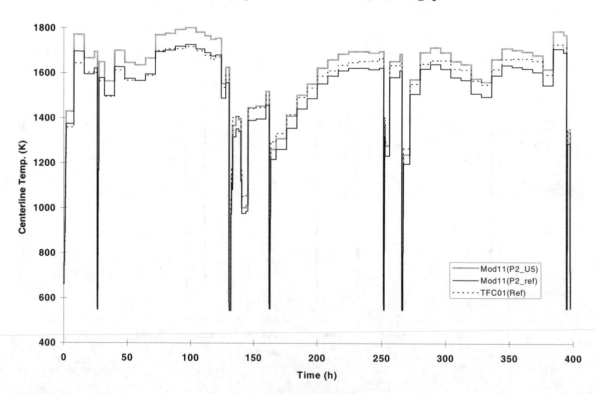

Figure 9. ε values for the example validation exercise

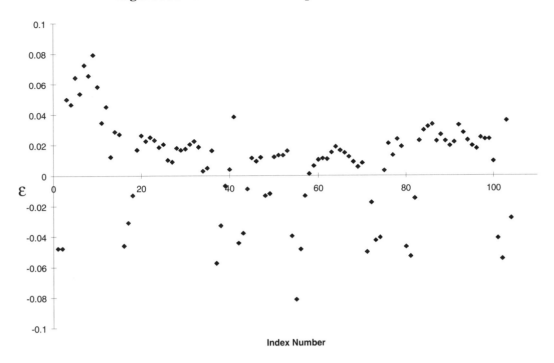

Figure 10. The theoretical and measured ε distribution

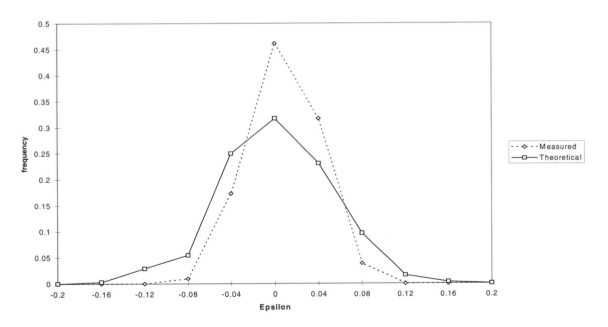

MODELLING PHENOMENA AFFECTING FUEL
ELEMENT THERMAL PERFORMANCE AT HIGH BURN-UP

V.I. Arimescu, M. Couture
Atomic Energy of Canada Ltd., Chalk River Laboratories, Canada

Abstract

The first part of the paper deals with burn-up degradation of UO_2 thermal conductivity. The processes affecting heat conduction in the fuel pellet are reviewed in terms of their impact at low and high burn-up. The main phenomena modelled presently are: solid fission product scattering effect on phonic term, gaseous fission product bubble porosity effect and low temperature irradiation damage effect.

Emphasis is placed on the first item, where data acquired at AECL for SIMFUEL have been used to derive a burn-up degradation factor. Different formulations of this correction factor are discussed and shown to provide very similar results. The effect on fuel performance is assessed in the framework of CANDU-type fuel.

The second part of the paper deals with another important contributing factor to heat conduction inside fuel pellet, namely flux depression and Pu build-up. In the previous versions of ELESIM/ELESTRES, the radial profile of the heat generation rate is modelled by a three-parameter correlation. Two databases exist for these parameters. Both were established by using the reactor physics code CE-HAMMER to calculate the local heat generation rate for fuel pins of various radii, enrichment and burn-up. The original database included fuel burn-ups in the range 0 to 480 MWh/kgU. The most recent database is an extension to burn-ups of 840 MWh/kgU. There exist, however, inconsistencies between the two databases that need to be resolved.

This motivated us to develop a new code, RAD_HEAT, whose evaluation of the local heat generation rate is based on a more mechanistic approach that does not rely on these databases. Two factors that contribute to changes in the heat generation are the thermal neutron flux depression due to self-shielding and a non-uniform build-up of ^{239}Pu. The radial profile of ^{239}Pu varies with burn-up and is characterised, at high burn-up, by a sharp increase in the concentration near the surface of the pellet. This non-uniform distribution is due to the resonance capture of epithermal neutrons by ^{238}U. In RAD_HEAT, our modelling of resonance capture phenomenon is based on Wagner's work which considers the fuel element as a pure absorber.

Before RAD_HEAT can be applied, a value for the parameter Σ_{av} must be determined. This was done by fitting the total (fast, epithermal and thermal) neutron capture rate in ^{238}U predicted by the code RAD_HEAT with experiment data obtained by A. Okazaki et al. at the ZEEP (Zero Energy Experimental Pile) reactor, Chalk River. In the end, the results of applying the new mechanistic model to CANDU-type fuel are outlined.

Introduction

In this paper we examine two phenomena that contribute to changes in thermal properties for UO_2 with burn-up. The first is burn-up degradation of UO_2 thermal conductivity. The processes affecting heat conduction in the fuel pellet are reviewed in terms of their impact at low and high burn-up. The main processes modelled presently are: solid fission product phonon, gaseous fission product bubble porosity formation and low temperature irradiation damage. Emphasis is placed on the first item, where data acquired at AECL for SIMFUEL have been used to derive a burn-up degradation factor. Different formulations of this correction factor are discussed and shown to provide very similar results. The effect of solid and gaseous fission products on fuel temperature is presented.

The second phenomenon that we examine is changes in heat generation rate due to flux depression and Pu build-up. In the previous versions of ELESIM/ELESTRES, the radial profile of the heat generation rate was modelled by a three-parameter correlation. Two databases exist for these parameters. Both were established by using the reactor physics code CE-HAMMER to calculate the local heat generation rate for fuel pins of various radii, enrichment and burn-up. The original database included fuel burn-ups in the range 0 to 480 MWh/kgU. The most recent database is an extension to burn-ups of 840 MWh/kgU. There exist, however, inconsistencies between the two databases that need to be resolved.

To address these discrepancies a new code, RAD_HEAT, was developed to evaluate the local heat generation rate with a more mechanistic approach, and accounts for thermal neutron flux depression due to self-shielding, and non-uniform build-up of ^{239}Pu. The radial profile of ^{239}Pu varies with burn-up and is characterised, at high burn-up, by a sharp increase in the concentration near the surface of the pellet. This non-uniform distribution is due to the resonance capture of epithermal neutrons by ^{238}U. In RAD_HEAT, modelling of resonance capture phenomenon is based on Wagner's work which considers the fuel element as a pure absorber. The theory behind RAD_HEAT and the derivation of key parameters are described. The results of applying RAD_HEAT to CANDU-type fuel are also outlined.

UO_2 thermal conductivity variation with burn-up

Theoretical background

Thermal conduction is a complex transport process, leading to a Boltzmann type equation, involving either phonon or electron migration and scattering through a lattice with point and extended defects. Analytical formulations are only possible by applying simplifying assumptions, which are valid only for limited ranges of relevant parameters.

Two different approaches have been used to model thermal conduction by phonons. A variational technique was used for lattice conduction first by Ziman [1] and by Leibfried and Schlomann [2], which is based on seeking the phonon distribution that satisfies a variational criterion equivalent to minimising the thermal resistivity. Phonon-phonon interactions are treated in terms of transition probabilities. Alternatively, a relaxation-time technique treats the problem in terms of restoring the phonon distribution to thermal equilibrium distribution at a rate proportional to the departure of the distribution from equilibrium. This technique was first proposed by Callaway [3] and later developed and applied by Klemens [4].

Both techniques arrive at the same result for pure material, namely a hyperbolic relation of the form C/T, where thermal conductivity is inversely proportional to temperature. The starting point to include the effects of lattice defects and impurities is the equation written by analogy with gas conduction:

$$k = 1/3 \int C(v)vl(v)dv \qquad (1)$$

where, $C(v)$ is the specific heat, v the group velocity, $l(v)$ the mean free path and, v the frequency of lattice oscillations. The dependence of C and l on v is called "dispersion relation". If different scattering processes are assumed independent, the relaxation-time method will lead to a harmonic mean rule of addition for the mean free path associated with these processes: $1/l = \Sigma 1/l_i$, where i numbers scattering processes. The main contributions come from Umklapp three-phonon interactions, and scattering by impurities and lattice defects(interstitials and grain boundaries).

Umklapp interactions occur in both pure and impure materials and their contribution to the mean free path, $1/l_u$, is proportional to v^2 and T and leads to the aforementioned C/T expression for thermal conductivity of pure materials. If the second process, scattering by defects, is assumed independent of phonon frequency, the following relation is obtained:

$$k = 1/(A + BT) \qquad (2)$$

where the BT is the intrinsic part, which comes from the anharmonicity of the lattice vibrations (Umklapp three-phonons collisions) and the A term represents the effect of impurities present in the lattice (such as fission products in irradiated material).

Alternatively, if the second process is treated according to Klemens' method for substitutional impurities, $1/l_i$, is proportional to v^4 and the final thermal conductivity expression is:

$$k = k_p X \operatorname{atan}(1/X) \qquad X = aT^{1/2} \qquad (3)$$

where and k_p is the pure material thermal conductivity, expressed as 1/BT in the previous relation, and X is the ratio of equivalent defect frequency (defined as the frequency at which defect mean free path equals intrinsic mean free path) and the Debye frequency.

The range of validity for both Equations (2) and (3) is limited to low and medium impurity concentrations, because it is only in this case that the assumption of different processes occurring independently holds. Numerically, this case corresponds to X >> 1 [6]. In this case, 1/X << 1 and atan(1/X) in Equation (3) can be expanded in power series, as follows:

$$k = k_p X (1/X)[1 - 1/(3X^2)+ \ldots] = 1/(BT) \, 1/[1 + 1/(3a^2T)] = 1/[BT + B/(3a^2)] \qquad (4)$$

where use was made of Equation (2) to express k_p, and the expansion $1/(1 + x) = 1 - x + \ldots$, was used for the second equality. Thus, Equation (3) is similar to Equation (2) for weak scattering from impurities.

Burn-up degradation correction from SIMFUEL data

Both formulations have been used at AECL to describe the burn-up degradation of UO_2 thermal conductivity due to solid fission product accumulation in the matrix as either precipitates or interstitials. SIMFUEL data [5] can be used to derive the unknown parameters for the effects of fission products on thermal conductivity. For example, the hyperbolic Equation (2) was used to derive a burn-up dependent correction relation, of the form:

$$k = 1/[A + A_b.Bu + BT], \quad A_b = 0.01436, \text{ for } A = 0.0378 \text{ and } B = 2.4E\text{-}04 \tag{5}$$

where, Bu, is the burn-up expressed in atom%*100, and k is expressed in W/mK.

The problem with applying Equation (3) is that fresh UO_2 is not a pure material and the fitting of this relation to experimental data is more difficult [6]. However, comparative studies proved that the results are very similar to those provided by Equation (5).

The effect of gaseous fission products which precipitate into bubbles, especially on the grain boundaries, is treated separately through the porosity correction term [7]. As irradiation progresses, this correction term becomes very important in the inner core of the pellet where gaseous swelling is significant.

Figure 1 shows the fractional degradation factor due to solid fission products for burn-ups up to 1440 MWh/kgU, while the combined effect of both solid and gaseous fission products is presented in Figure 2.

Another process that affects thermal behaviour and varies with burn-up is related to heat generation and its radial distribution, and the recently developed model is outlined below.

Radial profile of heat generation rate

In RAD_HEAT, the radial profile of the heat generation rate is calculated by dividing the fuel into a number, n_{an}, of annuli and solving for each annulus, J, the power density H^J. The local heat generation rate $H^J(t)$ in annulus J and at time t is defined as:

$$H^J(t) = \alpha_f \eta_{ff} \Phi_{ther}^J \sum_k N_k^J(t) \sigma_k^f \qquad \text{for } J = 1, n_{an} \tag{6}$$

where α_f is the energy release per fission (≈ 200 MeV), η_{ff} is the ^{238}U fast fission factor (1.08), Φ_{ther}^J is the time and volume averaged thermal flux in annulus J, $N_k^J(t)$ is the volume-averaged concentration of fissile nuclei k ($k = 1$ for ^{235}U, $k = 2$ for ^{239}Pu) in annulus J and at time t, and σ_k^f is the time and volume averaged weighted fission cross-section of nuclei k.

Neutron thermal flux contribution

The time-averaged radial distribution $\Phi_{ther}(r)$ of the thermal flux within the pellet is assumed to be a solution of the one-dimensional diffusion equation. For a solid pellet:

$$\Phi_{ther}(r) = I_0(\kappa r) \qquad 0 \leq r \leq R_{pin}$$

where I_0 is the modified Bessel function of the first kind and of order zero, R_{pin} is the radius of the pin and κ is the inverse of the thermal diffusion length. The value of κ is obtained by fitting $\Phi_{ther}(r)$ to the thermal fluxes (for various burn-ups) calculated with WIMS-AECL. By averaging over the J-th annulus, this thermal flux expression can be used in Equation (6) to define a relative power density factors $\Gamma^J(t)$ according to:

$$\Gamma^J(t) \equiv \frac{H^J(t)}{\overline{H}(t)} \qquad \text{for } J = 1, n_{an} \tag{7}$$

The fission cross-sections in Equation (6) are obtained through a WIMS-AECL calculation. The concentrations are the only remaining unknowns in Equation (6). They can be obtained by solving a set of depletion equations, the formulation of which requires a model for the radial profile of the resonance capture rate in ^{238}U, which is derived below.

Neutron resonance capture contribution

The radial profile of ^{239}Pu varies with burn-up and is characterised, at high burn-up, by a sharp increase in the concentration near the surface of the pellet. This non-uniform distribution is due to the resonance capture of epithermal neutrons by ^{238}U. In RAD_HEAT, our modelling of resonance capture phenomenon is based on Wagner's work [10] which considers the fuel element as a pure absorber.

An important feature of his model is that for most points within the fuel the radial distribution of resonance capture is expressed as a universal function independent of the resonance parameters. Comparisons [10] with experimental data published by Hellstrand [11] strongly suggest that this function could be a good approximation for the radial profile of the total (all resonances) capture rate. The model depends on the free parameter Σ_{av}, which can be regarded as an average resonance cross-section and can be determined by comparing code predictions and experimental data.

In the code RAD_HEAT, the fuel pin is divided into several annuli and Wagner's model is used to calculate the volume-averaged rate of resonance captures in each annulus J and in the pin:

$$C_{tot}^J = \frac{\int_J C_{tot}(r) r\, dr}{\int_J r\, dr} \quad ; \quad C_{tot}^{pin} = \frac{\int_{pin} C_{tot}(r) r\, dr}{\int_{pin} r\, dr}$$

The rate of resonance capture for the case of a pin with a hole is assumed to be identical to that of a pin with no hole. For each annulus the following ratio is defined for use in modelling the resonance capture term in the depletion equations:

$$\Lambda^J = \frac{C_{tot}^J}{C_{tot}^{pin}} \qquad \text{for } J = 1, n_{an} \tag{8}$$

Depletion equations

Since the percentage of depletion of ^{238}U is small, we consider its concentration as constant throughout burn-up and denote it $\overline{N}_{238}$. The depletion equations are formulated and solved for each annulus:

293

$$\frac{dN_{235}^J(t)}{dt} = -\sigma_{235}^a N_{235}^J(t)\Phi_{ther}^J \qquad \text{for } J = 1, n_{an} \tag{9}$$

$$\frac{dN_{239}^J(t)}{dt} = -\sigma_{239}^a N_{239}^J(t)\Phi_{ther}^J + \sigma_{238,res}^c \overline{N}_{238}\Phi_{epi}^J + \sigma_{238,ther}^c \overline{N}_{238}\Phi_{ther}^J \qquad \text{for } J = 1, n_{an} \tag{10}$$
$$+ \sigma_{238,fast}^c \overline{N}_{238}\Phi_{fast}^J$$

where $N_k^J(t)$ is the volume-averaged concentration for nuclei k (k = 235 for ^{235}U and k = 239 for ^{239}Pu) in annulus J at time t. Φ_{epi}^J and Φ_{fast}^J are respectively the time and volume-averaged epithermal and fast fluxes in annulus J. $\sigma_{238,res}^c$, $\sigma_{238,ther}^c$ and $\sigma_{238,fast}^c$ are the time and volume-averaged weighted resonance, thermal and fast capture cross-sections of ^{238}U. Finally, σ_{235}^a and σ_{239}^a are the time and volume-averaged weighted absorption cross-sections of ^{235}U and ^{239}Pu.

All the cross-sections can be evaluated using the reactor physics code WIMS-AECL and the thermal flux is given by Equation (7). What remains to be specified, in Equations (9) and (10), are the epithermal and fast fluxes. The goal is to express them in terms of the thermal flux. Using the model of resonance capture described before and Equation (8):

$$\frac{\Phi_{epi}^J}{\Phi_{epi}^{pin}} = \Lambda^J \tag{11}$$

The ratio of pin-averaged epithermal to thermal flux can then be calculated using WIMS-AECL:

$$\beta \equiv \frac{\phi_{epi}^{pin}}{\phi_{ther}^{pin}} \tag{12}$$

where the use of ϕ instead of Φ indicates that these are fluxes calculated by the code WIMS-AECL. Combining Equations (11) and (12) we get:

$$\Phi_{epi}^J = \beta \Lambda^J \Phi_{ther}^{pin} \tag{13}$$

A similar strategy is used for the fast flux:

$$\varsigma_{fast}^J \equiv \frac{\phi_{fast}^J}{\phi_{ther}^J} \quad \Rightarrow \quad \Phi_{fast}^J = \varsigma_{fast}^J \Phi_{ther}^J \tag{14}$$

where ϕ represents fluxes calculated using WIMS-AECL. The parameters β and ς_{fast}^J will vary with burn-up and fuel type.

The next step consists of formulating the depletion equations in terms of local burn-ups using the following relations:

$$\frac{dbu^J}{dt} = \frac{H^J(t)}{\rho_{fuel}} \qquad for \quad J = 1, n_{an} \quad ; \quad \frac{d\overline{bu}}{dt} = \frac{\overline{H}(t)}{\rho_{fuel}} \tag{15}$$

where ρ_{fuel} is the fuel density, bu^J the volume averaged burn-up in annulus J, and $\overline{bu}$ the volume-averaged burn-up in the pin. Recalling Equation (6), and combining the results expressed in Equations (13), (14) and the first relation of Equation (15), one gets:

$$\frac{dN_{235}^J(bu^J)}{dbu^J} = -l_0^J N_{235}^J(bu^J) \quad \text{for } J = 1, n_{an} \tag{16}$$

$$\frac{dN_{239}^J(bu^J)}{dbu^J} = -l_3^J N_{239}^J(bu^J) + \left(l_1^J + l_2^J + l_4^J\right)\overline{N}_{238} \quad \text{for } J = 1, n_{an} \tag{17}$$

where:

$$l_0^J = \frac{\sigma_{235}^a \rho_{fuel}}{\eta_{ff}\alpha_f \sum_k N_k^J(bu^J)\sigma_k^f} \quad ; \quad l_1^J = \frac{\beta \Lambda^J \rho_{fuel} \sigma_{238,res}^c}{\eta_{ff}\alpha_f \sum_k N_k^J(bu^J)\sigma_k^f} \frac{\Phi_{ther}^{pin}}{\Phi_{ther}^J}$$

$$l_2^J = \frac{\sigma_{238,ther}^c \rho_{fuel}}{\eta_{ff}\alpha_f \sum_k N_k^J(bu^J)\sigma_k^f} ; l_3^J = \frac{\sigma_{239}^a \rho_{fuel}}{\eta_{ff}\alpha_f \sum_k N_k^J(bu^J)\sigma_k^f} ; l_4^J = \frac{\sigma_{238,fast}^c \rho_{fuel} \varsigma_{fast}^J}{\eta_{ff}\alpha_f \sum_k N_k^J(bu^J)\sigma_k^f}$$

Since the coefficients l_i^J ($i = 0, 4$) are functions of the local concentrations, they are burn-up dependent. The task of solving these equations is simplified greatly if we seek solutions which are valid for a small local burn-up increment $\Delta bu^J = bu_f^J - bu_i^J$. Here bu_i^J and bu_f^J are the initial and final local burn-up values. If Δbu^J is small enough, a reasonable approximation is to consider the local power density to be constant during that burn-up interval. With such an approximation, the coefficients l_i^J can be considered constants whose values are determined by the concentrations at the initial burn-up value bu_i^J. Equations (16) and (17) are then first order linear differential equations which can be solved by standard methods:

$$N_{235}^J(bu_f^J) = N_{235}^J(bu_i^J)exp(-l_0^J\Delta bu^J) \tag{18}$$

$$N_{239}^J(bu_f^J) = exp(-l_3^J\Delta bu^J)\left[\left(\frac{(l_1^J + l_2^J + l_4^J)\overline{N}_{238}}{l_3^J}\right)\left(exp(l_3^J\Delta bu^J) - 1\right) + N_{239}^J(bu_i^J)\right]$$

In RAD_HEAT, for each global burn-up step $\Delta\overline{bu}$ the corresponding local burn-up step Δbu^J is evaluated using the following relation:

$$\Delta bu^J = \Gamma^J(t_i)\Delta\overline{bu} \quad \text{for } J = 1, n_{an}$$

where here $\Gamma^J(t_i)$, given by Equation (7), is the value of the power density factor in annulus J at the beginning of the burn-up step ($\Delta t = t_f - t_i$). The result is then substituted on the right hand side of Equation (18) to obtain the new concentrations. The process is repeated for each global burn-up step.

Model verification and tuning

Before RAD_HEAT can be applied, a value for the parameter Σ_{av} must be determined. This was done by fitting the total (fast, epithermal and thermal) neutron capture rate in ^{238}U predicted by the code RAD_HEAT with experimental data obtained by A. Okazaki *et al.* [12] at the ZEEP (Zero Energy Experimental Pile) reactor. The total capture rate in ^{238}U is modelled by the second term on the right hand side of Equation (17).

Neutron capture in ^{238}U results in the formation of ^{239}Pu. The transformation of ^{238}U into ^{239}Pu is accompanied by x-ray emissions that can be measured. Okazaki *et al.* obtained the radial profile of neutron capture in ^{238}U by measuring this activity in thin samples taken at various depths of ZEEP rods. The moderator in this experiment was heavy water. Figure 3 shows the total capture as predicted by RAD_HEAT (a burn-up of 0 is assumed) and as measured by Okazaki *et al.* [12] for a natural uranium metal rod of 3.26 cm in diameter with an aluminium cladding of 0.1 cm in thickness and a 0.015 cm gap. The rod was located at the centre of the core whose hexagonal lattice had a centre-to-centre spacing of 5.77 cm. The results are normalised to unity at the point nearest to the surface. The value that provided the best fit was $\Sigma_{av} = 230 \text{ cm}^{-1}$.

This value was then used to compare the power density factors (see earlier definition) predicted by RAD_HEAT with those of the existing correlation in the code ELESIM/ELESTRES using the initial (original) and extended databases. The case was that of a single uranium oxide rod of 12.15 mm in diameter with an enrichment of 0.71% and burn-up of 400 MWh/kgU. The rod was immersed in D_2O. Cross-sections used in this calculation were those associated with the outer pins of a 37-pin CANDU bundle and were obtained from a WIMS-AECL calculation. Power density factors are shown in Figure 4 as a function of normalised (fraction of pellet radius) radial position. The code CE-HAMMER could divide the fuel pin in at most 10 regions, while we have been using RAD_HEAT with over a thousand mesh points. The effect of a coarser mesh is to average the power density over a larger annulus, therefore decreasing its value, especially in the vicinity of the surface where the variation of the resonance capture rates is the greatest. This would explain why the difference between the power density factors calculated by RAD_HEAT and those evaluated by the existing correlation in ELESIM/ELESTRES is greater near the surface of the fuel pin, since the correlations are based on fitting CE-HAMMER calculated values.

Conclusions

The burn-up dependence of UO_2 thermal conductivity is modelled by accounting for three processes: phonon scattering by solid fission products, thermal resistance of fission gas bubbles and phonon scattering of low temperature irradiation damage. For a maximum CANDU burn-up of 450 MWh/kgU, the effect of this burn-up degradation of thermal conductivity is a 10% relative increase in central temperature.

The temperature distribution in the fuel pellet is affected by the radial variation of the heat generation rate which depends on thermal flux self-shielding and Pu build-up at the periphery of the pellet. A mechanistic model to describe both processes has been developed and verified.

REFERENCES

[1]   J. Ziman, "Electrons and Phonons", Oxford, Clarendon Press, 1960.

[2]   G. Leibfried and E. Schlomann, Nach. Akad. Wiss. Gottingen, 29, 1, 1954.

[3]   J. Callaway, *Phys. Rev.*, 113, 1046, 1959.

[4]   P.G. Klemens, *Phys. Rev.*, 119, 507, 1960.

[5]   P.G.Lucuta, B.J. Palmer, H.J. Matzke, D.S. Hartwig, Proc. 2nd Intl. Conf. on CANDU Fuel, CNS Toronto, 1989, 292.

[6]   P.G. Lucuta, H.J. Matzke, I.J. Hastings, *JNM*, 232, 166-180, 1996.

[7]   J.R. MacEwan, R.L. Stoute and M.J.F. Notley, *J. Nucl. Mat.*, 24, 109-112, 1967.

[8]   M.J.F. Notley, "ELESIM: A Computer Code for Predicting the Performance of Nuclear Fuel Elements", *Nuclear Technology*, Vol. 44, p. 445, August 1979.

[9]   H.H. Wong, E. Alp, W.R. Clendening, M. Tayal and L.R. Jones, "ELESTRES: A Finite Element Fuel Model for Normal Operating Conditions", *Nuclear Technology*, 57, 203-212, May 1982.

[10]   M. Wagner, "Spatial Distribution of Resonance Absorption in Fuel Elements", *Nuclear Science and Engineering*, Vol. 8, p. 278, 1960.

[11]   E. Hellstrand, "Measurements of Neptunium Activity in Irradiated Uranium Rods", Proceedings of the Brookhaven Conference on Resonance Absorption of Neutrons in Nuclear Reactors", p. 32, July 1956.

[12]   A. Okazaki, D.W. Hone and D.F. Allen, Internal AECL unpublished work, 1959.

Figure 1. Fractional burn-up degradation factor variation with burn-up

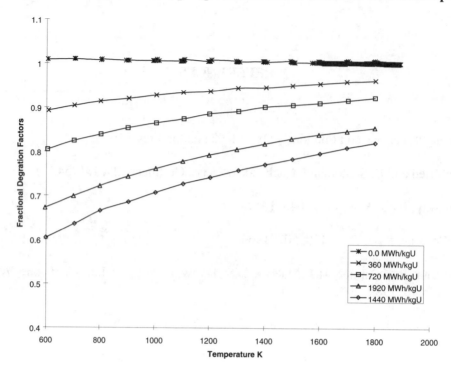

Figure 2. Relative increase in fuel centre temperature with burn-up

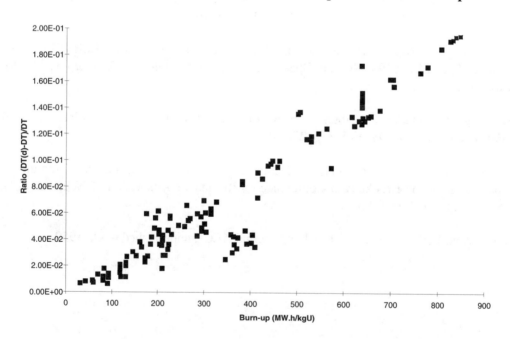

Figure 3. Relative total neutron capture rate

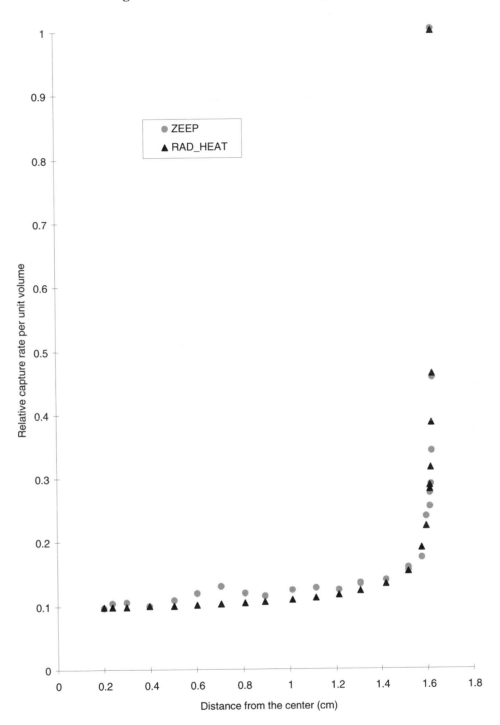

Figure 4. Radial profile of relative power density factor for a burn-up of 400 MWh/kgU; case of a natural uranium oxide rod of 12.15 mm in diameter

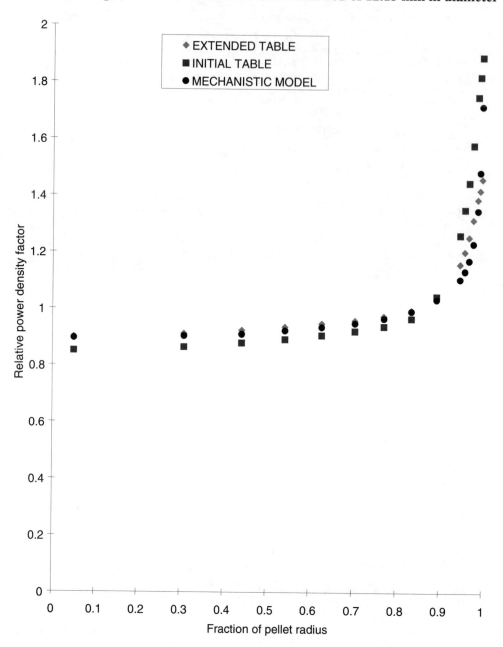

THERMAL PERFORMANCE MODELLING WITH THE ENIGMA CODE

G.A. Gates, P.M.A. Cook, P. de Klerk, P. Morris and I.D. Palmer
British Nuclear Fuels plc, Springfields, Preston, UK

Abstract

The ENIGMA fuel performance code has been used within BNFL to license UO_2, gadolinia and MOX fuels. The validation of ENIGMA has been extensive with over 500 rod irradiations covering thermal performance, fission product release, dimensional changes and clad corrosion to burn-ups of over 80 MWd/kgHM. The high burn-up thermal performance of the code is illustrated using a selection of validation data covering UO_2, gadolinia and MOX fuels. A brief description of the model changes made to address the thermal performance in these fuel types is presented with particular attention given to the modelling of the thermal conductivity degradation and "rim effect" in high burn-up fuel.

Introduction

The ENIGMA [1,2] fuel performance code has been used within BNFL to license UO_2, gadolinia and MOX fuels. The validation database for ENIGMA currently consists of roughly an equal mixture of commercial and test reactor data from various international projects. In total, the database includes over 500 rod irradiations from 46 different programmes, supplemented also by post-irradiation test data and PIE measurements. The total database covers thermal performance, fission product release, dimensional changes and clad corrosion with ample data for burn-ups up to 60 MWd/kgHM and a limited amount of data at burn-ups up to 80 MWd/kgHM. The present paper is limited to illustrating the high burn-up thermal performance of the code using a selection of validation data covering UO_2, gadolinia and MOX fuels.

Of particular importance at high burn-up is fuel thermal conductivity degradation and the change in pellet microstructure that occurs at the pellet periphery. This is characterised by a highly porous annulus of thickness typically 100-200 μm leading to an enhancement in the local fission gas release and a reduction in the local fuel thermal conductivity, the so-called "rim effect" [3]. A simple rim effect model is presented along with supporting validation data to account for this phenomenon. A brief description of other model changes made to address the thermal performance of the three fuel variants is also given.

Code overview

ENIGMA has been under development since the mid-1980s. Initially, the work was performed collaboratively by BNFL and CEGB (now Nuclear Electric), with support from UKAEA (now AEA Technology), and early versions of the code were jointly owned. Since 1991, BNFL and NE have pursued separate lines of development, reflecting different commercial and technical interests and priorities. For its part, BNFL has concentrated on modelling LWR fuel including UO_2, gadolinia and MOX fuel variants, with recent emphasis on high burn-up operation.

ENIGMA predicts the thermomechanical response of a fuel rod to a specified rating history using a 1½-dimensional representation in which the fuel rod is represented by a series of axial zones which are coupled by the coolant energy equation, rod internal pressure and gas transport. Within each axial zone, the code performs a one-dimensional axi-symmetric mechanical calculation under the assumption of generalised plane strain in both pellet and clad, with the pellet solution perturbed to take account of axial extrusion and pellet wheatsheafing (hourglassing). The fuel is treated as a non-isotropic but homogeneous material with directionally dependent elastic constants; these are functions of the state of cracking, which is constrained to occur only on the r-θ and r-z principal planes. Plasticity and creep in both fuel and clad are assumed to obey the Levi-Mises equations for an isotropic material. These are related to uniaxial material properties through the generalised stress. The stress equilibrium and strain-displacement relationships are approximated by a finite difference scheme in which deformations are treated by updating the fuel and clad deformed geometry at the end of each calculation step. An iterative marching procedure is used to solve the thermomechanical equations through successive clad and fuel annuli. A stable solution is obtained, irrespective of timestep length, using the virtual stress concept, which provides numerical stability and reasonable accuracy even for large history steps.

The code contains a set of self-contained modules to represent the various physical and mechanical properties of the fuel, clad and gaseous components of the rod, together with models representing the effects of irradiation, including fuel densification/swelling and fission product release. The output from the code is a set of calculated parameters describing the rod state at each point during the irradiation. The code can be used for the assessment of LWR fuel under all steady state and off-normal conditions, and provides calculations of all design and licensing related parameters including temperature, stress and strain distributions through pellet and clad, plus clad oxidation/hydriding and fatigue. The code also includes models for azimuthal and axial stress concentration and clad stress corrosion cracking associated with pellet-clad interaction.

The code currently consists of around 60 modules written to FORTRAN-77 and FORTRAN-90 coding standards, and has been developed and maintained using modern software quality assurance techniques. It is fast-running, readily portable, and provides a number of user convenience features including a graphical user interface. Most input parameters can be defaulted, leading to compact data files, and guaranteeing standardisation in validation and definitive design calculations. However, flexibility is retained to allow the user to investigate the sensitivity of code predictions to parameter uncertainties or different modelling assumptions. For code validation, a database of over 500 test reactor and commercial reactor rod irradiations has been assembled, which provides a thorough testing of all of the code's models over a very broad range of conditions. The validation procedure has been fully automated (taking around five hours on a modern PC) and generates some 1600 plots comparing predicted and measured values.

Modelling high burn-up fuel

UO_2 fuel

The ENIGMA validation database for standard UO_2 fuel consists of some 414 rod irradiations from 33 different programmes to burn-up of over 80 MWd/kgHM. Using this database the fuel thermal conductivity has been fitted to the following standard equation:

$$k = \frac{1}{a + bT} + electronic\ term \tag{1}$$

$$a = a_0\left(1 + \alpha\ Burnup\right)$$

where T is the fuel temperature, a is the phonon scattering term caused by the scattering from lattice defects and impurities and b describes the phonon-phonon scattering which is a characteristic of the host material. Burn-up increases the impurity content, specifically fission products, and hence leads to an increase in the value of the phonon scattering term, a, as shown in Equation (1) where α is the rate of thermal conductivity degradation. Values for a_o and b are derived from out-of-pile conductivity measurements on unirradiated fuel, $a_0 = 37.5$ and $b = 0.2165$ when k is in units of kW/m/K.

The full conductivity model contains correction terms to account for the presence of as-fabricated and gas bubble porosity and deviations from stoichiometry. At the present time no attempt has been made to model the transient thermal conductivity recovery effect observed in out-of-pile tests on irradiated fuel. This effect is attributed to irradiation damage annealing and gas bubble precipitation and growth, but under in-pile conditions the kinetics are considered to be too rapid for the effect to be observed. However, the model does include an upper limit on conductivity, due to the absence of damage annealing, for fuel at low temperatures.

A number of experiments have been conducted in Halden and Risø to assess the magnitude of thermal conductivity degradation with burn-up. Data at over 80 MWd/kgHM burn-up have been collected and these have been used to determine a value of 0.08 $(MWd/kgUO_2)^{-1}$ for the degradation parameter, α, in Equation (1).

An example of the ENIGMA predictions of fuel centre temperature at high burn-up can be seen in Figure 1. The data in this figure is from the ultra-high burn-up Halden rig, IFA-562. This rig was loaded with rods of small diameter and high enrichment allowing a very high rate of burn-up accumulation. From this figure it is clearly seen that ENIGMA predicts the correct burn-up dependence of the thermal conductivity degradation of UO_2 to burn-ups of over 80 MWd/kgHM. The early in life fuel centre temperatures are over predicted by about 100°C which may be due to an overestimate in the rod powers.

**Figure 1. Comparison of ENIGMA fuel centre temperature
predictions with data from Halden ultra-high burn-up rig, IFA-562 Rod 16**

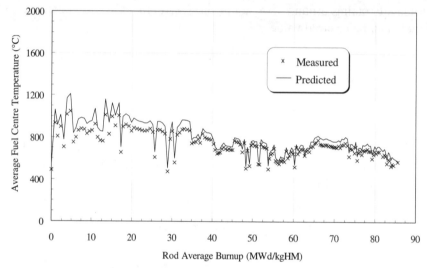

For a number of years it has been known that at high burn-ups (≥45 MWd/kgHM radial average) a change in the fuel microstructure occurs at the periphery of the fuel pellet, termed the rim effect [3]. This rim region is characterised by a loss of definable grain structure and a depletion of matrix fission gas which leads to an increase in the local fission gas release rate and a reduction in the local fuel thermal conductivity. This effect is attributed to strong resonance absorption of neutrons in ^{238}U, leading to an enhancement in the level of ^{239}Pu at the pellet surface which leads to the changes in the pellet microstructure. The formation of a fuel rim has been observed in fuel irradiated to a mean burn-up of ≥45 MWd/kgHM. At the present moment there is limited understanding of the detailed mechanism underlying the formation of this rim region.

The rim model used in ENIGMA is based on the empirical correlations reported by Barner *et al.* [4] and is basically an enhancement to the athermal fission gas release caused by the formation of a new microstructure at the pellet surface. The rim width is correlated with the pellet edge burn-up beyond a certain threshold value, that is:

$$w_{rim} = r_{grow} \sqrt{bu_{rim} - bu_{thres}} \tag{2}$$

where w_{rim} is the rim width, r_{grow} is the rim growth rate, bu_{rim} is the pellet edge burn-up and bu_{thres} is the local burn-up threshold for rim formation. The local fission gas release within the rim layer is taken as a linear function of the excess burn-up beyond the threshold:

$$f_{rim} = r_{fr}\left(bu_{rim} - bu_{thres}\right)$$
(3)

where f_{rim} is the fraction of gas release from the rim region and r_{fr} is the rim fractional release factor. In [4], slightly different values of threshold burn-ups are used in Equations (2) and (3), but in the ENIGMA model a single value has been chosen which has been fitted to fission gas release data. Also, the values for r_{grow} and r_{fr} have been correspondingly modified slightly from the values used in [4]. The highly porous nature of the rim region leads to a reduction in the local fuel thermal conductivity; this is reflected in the model by reducing the local fuel conductivity by an empirically derived factor.

The improvements in modelling predictions of fission gas release by including this simple rim release model can be seen in Figure 2. In his figure the target P/M=1 (predicted-over-measured) line together with the ×/÷ 2 band are shown. This band is generally considered as being equivalent to a rating uncertainty of about ±5%, and hence represents the tightest grouping of predictions that can realistically be expected. With the rim model switched off (×) the P/M fission gas release value is 0.818, this is dramatically improved to a P/M value of 1.09 when the rim model is included ($\square$). The burn-up dependence of the P/M gas release values is shown in Figure 3. Here it is seen that the inclusion of the rim model improves the fission gas release predictions for rod average burn-ups greater than 40 MWd/kgHM. The model performs well up to ~65 MWd/kgHM.

Another example of the improvements in model predictions gained by using this simple rim model is shown in Figure 4. Here pressure data from Rod 13 of IFA-562 is plotted along with the ENIGMA predictions with and without the rim model included. It can clearly be seen that to burn-ups of up to ~90 MWd/kgHM the ENIGMA predictions are dramatically improved with the inclusion of this new rim model even though the softer neutron flux spectrum in the Halden reactor leads to a much less pronounced rim structure when compared to that produced in a commercial reactor at the same equivalent average burn-up.

Figure 2. Fission gas release predictions of commercially irradiated fuel, with and without the rim model included

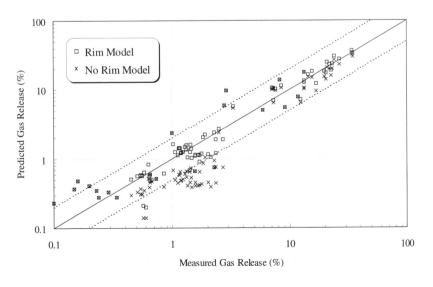

**Figure 3. Burn-up dependence of the fission gas release predictions
of commercially irradiated fuel, with and without the rim model included**

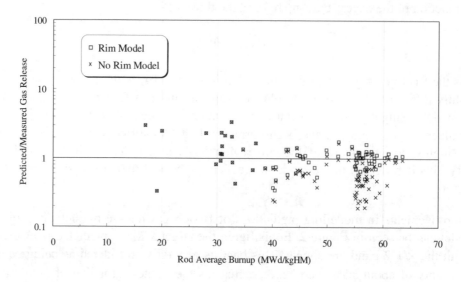

**Figure 4. Comparison of ENIGMA rod pressure predictions
with data from the Halden ultra-high burn-up rig, IFA-562 Rod 13**

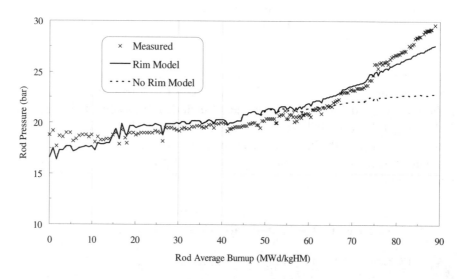

Gadolinia-doped fuel

The addition of gadolinia to uranium fuel has significant effects on the physical and thermal properties of the fuel [5,6], and in particular has a negative effect on the thermal conductivity, resulting in higher fuel temperatures. Therefore it is important to be able to model these effects within any fuel performance code. Within ENIGMA, gadolinia-doped fuel is modelled as standard UO_2 fuel apart from three basic model changes:

- The thermal conductivity of gadolinia-doped fuel is lower than that of UO_2, this is reflected in a change to the fuel conductivity equation, see below.

- The fission gas diffusion rate is suppressed in line with diffusivity measurements.

- The highly absorbing isotopes of gadolinia cause a severe and rapidly varying radial power profile across the fuel pellet; the RAting Depression Analysis Routine (RADAR) [7] used within ENIGMA, which calculates the radial power profile across the fuel pellet, has been rewritten to account for this. This model has been validated against the WIMS and CASMO neutronics codes.

The above model changes have been validated against the ENIGMA gadolinia database which contains data from seven different programmes to a maximum burn-up of ~60 MWd/kgHM.

To model the effect of the reduction in the thermal conductivity of UO_2 with the addition of gadolinia an additional degradation term is added to the phonon scattering term in Equation (1) to give the following equation,

$$a = a_0\left(1 + \alpha\, Burnup + \beta f_{gad}\right) \tag{4}$$

where f_{gad} is the fractional gadolinia content, and β is an empirical parameter derived from thermal conductivity measurements made as part of the NFIR programme [6]. The NFIR tests included samples of fuel doped with 4, 8 and 12% gadolinia for temperatures up to 1500°C.

An example of the ENIGMA predictions of fuel centre temperature for gadolinia-doped fuel is given in Figure 5. This figure shows data for the two standard grain size rods in the Halden gadolinia experiment IFA-515.10 [9]. This rig contains six small diameter rods, three UO_2 and three UO_2-Gd, of high enrichment to achieve a high rate of burn-up accumulation. The gadolinia dopant is at the 8% level, however the source gadolinium is not of natural composition, but rather consists of more than 98% of the non-absorbing [160]Gd. This approach gives an experiment in which the impact of gadolinia doping on fuel thermal properties can be isolated from the usual severe impact that the dopant has on early-in-life powers and the radial power profile. From Figure 5 it is seen that the ENIGMA predictions of the through-life temperatures, at a constant rating of 16 kW/m, for both the gadolinia and UO_2 rods are well predicted to a burn-up of over 50 MWd/kgHM. The measurements display a distinct shift at a burn-up of ~25 MWd/kgHM; this is believed to be due to instrumentation drift in the expansion thermometer.

**Figure 5. Comparison of ENIGMA predictions of fuel centre
temperature with data from the Halden gadolinia experiment, IFA-515.10**

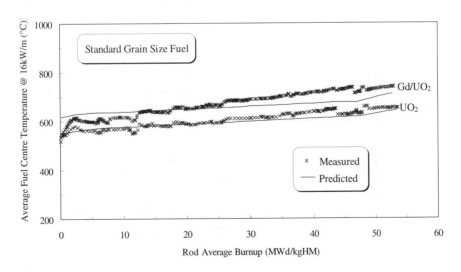

Figure 6 shows the predicted versus measured fission gas release values derived from the puncture tests of gadolinia-doped rods. Overall for burn-ups of up to ~60 MWd/kgHM the ENIGMA predictions are within the ×/÷2 band. However, three less satisfactory predictions are seen, these all being overpredictions. All these rods have quoted releases that appear very low for the power experienced during their irradiation, and hence the accuracy of the fission gas release for the rods is open to question.

Figure 6. Predicted versus measured gas release values for gadolinia-doped fuel

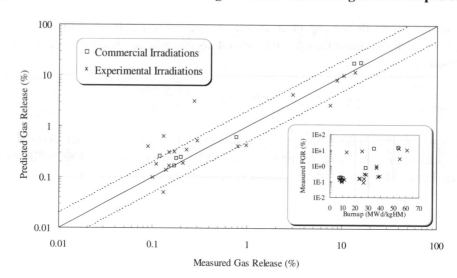

MOX fuel

As for gadolinia-doped fuel mixed oxide (MOX) fuel is modelled within ENIGMA as standard UO$_2$ fuel apart from four basic model changes:

- The thermal conductivity of MOX fuel is lower than that of UO$_2$, this is reflected in a change to the fuel conductivity equation, see below.

- Based on out-of-pile measurements the thermal creep rate of MOX fuel is enhanced compared with that of UO$_2$ under the conditions of most relevance to pellet behaviour in normal and transient conditions, although in other, less common, regimes of stress and temperature its creep rate can be lower.

- The presence of the plutonium isotopes increases neutron absorption at the start of life, giving a larger radial power depression and a different evolution with burn-up compared to standard UO$_2$ fuel, the RADAR model has been tuned to model this.

- The high plutonium isotopes lead to earlier and larger generation of curium-242 which is strongly alpha-active and can produce significant amounts of helium at high burn-up; a model is incorporated in ENIGMA to treat this.

The above model changes have been validated against the ENIGMA MOX database which contains data to a maximum burn-up of ~55 MWd/kgHM.

The fuel thermal conductivity relationship for MOX fuel is that used for UO_2, see Equation (1), with a reduction of 8% to account for the lower thermal conductivity of MOX. This value of 8% is based on in-pile fuel centre temperature measurements from a number of MOX irradiation tests. An example of ENIGMA centre temperature predictions for a Short Binderless Route (SBR) [9] MOX rod is shown in Figure 7. From this figure it is clearly seen that the ENIGMA predictions of fuel centre temperature are good to burn-ups of ~55 MWd/kgHM.

Figure 7. Comparison of ENIGMA fuel centre temperature predictions with data from an SBR MOX fuel rod

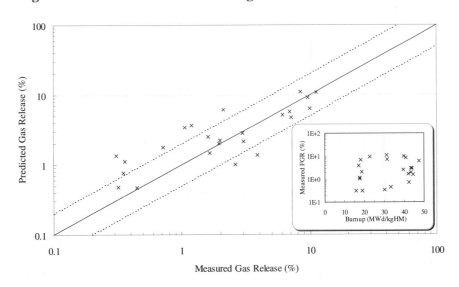

Figure 8 shows the predicted versus measured fission gas release values from the MOX validation database. The code predictions are reasonable to a maximum burn-up of just under 50 MWd/kgHM.

Figure 8. Predicted versus measured gas release values for MOX fuel

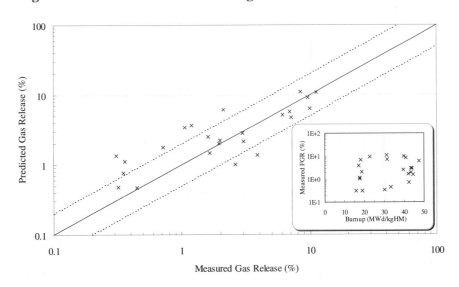

309

Conclusions

It has been shown that the ENIGMA fuel performance code has the capability of modelling the thermal performance of UO_2, gadolinia-doped and MOX fuels to burn-ups in excess of 55 MWd/kgHM.

It has been found that the inclusion of a model for the high burn-up rim effect, together with a simple treatment of fuel thermal conductivity degradation, are sufficient to provide good agreement on both fuel temperatures and fission gas release through to very high burn-ups. The modelling of these effects will continue to be reviewed as new data becomes available.

REFERENCES

[1]   P.A. Jackson, J.A. Turnbull and R.J. White, "ENIGMA Fuel Performance Code", *Nucl. Energy*, 29, 107-114, April 1990.

[2]   W.J. Kilgour, J.A. Turnbull, R.J. White, A.J. Bull, P.A. Jackson and I.D. Palmer, "Capabilities and Validation of the ENIGMA Fuel Performance Code", ANS Avignon, April 1991.

[3]   Hj. Matzke and M. Kinoshita, "Polygonization and High Burn-up Structure in Nuclear Fuels", *J. Nucl. Mater.*, 247, 108-115 (1997).

[4]   J.O. Barner, M.E. Cunningham, M.D. Freshley and D.D. Lanning, "Relationship Between Microstructure and Fission Gas Release in High Burn-up UO_2 with Emphasis on the Rim Region", ANS Avignon, April 1991.

[5]   R.A. Busch, "Properties of the Urania-Gadolinia System (Part 1)", NFIR Report, EPRI NP-7561-D, January 1992.

[6]   R.H. Watson, "Properties of the Urania-Gadolinia System (Part 2)", NFIR Report, EPRI NP-5861-LD, June 1988.

[7]   I.D. Palmer, K.W. Hesketh and P.A. Jackson, "A Model for Predicting the Radial Power Profile in a Fuel Pin", Proceedings IAEA Specialists' Meeting, Preston, March 1982.

[8]   M.T. Alvarez, M. Hirai and W. Wiesenack, "Analysis of the Thermal Behaviour of Gd-bearing Fuel in IFA-515.10", Halden Report, HWR-470, April 1996.

[9]   J. Mullen, C. Brown, I.D. Palmer and P. Morris, "Performance of SBR MOX Fuel in the Callisto Experiment", TopFuel '97, Proceedings of the Conference organised by the British Nuclear Energy Society and held in Manchester on 9-11 June 1997: Vol. 1., June 1997.

ANALYSES OF FUEL ROD THERMAL PERFORMANCE
AND CORRELATIONS WITH FISSION GAS RELEASE

L.C. Bernard, P. Blanpain, E. Bonnaud, E. Van Schel
Framatome Nuclear Fuel, Lyon, France

Abstract

The accurate evaluation of temperature distribution inside the fuel rod is most important for good fuel rod modelling as key phenomena, like Fission Gas Release (FGR), are strongly dependent on temperature. The strong coupling between temperature and FGR implies correlations, useful to provide bounds on thermal effects like degradation of fuel thermal conductivity with burn-up. This paper presents thermal and FGR analyses done with COPERNIC, the FRAMATOME fuel rod performance code, based on the TRANSURANUS code. COPERNIC was developed to accurately predict extended-burn-up fuel performance while including advanced material models.

COPERNIC thermal calculations include new or improved models: porosity increase in the rim region, heat transfer multiplier for the pellet-cladding gap, degradation of fuel thermal conductivity with burn-up, gaseous and contact conductances, validated separately. The thermal model is benchmarked with over 1800 fuel centreline temperatures up to a rod average burn-up of 80 GWd/tM. UO_2, UO_2-Gd_2O_3, and MOX data have been obtained through French or international R&D programmes, and, in particular, the use of high burn-up irradiated fuels.

The key feature of FGR is the presence of an *Incubation Threshold* (IT): FGR occurs above a threshold temperature, which decreases as burn-up increases. The COPERNIC model describes FGR as a two-step diffusion process: the gas atoms diffuse first from inside the grain to the grain boundaries where they accumulate until a saturation threshold is reached. The FGR model predicts well the IT curve, usually derived numerically. An excellent Analytical IT expression (AIT) is found and has the merit of isolating the parameters pertaining to the mechanisms which come into play. The AIT expression includes one term similar to that of the Empirical IT (EIT) and additional terms with parameters sensitive to temperature. The AIT parameters have been fitted on close to 200 FGR measurements from UO_2 fuel rods in steady-state operations, up to a rod average burn-up of 67 GWd/tM and an LHGR up to 50 kW/m. As a result, the AIT temperature is comparable to the EIT temperature at low burn-ups but it is lower at high burn-ups. The origins of this difference are discussed.

Introduction

It is important to accurately predict the temperature distribution inside the fuel rod as many mechanisms are critically dependent on temperature. Moreover, at extended burn-ups, new effects take place like the rim effect and the degradation of fuel thermal conductivity with burn-up. Improved modelling is needed to take these effects into account.

A new fuel performance code, COPERNIC, has recently been developed at FRAMATOME [1]. This code, based on the TRANSURANUS code [2], includes advanced models and has been extensively benchmarked taking into account newly available high burn-up and transient experimental data. Also, COPERNIC predictions apply to UO_2, UO_2-Gd_2O_3 and MOX fuel types.

This paper presents, first, the thermal model of COPERNIC and its benchmark. Then, we present the Fission Gas Release (FGR) model as FGR is strongly dependent on temperature and the presence of a thermally activated FGR threshold [3] gives indications on temperature. Last, these results are discussed.

Thermal model

The model

The model contains several sub-models: gap conductance, gap closure, fuel thermal conductivity, radial power profile through the pellet, and fuel rim.

The gap conductance model includes a description of gap conductance [4] and an adjustment on contact conductance data [5].

The degradation of fuel conductivity with burn-up is the main issue one must address to accurately predict high burn-up fuel performance. Even if questions remain to fully explain some experimental results, the data available today allow one to quantify this effect reasonably well. These data are of three types:

- thermal diffusivity measurements on simulated fuel (SIMFUEL [6]);

- thermal diffusivity on irradiated fuel;

- central temperature measurements on instrumented fuel rods.

A detailed interpretation of these data has been performed. As proposed by several authors, the burn-up appears to mainly act on the phonon-impurity interaction term. Other phenomena, such as irradiation damage, are of second order magnitude for power reactor conditions.

The following, now classical form, was chosen to predict the effect on irradiated fuel conductivity:

$$\lambda_{100\%} = (A_0 + A_1.B + A_2.T)^{-1} + f(T)$$

where $\lambda_{100\%}$ is the 100% dense UO_2 fuel conductivity, B is burn-up, T is temperature and $f(T)$ is the electronic contribution which is small up to high temperatures. The porosity correction to fuel conductivity is taken from [7]. The A_I coefficient, which was first derived from SIMFUEL data, has been slightly increased to better fit thermal diffusivity measurements on irradiated fuel and above all, central temperature measurements for high burn-up fuel rods.

The use of accurate radial profiles calculated with the advanced APOLLO2 neutron physics code and the addition of a rim porosity model also improves the thermal predictions at high burn-ups, mainly for thermocoupled fuel rods which were first pre-irradiated in a commercial PWR.

Qualification

The thermal model has been benchmarked with over 1800 centreline temperature data up to a rod average burn-up of 80 GWd/tM. Figure 1 shows the Predicted (*P*) vs. Measured (*M*) values of fuel centreline temperatures. The average value of *P/M* is 1.02 and the 2-standard-deviation limit is ±12.8%. Figures 2 and 3 show the ratios *M/P* vs. burn-up and LHGR, respectively. There is a slight trend of calculated temperatures with burn-up. The trend with LHGR is excellent.

As high burn-ups are of most interest, we present in more detail the comparison of calculations with high burn-up data. Figure 4 shows such a comparison for the EXTRAFORT experiment [8] which reached 62 GWd/tM and Figure 5 shows the results for the Halden IDA 562.2-16 experiment which reached 80 GWd/tM. Even if slight trends are observed with either temperature or burn-up, the difference between predicted and measured temperatures is below 100°C. More discussion on the thermal model is found in the *Conclusion*.

Fission gas release

The Fission Gas Release (FGR) model includes an athermal and a thermal mechanism.

Athermal release

Athermal release arises from the contributions of the recoil/knockout mechanisms and of the rim effect. Neglecting the recoil contribution, the FGR fraction, F_{ATH}, is simply of the form [9]: $F_{ATH} = C_I(S/V)B$ where C_I is a model parameter and S/V is the specific surface of the fuel. Contributions from fuel open porosity and from the rim are included in the specific surface. Parameters of the athermal model were adjusted on experimental data where no thermal release was observed (absence of a radius of inter-granular bubble precipitation).

Thermal release: formulation

The thermal release model is based on previous closely related models (see for example [10,11]). Gas release is calculated by modelling one UO_2 grain idealised as a sphere of radius *a*. Gas release is a two-step diffusion process from the interior of the grain to the grain boundary. In the first step, gas atoms accumulate at the grain boundary without release. In the second step, gas release is activated when the gas density at the boundary is bigger than a saturation density, N_S.. The phenomenon of *re-solution* is also taken into account. Fission spikes bring back into solution a fraction of inter-granular atoms. This induces a re-solution flux which counteracts in part the diffusion flux. Re-solution is also acting inside the grains where atoms cluster into fine bubbles. These bubbles are brought back into solution, recombination and re-solution balancing out. The problem is then to solve a diffusion equation inside the grain with a time-varying condition at the grain boundary.

The diffusion equation is expressed as:

$$\frac{\partial c}{\partial t} = \beta + \div (D \cdot \Delta(c))$$

where $c(r,t)$ is the local concentration, $r(t)$ is the space (time) variable, β is the creation rate, and D is the diffusion coefficient. The boundary condition is given by:

$$c(a) = (bN\delta) / (2D)$$

where b is the probability of re-solution on the inter-granular atoms, N is the density per unit area at the grain boundaries, and δ is the width of the inter-granular re-solution layer. The following balance equation closes the problem when $N < N_s$:

$$\int cdVdt + \frac{1}{2} \frac{N}{a} = \int \beta dVdt$$

Several numerical algorithms have been devised to solve the above-mentioned problem [11-13]. We have recently developed an efficient and accurate algorithm based on finite volumes [14]. However, in the case of steady-state operation, we have found a simple analytical approximation described below in the section entitled *The incubation threshold*.

Thermal release: analytical approximation

The incubation threshold

Following [15], D is chosen as the sum of three components D_1, D_2, and D_3. The D_1 term represents the intrinsic diffusion in the absence of irradiation. The D_2 and D_3 terms represent the thermal and athermal contributions induced by irradiation, respectively. A correction is added due to the irradiation-induced re-solution of intra-granular bubbles is [16]:

$$D = \frac{1}{\dfrac{1}{D_{1+2+3}} + \dfrac{1}{L^2 b'}}$$

where $D_{1+2+3} = D_1 + D_2 + D_3$, b' is the probability of re-solution of intra-granular atoms by fission spikes, and L is the mean free-path between two bubbles.

$$D_1 = D_{01} \exp\left(\frac{-T_{01}}{T_K}\right)$$

$$D_2 = D_{02} \exp\left(\frac{-T_{02}}{T_K}\right)\sqrt{\frac{P'}{20}}$$

$$D_3 = D_{03} \frac{P'}{20}$$

where T_K is the local temperature (K) and P' is the Linear Heat Generation Rate (LHGR) expressed in kW/m units. The following values were chosen: $D_{01} = 3.9 \times 10^{6}$ m²/s, $T_{01} = 45275$ K, $D_{02} = 1.77 \times 10^{-15}$ m²/s, $T_{02} = 13800$ K, $D_{03} = 2 \times 10^{-21}$ m²/s, $L^2 b' = 10^{-15}$ m²/s.

The expression of the incubation burn-up $B_i \propto \beta t_i$ can be deduced following [14]. At low temperatures:

$$B_i \propto \frac{bN_S\delta}{2(D_2 + D_3)}$$

and at high temperatures:

$$B_i \propto \left(\frac{9}{8\pi}\right)^{1/3} (N_S)^{2/3} \beta^{1/3} \left(\frac{1}{D_1} + L^2 b'\right)^{1/3}$$

Two important remarks enable one to combine these two last expressions into one single expression. First, $T_{02} \sim T_{01}/3$. Second, there are great uncertainties for the values of the parameters b, δ et N_s and, *a fortiori*, on the combination of these parameters. We therefore seek a solution for the incubation threshold of the form:

$$B_i = \frac{B_1}{exp\left(-\frac{T_1}{T_K}\right) + \frac{T_K - T_2}{T_3}} + B_2$$

The parameters B_1, B_2, T_1, T_2, and T_3 are adjustable parameters of the model, in agreement with experimental fission gas release results (see section entitled *Qualification*). The expression has similarities and differences with the Halden threshold [3] which are discussed in the *Conclusion*.

The gas release fraction

By comparison with numerical results [14] and in the case of steady-state operations, we have found an excellent analytical approximation for the release fraction, F, namely:

$$F = (1-t/t_i) \, F_B(\tau)$$

where F_B is the Booth [17] solution evaluated at the reduced time $\tau = D(t-t_i)/a^2$:

$$F_B(\tau) = 1 - \frac{6}{\tau} \left[\sum_{n=1}^{\infty} \frac{1 - exp\,(n^2\,\pi^2\,\tau)}{n^4\,\pi^4} \right]$$

Qualification

The parameters of the incubation threshold have been fitted on the FRAMATOME UO$_2$ steady-state database which includes close to 200 fuel rods irradiated in commercial or experimental reactors up to a rod average burn-up of 67 GWd/tM and an LHGR up to 50 kW/m. The result is shown on Figure 6, together with numerical results obtained with the method [14] where now the fitting parameters are N_s and b. Choosing $\delta = 8 \times 10^{-8}$ m, a best fit is found for $N_s = 8 \times 10^{-19}$ m^{-2} and $b(s^{-1}) = 3 \times 10^{-6}$ (P/20). There is an excellent agreement between the analytical and numerical threshold curves.

Figure 7 shows the comparison between measured (*M*) and predicted (*P*) gas release fractions, together with 95% uncertainty levels. The average value of *P/M* is 1.04. The uncertainty spread is equal to times/divide 2.36.

Discussion

Thermal model

Based on our experience, the value of the coefficient A_l which indicates the rate of the fuel thermal conductivity degradation with burn-up is higher than SIMFUEL's [6] (A_l ~ 15 KmtM/WGWd for 95% dense fuel but lower than Halden's recommendation [18] (A_l ~ 26). We tried additional corrections to the fuel conductivity like the contribution at low temperature of irradiation defects [19] or of volatile fission products [20]. These corrections did not improve significantly the predictions. Also, we notice that the correction [20] should apply when the correction [16] to the diffusion coefficient does not apply. We found that the range of temperatures for these corrections are not quite consistent.

Fission gas release model

The FGR threshold is quite similar to the empirical Halden threshold, B_{iH}:

$$B_{iH} = 5.67 \exp\left(\frac{9800}{T}\right)$$

In fact, the Halden threshold is given for $F_{av} > 1\%$ where F_{av} is the pellet average value of the release fraction. Figure 8 compares the Halden threshold with F_{av} values = 1 and 3% for COPERNIC and FRAMATOME fuels in PWRs. It is seen that the COPERNIC threshold matches the Halden threshold at low burn-ups but is lower at higher burn-ups. This was also observed before for PWR fuels [21,22]. As the thermal model has been fitted on many experimental data including ultra high burn-ups, we look at other factors to explain this difference.

First, the Halden threshold was derived up to a burn-up below 40 GWd/tM. Recent data [23] suggest that the temperature threshold is lower at higher burn-up (see Figure 8). This, most likely, translates effects of rim and fuel conductivity degradation with burn-up. However, there is still a difference of ~200°C between the COPERNIC and Halden threshold temperatures.

Second, a threshold is difficult to accurately predict. This is reflected on Figure 7 where the scatter of the data is highest in the region around 1%. Figures 6 and 8 also show that the COPERNIC threshold temperature varies significantly going from the 0% level to the 1% level. However, the 3% level is still below the Halden threshold.

Last, a combination of other small effects could play a role. We may mention the power profile through the pellet which could be significantly different for PWR reactors and the Halden reactor. This could lead to a difference between pellet central temperatures. We also notice a great uncertainty in the diffusion coefficient, particularly at low temperatures and high burn-ups where the D_3 term dominates. White [24] suggests this term should not be taken into account for stable gas release. As it corresponds to an athermal release, it could be included into the athermal release formulation. This could result in a slightly different threshold.

Conclusion

We have analysed the thermal behaviour of PWR fuel rods with COPERNIC, the FRAMATOME fuel rod performance code. The thermal model was extensively benchmarked on experimental data up to ultra high burn-ups, as well as the fission gas release model because fission gas release is strongly correlated to temperature. The predictive performance of both models has been shown to be good.

The rate of the degradation of fuel thermal conductivity with burn-up was found to be intermediate between the values derived from SIMFUEL [6] and Halden's recommendation value [18]. Also, the threshold temperature for thermally activated fission gas release was found smaller than the Halden threshold temperature at high burn-ups. These differences may come from differences between PWR reactors and the Halden reactor such as the power radial profile through the pellet.

REFERENCES

[1]   E. Bonnaud *et al.*, "COPERNIC: A State-of-the-Art Fuel Rod Performance Code", ANS Winter Meeting, Albuquerque, NM, USA (1997).

[2]   K. Lassmann, *J. Nuclear Mater.*, 188 (1992) 295.

[3]   C. Vitanza, E. Kolstad and U. Graziani, "Fission Gas Release from UO_2 Pellet Fuel at High Burn-up", ANS Topical Meeting on LWR Fuel Performance, Portland, USA (1979) 361.

[4]   D.A. Wesley and M. Yovanovich, *Nuclear Technology*, 72 (1986) 70.

[5]   J.E. Garnier and S. Begej, "Ex-Reactor Determination of Thermal Gap Conductance Between Uranium Dioxide-Zircaloy 4 Interfaces. Stage I: Low Gas Pressure", NUREG/CR-0330 (1979).

[6]   P.G. Lucuta *et al.*, *J. Nuclear Mater.*, 188 (1992) 198.

[7]   M.J. Notley and J.R. McEwan, *Nucl. Appl. Technol.*, 2 (1966) 117.

[8]   S. Bourreau *et al.*, "Temperature Measurements in High Burn-up UO_2 Fuel: EXTRAFORT Experiment", this conference.

[9]   D.R. Olander, "Fundamental Aspects of Nuclear Reactor Fuel Elements", ERDA, USA (1974).

[10] R.J. White, M.O. Tucker, *J. Nuclear Mater.*, 118 (1983) 1.

[11] K. Forsberg and A.R. Massih, *J. Nucl. Mater.*, 135 (1985) 140.

[12] D.M. Dowling, R.J. White and M.O. Tucker, *J. Nucl. Mater.*, 110 (1982) 37.

[13] K. Ito, R. Iwasaki and Y. Iwano, *J. Nucl. Sci. Tech.*, 22 (1985) 129.

[14] L.C. Bernard and E. Bonnaud, *J. Nucl. Mater.*, 244 (1997) 75.

[15] J.A. Turnbull, R.J. White and C. Wise, "The Diffusion Coefficient for Fission Gas Atoms in Uranium Dioxide", IAEA Tech. Com. Meeting on Water Reactor Fuel Element Computer Modelling in Steady State, Transient and Accident Conditions, Preston (1988).

[16] M.V. Speight, *Nucl. Sci. Eng.*, 31 (1969) 180.

[17] A.H. Booth, "A Method of Calculating Fission Gas Diffusion from UO_2 Fuel and its Application to the X-2-f Loop Test", Report Atomic Energy of Canada Limited CRDC-721 (1957).

[18] W. Wiesenack, "Assessment of UO_2 Conductivity Degradation Based on In-Pile Temperature Data", International Topical Meeting on LWR Fuel Performance, Portland (1997).

[19] P.G. Lucuta, Hj. Matzke, and I.J. Hastings, *J. Nuclear Mater.*, 232 (1996) 166.

[20] J.A. Turnbull, private communication.

[21] R. Manzel and R. Eberle, "Fission Gas Release at High Burn-up and the Influence of the Pellet, Rim", ANS/ENS International Topical Meeting on LWR Fuel Performance, Avignon (1991).

[22] C. Forat *et al.*, "Fission Gas Release Enhancement at Extended Burn-up: Experimental Evidence from French PWR Irradiation", IAEA Technical Committee Meeting, Pembroke (1992).

[23] OECD Halden Reactor Project, Report HP-1005, (1996).

[24] R.J. White, "A New Mechanistic Model for Calculation of Fission Gas Release", International Topical Meeting on LWR Fuel Performance, West Palm Beach (1994).

Figure 1. Qualification of the COPERNIC thermal model

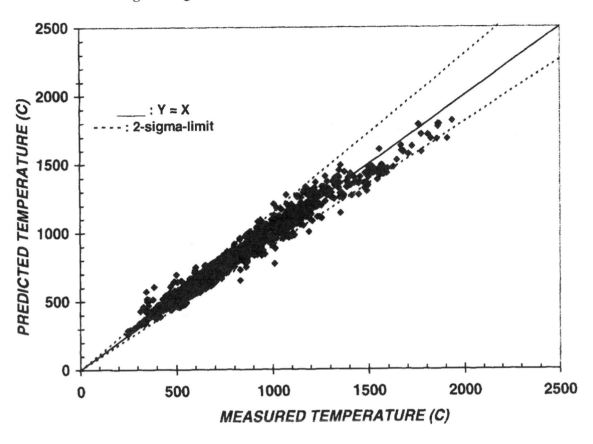

Figure 2. Measured over predicted fuel temperature vs. burn-up

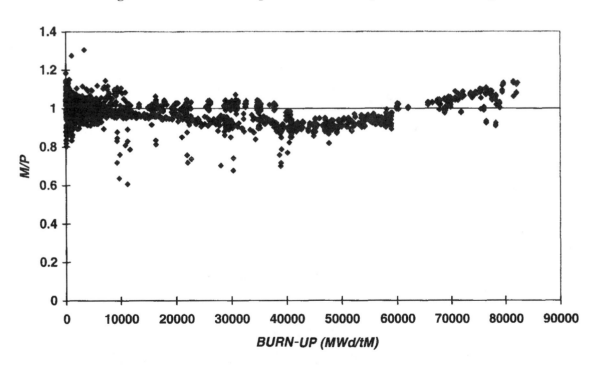

Figure 3. Measured over predicted fuel temperature vs. LHGR

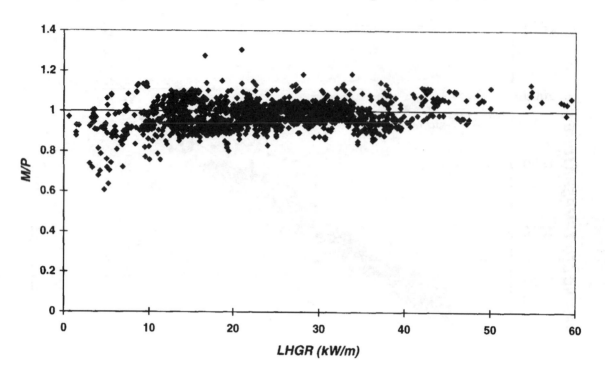

Figure 4. EXTRAFORT experiment at 62 GWd/tM

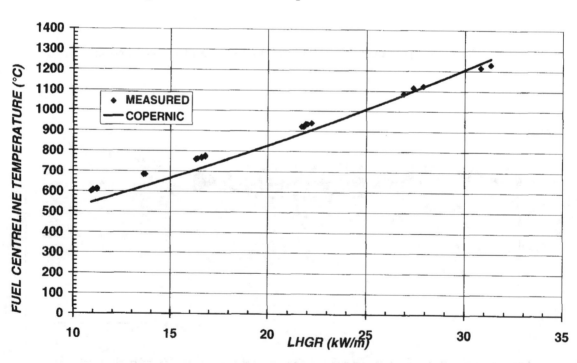

Figure 5. IFA 562.2-16 experiment

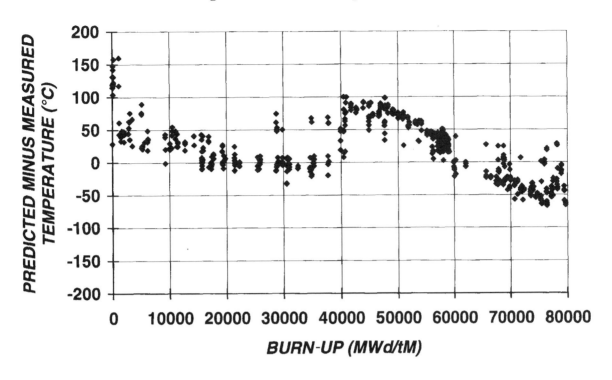

Figure 6. COPERNIC fission gas release model: incubation threshold

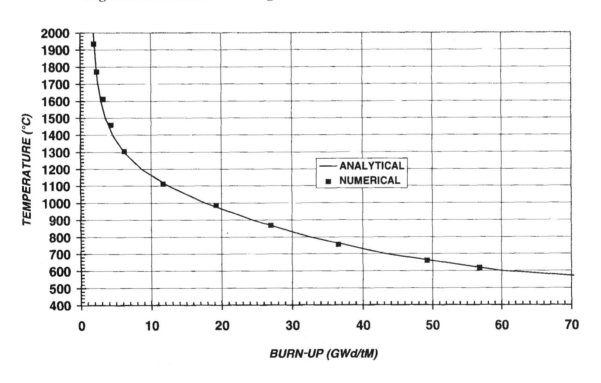

Figure 7. Qualification of the COPERNIC fission gas release model

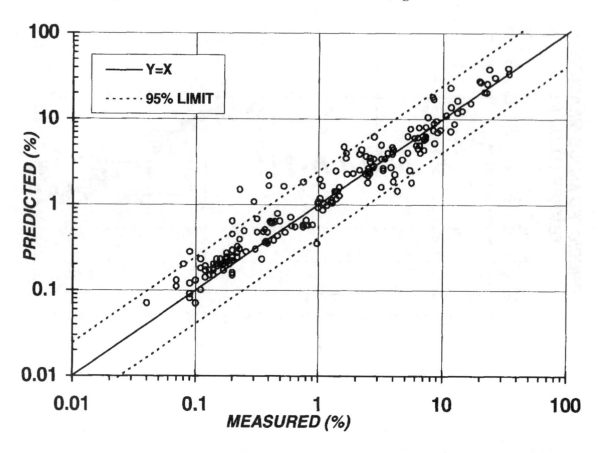

Figure 8. Comparison between COPERNIC and Halden FGR thresholds

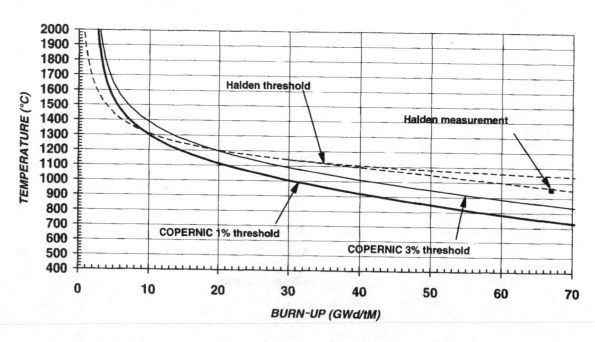

CYRANO3: EDF'S FUEL ROD BEHAVIOUR CODE: PRESENTATION AND OVERVIEW OF ITS QUALIFICATION ON HRP AND VARIOUS OTHER EXPERIMENTS

Nicolas Cayet (EDF/SEPTEN France)
Daniel Baron (EDF Études et Recherches, France)
Stéphane Beguin (EDF/SEPTEN, France)

Abstract

CYRANO3 is the new code developed and used by the EDF as a means of simulating the thermomechanical behaviour of the fuel rod. The code aims to verify the integrity of the fuel cladding in a PWR and enables the evaluation of parameters subjected to safety criteria such as the internal pressure, the corrosion layer, the central temperature or cladding stresses.

Its development has benefited from the latest R&D results in areas such as fuel conductivity degradation with burn-up, fragmentation-relocation or corrosion. Because of the new operating conditions, emphasis is put on the modelling of PCMI, the rim effect and high exposure behaviour. The fuel conductivity and its degradation with burn-up are modelled according to a unified modelling, able to simulate UO_2, MOX and gadolinia fuels.

Because it is a driving parameter of the overall behaviour of the fuel rod, the thermal aspects have been extensively validated and qualified over various French and international experiments. By choosing analytical experiments, it has been possible to validate the different models acting directly in the thermal chain.

Regarding the thermal aspects, CYRANO3 is validated for UO_2 fuels irradiated up to a burn-up of 100 MWd/kgU.

Introduction

CYRANO3 is the new code used by the EDF to simulate the LWR fuel rod thermal/mechanical behaviour under irradiation. EDF needs such a code to check the respect of the different design and safety criteria before licensing any kind of fuel. This tool integrates the entirety of the knowledge gathered in the previous code CYRANO2 but was also designed to cope with the increasingly descriptive modellings expected over the next decade. Convivial, this code is suitable for design analysis and for the interpretation of experiments as well. It accounts for any kind of fuel actually loaded in the French reactors: UO_2, UO_2 with gadolinium and MOX fuels.

CYRANO3 is built around an accurate finite element kernel. The first part of this paper describes the code architecture.

Major improvements concerning fuel rod modelling have been introduced in CYRANO3. These developments respond to an industrial need: to improve the safety and the competitivity of the French reactors. The models presented in the second part reflect the answers given to the industrial questions. The fuel thermal conductivity degradation model has been improved to cope with high exposure irradiation campaigns. An original relocation model had been developed to simulate a better PCMI behaviour for an increased manoeuvrability. The validation of the presented models is illustrated in the third part over selected HRP experiments and some others.

CYRANO3, however, is not the final response of the EDF in the domain of fuel modelling. Several research programmes have already been engaged on topics such as pellet viscoplasticity, a descriptive mechanical behaviour of the cladding or fission gas release. Their results will be included in subsequent versions of CYRANO3. The fourth part of this paper details the main programmes engaged.

Description of the code architecture

The CYRANO3 code was developed to simulate the overall fuel rod thermal/mechanical behaviour for design and safety analysis. Several rod power histories can be run in the same time for a common initial geometry. The main choices have been determined to cope with an acceptable running time, a compatibility with different computers, a convivial treatment of maximum informations stored during calculations and a good suitability with the different types of LWR fuel rods expected to appear over the next ten years. For example it is possible to simulate a fuel rod with different fuel pellet geometries and contents.

Geometrical configuration

Standard geometrical structure

Due to the disproportion between radial and axial gradients, a 1-D thermal/mechanical finite element kernel has been retained to perform the fuel rod simulations. However, some 2-D and 3-D effects, such as axial binding between the cladding and the fuel stack, cladding ridging and pellet fragments relocation are taken into account, leading the present CYRANO3 code to be characterised as a "1.5-D" code.

Figure 1 shows how the fuel rod is depicted by finite elements in the CYRANO3 code. The fuel rod is split into axial zones. Each one is divided into radial ring elements. The radial and axial discretisations are free (only limited by the memory of the computer) and the mesh can be equivolume, equidistant or arbitrary. The thermal and mechanical equations are solved on the radial mesh, where the pellet-to-clad gap is considered as a coupling element. The free volume of the plenum is treated as a specific axial slice. These principles are valid for both solid and annular fuel pellets.

Ridging geometrical structure

In addition to the standard axial slices describing the mean pellet radius, a specific option is available to deal with the ridging effect. The thermal and mechanical balance equations are solved at pellet-pellet interface, assuming the same temperature gradient but accounting for an extra deformation of the pellet ends, due to the thermal hourglassing effect. The calculated axial profilometry then reflects the ridging of the cladding.

Time discretisation

CYRANO3 has been developed to handle normal and off-normal conditions (up to Class 2 transients). The power history is given by the user as an input by means of the initial and final power levels and of the corresponding time step. An axial flux profile function is applied to these power levels to define the local axial power level at each axial position during a time interval. The time step is subdivided in an *a priori* manner so as to ensure the convergence of the calculations in any case. However if there are still troubles because of the strong non-linearities of the problem, the code will again subdivide the time step during the calculation itself.

Solving scheme

The fuel rod is assumed to be axially symmetric in order to solve the thermal and mechanical equations. Two nodes are considered in each element of the mesh, each node having two degrees of freedom. The integration is made both along the length of the elements and their thickness. The non-linear heat equation is solved in the finite element module accounting for the non-linearities of the materials properties, the heat transfer through the gap and the boundary conditions.

The same module also solves the classical 1-D axisymmetric mechanical equations with their boundary conditions (pressure, temperature, stresses and displacements). An option can be activated to estimate the axial and shear stresses by means of a partial dualisation technique.

The axial mechanical balance between the fuel stack and the cladding tube is evaluated at each time step. The strains obtained are then introduced in the radial mechanical balance as specific boundary conditions.

The overall flow diagram of the code is given on Figure 2. After completion of a detailed check of the code input data and the specification of the power history, the CYRANO3 code repeats the finite elements assembling and the global equations solving procedure, and yields step-by-step results for the entire irradiation history.

All the calculation results (more than 100 tables) are stored in a binary form file at the end of the run, usable by the different post-processors, as far as no writing is accepted during the run. That means that all the results are progressively stored during calculations in large tables controlled by an optimised dynamic memory management.

Other possibilities

Refabrication

CYRANO3 offers the possibility to simulate the refabrication of a sibling fuel rod segment cut from a mother rod after its base irradiation. The fuel rod characteristics are saved after the base irradiation in a specific file at the time step chosen by the user. This file will be the input of the restart run to initiate the geometry of the sibling rod segment. The user then specifies the spatial discretisation of the sibling rod, its power history, the new filling gas composition, the new gas pressure and the new plenum size.

Coupling with other codes

CYRANO3 can be coupled with other codes for specific needs.

The coupling of CYRANO3 with core physics and accident analysis codes has been done and is operated as a mock-up [1].

For specific ramp test analysis, CYRANO3 can be coupled with the EDF's structure mechanics code, *Code_Aster*[TM] [2]. CYRANO3 generates input data for more precise thermal/mechanical calculations performed by *Code_Aster*[TM]. A more detailed and multi-dimensional modelling of part of the fuel rod (two pellets) has been developed. This modelling contains the short term effect fuel rod specific models included in CYRANO3 such as the clad creep model. Moreover the meshes of the pellet and the cladding tube are refined and the simulation does not have to be axially symmetric. 2-D(r,q) and (r,z) calculations have already been performed. This tool can then be used to assess more accurately the mechanical state of the fuel rod but also to develop and validate new simplified mechanical models to be included in CYRANO3 later on.

CYRANO3, an element of response to EDF's main future industrial targets

The new fuel rod behaviour code CYRANO3 has to respond to the EDF's future operating conditions. To make nuclear energy safer and more competitive, the EDF has orientated its policy in different areas. The extension of the irradiation periods and the increase of the reactors' manoeuvrability (including MOX fuel reloadings) are among the most central. The simulation tools and the models have to be improved to reach these goals.

For the thermal aspect, the validity domain of the fuel thermal conductivity model has to be extended. Moreover this model must be able to simulate MOX and gadolinia fuels which will be loaded more and more into the French reactors.

More accurate calculations of the clad stress state are needed to allow an increase of the reactors' manoeuvrability. Basically one has to prove that the increased manoeuvrability will not lead to a PCMI cladding failure. Several original mechanical models, such as a fragmentation-relocation model and a phenomenological clad creep model, have thus been developed and introduced (partly for the clad creep model) in CYRANO3 to answer this question.

Linked with the extension of the irradiation periods but also with the coming of new cladding alloys, the corrosion model has been improved. The internal pressure is a safety parameter which will also be effected by the increased discharged burn-ups. The fission gas release model has thus been reviewed as well.

The models presented hereafter are the major advances introduced in CYRANO3 to meet the industrial needs of the EDF.

Development of a unified fuel conductivity model

The fuel conductivity and its degradation with burn-up can be modelled in various ways. CYRANO3 contains several fuel thermal conductivity models such as the Stora model and the HRP correlation [3].

The Stora fuel conductivity model is based on the following equation:

$$K_c = K_p \cdot \left(5.8915 \cdot 10^{-2} - 4.3554 \cdot 10^{-5} \cdot T + 1.3324 \cdot 10^{-8} \cdot T^2 \right)$$

where T is the temperature in °C and K_p, the porosity correction given by:

$$K_p = \frac{1 - (2.58 - 0.58 \cdot 10^{-3} \cdot T) \cdot p}{1 - (2.58 - 0.58 \cdot 10^{-3} \cdot T) \cdot 0.036}$$

p being the initial porosity.

But CYRANO3 has also benefited from the introduction of a fuel conductivity model of a new type. This unified model has the advantage of treating UO_2, MOX and gadolinia fuels [4,5].

The unified Baron-Hervé fuel conductivity model presented in [4] has been improved for high temperature simulations (>2000 K). It is nevertheless still based on the following equation for stoichiometric UO_2 fuel, 95% TD:

$$K_0(T) = \frac{1}{A + B \cdot T} + \frac{C}{T^2} \cdot \exp\left(-\frac{W}{kT}\right)$$

where K_0 is the fuel thermal conductivity in $W.m^{-1}.K^{-1}$, and T the temperature in K.

To take into account the stoichiometric ratio, the gadolinia and plutonium contents, the previous equation is modified as follows:

$$K_0(T) = \frac{1}{A_0 + 4x + A_1 g + A_2 g^2 + \left(B_0 (1 + B_1 q) B_2 g + B_3 g^2 \right) T} + \frac{C + Dg}{T^2} \exp\left(-\frac{W}{kT}\right)$$

where K_0 is the fuel thermal conductivity in $W.m^{-1}.K^{-1}$, T the temperature in K, x the stochiometric ratio (2-O/M), q the plutonium content and g the gadolinia content. These correlations are fitted and then available for fuels with 5% porosity (95% of the theoretical density).

To account for different porosities, the Loeb and Ross correction is applied:

$$K_1(T, p) = K_0(T) \times \frac{1 - p \cdot \alpha(T)}{1 - 0.05 \cdot \alpha(T)}$$

where $K_0(T)$ is the fuel thermal conductivity at 95% TD in $W.m^{-1}.K^{-1}$, T the temperature in K and a is given by:

$$\alpha(T) = 2.7384 - 0.58 \cdot 10^{-3} \cdot T \quad \text{for } T < 1273 \, K$$
$$\alpha(T) = 2 \quad \text{for } T \geq 1273 \, K$$

This model has been fitted on the NFIR1 data [6,7] and validated on a wider database.

The degradation of the fuel conductivity with burn-up has been investigated over the last ten years. Nevertheless, until now, the Lokken and Courtright correlation [8] was used to take into account the fuel conductivity degradation with burn-up as far as its validity was confirmed for the standard operating conditions:

$$K_2(T, p, bu) = \frac{1}{\dfrac{1}{K_1(T, p)} + E \cdot \dfrac{bu}{T}}$$

where T is the local fuel temperature in °C, bu the local burn-up in MWd/tU, K_1 the thermal conductivity of the fresh fuel and E a constant. Another equivalent correction was proposed, validated on the NFIR2 data [9]. The burn-up dependence of the fuel conductivity is then formulated as:

$$K_2(T, p, bu) = \frac{K_1(T, p)}{1 + A_1 \cdot \dfrac{\sqrt{bu}}{T^{A_2}}}$$

where K_1 is the thermal conductivity of the fresh fuel, bu the local burn-up in GWg/tU and T the local temperature in K. A new correlation is in preparation for the next future [5] accounting for the burn-up degradation, but also for restoration effects with temperature, directly on the phonon term of the fuel conductivity correlation.

The calculation of the local temperature not only depends on an accurate thermal conductivity expression but also on the evaluation of a correct radial power profile, accounting for the ^{238}U resonance absorptions and the plutonium formation, mainly in the periphery (neutronic self shielding effect). CYRANO3 proposes two simulations. The first one is based on an interpolation in an abacus system calculated with the core physics code APOLLO. It is valid for standard operating conditions. The second one is a call to the RADAR routine, developed in the TUI for the TRANSURANUS code by Doctor Lassmann [21], and integrated in CYRANO3. The heat transfer through the gap accounts for the fuel fragment relocation, using the Weissman correlation [29].

Concerning the boundary temperature conditions on the cladding, a simplified thermohydraulic routine computes the axial heat transport by the coolant and the heat transfer between coolant and cladding surface. However a coupling is already foreseen with the EDF's thermohydraulic code, THYC [1] to obtain more accurate boundary conditions between the cladding and the coolant.

A specific PCMI treatment

The general treatment in CYRANO3 of the basic mechanical phenomena is rather classical compared to any other fuel rod behaviour code. However the mechanical developments have been focused on the PCMI treatment.

The need was a better assessment of the cladding stress state. A major step has been made with the introduction of a fragmentation-relocation model in CYRANO3. Moreover the code estimates the axial stress equilibrium within the fuel rod. A specific contact model between the pellet and the cladding has been introduced in CYRANO3 to improve these calculations (on the stresses aspect but also on the cladding elongation and radial deformation aspects). As mentioned above, the pellet wheatsheaf effect is considered with a special treatment at the inter-pellets location.

Fragmentation-relocation model

The fragmentation of the fuel pellet occurs after the first power increase because of the rather large radial thermal gradient in the pellet producing thermal stresses higher than the UO_2 rupture stress (about 100 MPa). Once the contact between the pellet fragments and the cladding is established, the fragments formed tend to relocate.

The model introduced in CYRANO3 to simulate this phenomenon [10] is based on the fragment displacement and the solving at each time step of the quasi-static equilibrium of the fragment to determine its position. An initial roughness is added to simulate the friction between two fragments. A progressive wear of the roughness simulates its decrease with burn-up.

On one hand, because of the pressure gradient between its inner and outer parts, the cladding creeps down and constrains the fragments. On the other hand, because of the interaction between the fragments, a resistance force avoids the direct relocation of the fragment. Figure 3 illustrates the mechanical equilibrium of the fragment, the different forces being schematised. The kinetics of the phenomenon are controlled by the displacement speed of the fragment. The initial pattern of cracks is inspired by Oguma's modelling [11].

Contact between the pellet and the cladding

The axial contact between the fuel stack and the cladding depends on the local thermal and mechanical conditions. CYRANO3 takes these contact conditions into account and calculates the axial equilibrium. The calculated axial stresses and displacements are then introduced as boundary conditions in the radial equilibrium calculation.

The wheatsheaf effect

The pellet tends to have a wheatsheaf shape because of its strong radial thermal gradient. The deformations are then more important at the pellet's extremities than at its mid-plane. Figure 4 illustrates the geometry of the pellet resulting from this phenomenon.

To take account of this phenomenon in CYRANO3, two mechanical calculations are performed, one at the pellet mid-plane, and the other at the pellet-pellet interface.

The magnitude of the wheatsheaf effect depends on the number of radial and axial cracks, which is a function of the maximum local power achieved since the beginning of life. This pellet deformation results in:

- differential axial dilatation between the centre and the periphery of the pellet;

- fragment elastic flexion induced by the thermal gradient.

These two contributions are separately calculated and added to give the value of the end pellet extra strain.

The calculation of the cladding ridges induced by this hourglass effect can be directly activated in CYRANO3. For specific calculations – power ramps tests for example – CYRANO3 can be coupled with *Code_Aster*™ (see section above entitled *Coupling with other codes*). CYRANO3 is then used to evaluate the fuel rod geometry previous to the ramp test.

Extension of the fission gas release modelling to high burn-up

No fundamental developments have yet been included in CYRANO3 concerning the fission release modelling. The existing models have been tested on different fuels and at different burn-ups. The extension of their validity domain has been proved to be sufficiently accurate.

Two fission gas release models are available in CYRANO3. They are essentially built upon empirical bases. They were initially developed for UO_2 fuels but they have also been validated for MOX fuels. It is currently supposed that they are also valid for gadolinia fuels. We briefly present these two models hereafter.

The Capart model

The empirical Capart model [12] is based on an equivalent diffusion D_{eff} coefficient which applies to the total quantity of gas present in the matrix. D_{eff} is given by:

$$D_{eff} = \frac{D \cdot b}{b + g}$$

where b is the probability of resolution, g is the probability of intragranular trapping and D is given by:

$$D = D_t + D_{irr} = D_0 \cdot \exp\left(-\frac{4.18A}{RT}\right)\gamma \dot{F}$$

$D_0 = 0.1$ cm²/s, $R = 8.314$ J/mol.K, $g = 2.10^{-30}$ cm^5, T is the temperature (K) and $\dot{F}$ is the fission rate (fissions/cm^3.s). A is an activation energy which depends on the burn-up and temperature levels.

This model has been extensively used in the EDF's previous thermo-mechanical code, CYRANO2.

The empirical Baron-Maffeis model

The empirical Baron-Maffeis fission gas release model [13] splits the fission gas release fraction into an athermal part and a thermal part.

Thermal fission gas release is supposed to be the result of a diffusion process within the grains and a recirculation within the grain boundaries. The diffusion process is modelled according to the Booth approach [14]. The recirculation at the grain boundaries is modelled by a function depending on burn-up and temperature. This function simulates the evolution of the fission gas release temperature threshold with burn-up (it has a very similar approach to the well-known Vitanza curve).

The athermal release is mainly due to fission fragment collidings and ejection process but also to direct fission recoil energy. Its modelling is based on the following expression:

$$F_{ath} = a_1 \cdot \exp(a_2 BU^2) + a_3 \cdot ln(0.9 + poro)$$

where BU is the local burn-up (MWj/tU), poro the initial open porosity volume (%TD) and a_1, a_2 and a_3 are constants.

Corrosion

The current model used in CYRANO3 is based on the EPRI modelling [15]. The COCHISE corrosion model [16] has also been coupled to CYRANO3. This modelling takes into account the lithium concentration in the coolant. A specific lithium concentration history is then added to the classical power history. A first study has shown that the corrosion layer is more accurately predicted with this model.

Validation and qualification

Fuel thermal conductivity degradation with burn-up

UO$_2$ fuel

Fuel thermal conductivity degradation with burn-up has been qualified essentially on data from the Halden Reactor Project, IFA-562 and IFA-597. The EXTRAFORT experiment, a test similar to IFA-597.2 and performed by the CEA in the OSIRIS reactor, has also been used. The main experiments used to qualify the conductivity degradation model are listed in the Table 1.

Table 1

	Maximum burn-up (MWd/kgUO$_2$)
IFA-562.2, .3, .4, .5, .6	88
IFA-597.2, .3	61.5
EXTRAFORT	55

IFA-562

Some of the rods of IFA-562 were equipped with a fuel centre thermocouple whereas the others were equipped with pressure transducers. We focused mainly on Rod 16, for which the instrumentation was not damaged throughout the whole irradiation and which reached the highest burn-up (88 MWd/kgUO$_2$). The fuel rods were irradiated at rather low power levels to avoid fission gas release. Rod 16 is then completely dedicated to the study of fuel conductivity degradation with burn-up [3].

Figures 5 and 6 show the evolution of the fuel centre temperature as a function of burn-up for Rod 16. Two different fuel conductivity models are used, respectively the Stora model and the unified Baron-Hervé model, associated to the Lokken and Courtright burn-up degradation formulation. There is a good agreement between the measured and the calculated fuel centre temperatures for both models, up to 100 MWd/kgU. The difference between the calculated and measured temperatures is less than 40°C for the best modellings.

IFA-597.2 and .3

Three BWR fuel rods irradiated in a commercial reactor have been re-instrumented and re-irradiated in the HBWR for two loadings of IFA-597. Rod 8 was equipped with a centre thermocouple and provided valuable data on the thermal conductivity degradation since it has been irradiated to a burn-up of 61.5 MWd/kgUO$_2$ [17].

Calculations have been performed on this rod. The refabrication option of CYRANO3 has been typically used to simulate this rod. Figure 7 compares the measured and calculated temperatures for the two loadings. The observed differences (approximately 100-150°C) are attributed to the high conductivity degradation in the rim region which has proven to be important. An extra conductivity degradation is added to the regular conductivity degradation model to amplify this phenomenon in the rim region. The calculated temperature is closer to the measured temperature.

EXTRAFORT

This experiment was specifically dedicated to check the fuel thermal conductivity degradation modelling [18] at a high burn-up level. The EXTRAFORT fuel rod is a typical high burn-up fuel rod irradiated in a French PWR. It was instrumented with a centreline thermocouple after having been irradiated during five cycles in a commercial reactor. The rod was subjected to several power ramps between 10 and 35 kW/m.

The base irradiation of the initial fuel rod, the refabrication of the EXTRAFORT rod and its irradiation in the OSIRIS reactor have been simulated with CYRANO3. Figure 8 shows the good match between the measured and calculated temperatures obtained for this rod.

Gadolinia fuel: IFA-515.10

There are few data available to assess the thermal conductivity degradation of gadolinia fuels. One UO_2 fuel rod and one gadolinia fuel rod have been irradiated in IFA-515.10 up to a burn-up close to 55 MWd/kgUO_2 [19]. The unified Baron-Hervé fuel conductivity model associated with the Lokken and Courtright formulation has been tested on these rods.

The measured and calculated fuel centre temperatures are compared in Figure 9 for the UO_2 rod and in Figure 10 for the gadolinia rod. The unified Baron-Hervé fuel conductivity degradation model is used for both rods. The predictions are good for both rods. This experiment shows the ability of CYRANO3 to simulate gadolinia fuel.

Fission gas release at high burn-up

IFA-519.9

Three fuel rods have been irradiated in IFA-519.9 up to a burn-up of 90 MWd/kgUO_2 [20]. Two out of the three pressure transducers worked satisfactorily throughout the whole irradiation. This IFA thus gives valuable data on fission gas release at high burn-up.

Figures 11 and 12 show the predictions in fission gas release respectively for Rod DH and Rod DK, together with the experimental data. For each rod, the calculations were performed with the Capart model and the Baron-Maffeis model. Taking account of the uncertainties in estimating the FGR from the pressure measurements, one can observe the satisfactory behaviour of CYRANO3, despite the overestimation at the end of life for the large gap Rod DK.

IFA-562

Rods in IFA-562 were irradiated at rather low power in order to avoid fission gas release. However, a significant pressure increase in Rod 13 was observed since it reached a burn-up of 60 MWd/kgUO_2. This rod was simulated by CYRANO3, using the Baron-Maffeis model. Figure 13 shows the calculated and measured rod pressures, which were normalised to zero power. A good agreement is found between the experimental data and the predictions.

A FGR ratio of 19.5% was predicted for Rod 13 at the end of its irradiation in 1997. Only the athermal part of the Baron-Maffeis model was activated during the simulation, due to the low fuel temperatures.

Simulations of IFA-519.9 and IFA-562 confirm the validity of the Baron-Maffeis model on a large range of burn-ups and fuel temperatures.

Cladding creep: IFA-585.1

IFA-585.1 was designed to study cladding creep [22]. The code was run on the two Zr-4 PWR fuel rods L and U. Rod L was stressed permanently under tension loading, whereas Rod U was equipped with an external pressure line and was thus irradiated under both compression and tension stresses.

Figure 13 shows a good prediction of the outer cladding diameter deformations for Rod L. The tension creep is thus well modelled. Whereas, as shown in Figure 14, the creep deformations under compression are overestimated. The creep model in CYRANO3 considers basically the same behaviour for a tube under compressive stress as for a tube under tensile stress, whilst experimental data in IFA-585 have shown a much lower creep rate in compression.

Fragmentation-relocation

The RECOR experiment

The fragmentation-relocation model was validated on the RECOR experiment [23]. This experiment was specifically dedicated to the study of the fragmentation-relocation phenomenon. A fuel rod was irradiated in the OSIRIS reactor according to the power history depicted in Figure 16. The external diameter was measured continuously. This experiment was used to calibrate the different parameters of the model. Figure 17 makes the comparison between the measured and calculated external diameters. This ensures the correct modelling of the fragmentation-relocation phenomenon. The model was also validated against diameter measurements performed in commercial reactors.

IFA-504

The rods in the Gas Flow Rig IFA-504 can be flushed with gas from a supply external to the reactor. The hydraulic diameter is deduced from the measurement of the resistance to the axial flow. Gas flow measurements were performed since the beginning of the experiment which has now reached a burn-up of around 60 MWd/kgUO$_2$ [24]. IFA-504 thus provides valuable data for the study of pellet cracking and relocation of the fragments.

Rods 1, 2 and 3 were simulated with CYRANO3 in order to qualify the fragmentation-relocation modelling and its calibration. The calculated pellet-clad gap at zero power for Rod 2 is shown in Figure 18, together with the measured hydraulic diameter. The initial drop in hydraulic diameter due to pellet cracking is well reproduced, as well as the gradual reduction due to solid fission product swelling at a rate of 0.7% per 10 MWd/kgU. A progressive reduction in the calculated gap decrease rate is also observed: CYRANO3 takes into account the stress of the cladding on the pellets, which tends to relocate the fragments. At high burn-up, the calculated gap is slightly smaller than the measured hydraulic diameter. The latter does not reflect exactly the pellet-clad gap when it becomes small, because the gas flow is diverted within the fragments.

Cladding lift-off: IFA-610

IFA-610 was designed to study the cladding lift-off phenomenon. A high burn-up PWR fuel rod was irradiated in the HBWR. The rod was connected to an external pressure line and it was pressurised with argon up to 460 bar in steps of 50 bar. Cladding lift-off was supposed to be detected with a centreline line thermocouple and a clad elongation detector.

This IFA was simulated with CYRANO3 before its loading to compare different experimental procedures: external pressurisation or natural pressurisation by release of fission gas. It was concluded to be a more representative test with the former possibility. The code was also used to assess the required pressure level for the cladding lift-off to occur. An overpressure greater than 200 bar with a coolant temperature of 300°C and a linear heat rate of 17 kW/m was found to be needed in order to induce a gap opening and thermal feedback.

IFA-610 was recalculated after the first cycle of irradiation. The hypothesis for the simulation of the thermal and cladding creep aspects are the same as those for the EXTRAFORT rod and for IFA-585.1 (same type of cladding under tension), whose simulations with CYRANO3 were very satisfactory. Figure 19 shows the rod pressure history which has been applied to the fuel rod. Two calculations were performed: one for a fuel rod with hollow pellets subjected to the linear heat rate at the TF position, one for a fuel rod with solid pellets subjected to the rod average linear heat rate. The calculated and measured temperatures during the overpressure test (after the first 100 hours of operation) are depicted in Figure 20. The good agreement between the calculations and measurements corroborates the satisfactory behaviour of CYRANO3 in terms of fuel conductivity degradation. A slight increase in fuel temperature is observed during the last 300 hours of the test, whilst the rod power is slightly decreasing; this is attributed to the thermal feedback of the gap opening on the fuel temperature.

The predicted pellet-clad gap is closed or very small during the first two stages at 50 bar and 100 bar of overpressure. It appears to open clearly after the step of 150 bar overpressure and a temporary power reduction. From the period at +200 bar, the outward creep of the cladding is found to dominate the gap opening. It is noticed that the evolution of the gap at the TF position is consistent with the evolution of the normalised fuel temperature given in [25].

Power ramps

No real ramp tests have been performed in the Halden Boiling Reactor. So CYRANO3 has been tested on other ramps experiments carried out in French and international programmes such as OVERRAMP, SUPERRAMP and TRIBULATION.

In accordance with the fuel licensing, one has to assess the behaviour of a fuel rod in transient conditions. The EDF PCMI methodology is based on a cladding hoop stress criterion which defines the lower admissible stress boundary before rod failure.

The base irradiation of the mother rod is simulated and the refabrication option is used for the sibling rod. The ramp test is precisely simulated and the cladding stresses are evaluated. If the cladding experiences a hoop stress greater than the criterion, it will supposedly fail. A code is judged to properly simulate a ramp test if, over a large number of experiments, it can distinctively discriminate between failed and unfailed fuel rods.

The EDF's ramp tests database contains numerous experiments. CYRANO3 has shown over it its ability to simulate the PCMI phenomenon (see Figure 21).

Corrosion

CYRANO3 has not been specifically tested on HRP corrosion data. The qualification of the EPRI model has been carried out on in-reactor measurements for fuel rods irradiated up to 60 MWd/kgU.

The calculated oxide layer is compared to the measurements in Figure 22 for low-tin Zy-4 claddings. The EPRI model coefficients have been calibrated to get an envelope estimation of the oxide layer with a 95% confidence.

Future developments

Fission gas release

The CEA is developing a phenomenological modelling of the fission gas release within the fuel rod [26]. This model aims to treat each step of the fission gas release process, from the intergranular gas migration to the release in the rod free volume. It will be soon included in the CYRANO3 code.

Viscoplastic modelling of the pellet

The EDF has started a vast research programme to study the viscoplastic behaviour of the fuel pellet. Most of the mechanical modellings used in fuel rod behaviour codes extend creep law to any kind of loading. Stress relaxation and strain hardening are thus not correctly treated.

In the current version of CYRANO3, the pellet is supposed to be elastic. But recent ramp calculations have been perfomed by means of the coupling between CYRANO3 and *Code_Aster*™ to assess the stress concentration on the cladding due to the pellet radial cracks, the pellet having a viscoplastic behaviour in the rim region. They have shown that an eventual viscoplastic behaviour of the rim region changes drastically the cladding stresses in case of PCMI as the stress concentration in front of the cracks decreases significantly.

Thus specific tests such as microindentation [27] are planned to obtain a phenomenological viscoplaticity modelling of the irradiated fuel material. This modelling, after having been simplified and adapted to the code, will be introduced in CYRANO3.

Viscoplastic modelling of the cladding

The cladding deformations are commonly considered as the sum of elastic deformations, of irradiation growth deformations, of irradiation creep and of thermal creep. The last two parts cover the deformations due to:

- High stresses or high strain rates for the thermal creep. The instantaneous neutron flux is then supposed to have no effect whereas the irradiation effect is considered by the fluence parameter.

- Low mechanical loadings for the irradiation creep. The instantaneous neutron flux has to be considered in that case.

A cladding elasto-visco-plastic modelling has been developed [28] to replace the previous thermal creep. Its development has been focused on the PCMI behaviour, most of the tests having been done at high stress levels. The modelling will be introduced in CYRANO3. The coupling between CYRANO3 and *Code_Aster*™ (see section above entitled *Coupling with other codes*) has already been used to test this new cladding modelling. The test has shown real benefits for ramp test calculations.

Further experiments will be carried out in CEA laboratories to prove the assumption that the instantaneous flux has no impact. This will also validate the so-called irradiation creep modelling to be introduced in CYRANO3.

Conclusion

Major fuel modellings improvements have recently been made. In order to match its industrial needs, the EDF has developed an original fragmentation-relocation model and a unified fuel thermal conductivity model, able to simulate not only UO_2 fuel type but also MOX and gadolinia fuels.

These models and others have been introduced in CYRANO3, the EDF new fuel rod overall behaviour code. If CYRANO3 includes many models already used in the previous code (CYRANO2), it also proposes many new functionalities to improve the PCMI simulation. It has been developed in accordance with quality assurance rules. CYRANO3 is integrated in a global environment, including post-processor and optional couplings with other codes. The code architecture makes it very modular and convivial. The user, depending on his access rights, can eventually choose between different modellings. Some specificities have been added. For example, the simulation of a re-instrumented fuel rod is possible.

The validation of CYRANO3 different modellings has been completed on numerous French and international experiments, among them some selected Instrumented Fuel Assemblies experiments of the Halden Reactor Project. CYRANO3 is able to simulate highly irradiated fuels, up to 100 MWd/kgU, with respect to centre temperature and fission gas release. The corrosion modelling has been qualified over Zy-4 low tin cladding rods, irradiated five cycles in commercial French reactors. The introduction of a fragmentation-relocation model enables a better calculation of the fuel rod diametral deformation. The modelling has been validated over a specific experiment and it has been qualified over a large database of fuel rods irradiated in commercial reactors up to five cycles. The qualification of the fragmentation-relocation model is linked with the cladding creep modelling qualification. The ramp tests calculations confirm the accuracy of CYRANO3 PCMI modelling.

By developing CYRANO3, the EDF aimed to improve the simulation of the fuel rod behaviour at high burn-up for normal and transient operating conditions. This tool has already reached most of the initial goals. Several more predictive modellings have been developed and will be introduced in the next code release to achieve the development of this ten-years programme, allowing finally to give more valuable responses to the EDF's industrial questions. Emphasis is put on cladding and pellet visco-plastic behaviour, and on fission gas behaviour.

REFERENCES

[1]   S. Marguet, "A New Coupling of the 3-D Thermal-Hydraulic Code THYC and the Thermomechanical Code CYRANO3 for PWR Calculations", NURETH 8, October 1997, Kyoto, Japan.

[2]   EDF Internal Note, HI-75/96/073 (in French).

[3]   W. Wiesenack, R. Thankappan and M. Vankeerberghen, "Assessment of UO_2 Conductivity Degradation Based on In-Pile Temperature Data", HWR-469, May 1996.

[4]   S. Hervé and D. Baron, "Baron-Hervé Thermal Conductivity Modelling Available for UO_2, MOX and Gadolinia PWR Fuels and Extending to Irradiated Fuels", HT/B2/95/040/A.

[5]   D. Baron, "A Proposal to Simulate the Fuel Thermal Conductivity Degradation at High Burn-up and the Recovering Processes with Temperature", HBRP 4th Steering Committee, Karlsruhe, November 1997.

[6]   R.A. Bush, "NFIR-Properties of the Urania-Gadolinia System (Part 1)", NFIR-RP03 (Final Report) December 1986.

[7]   R.H. Watson, "NFIR-Properties of the Urania-Gadolinia System (Part 2)", NFIR-RP04 (Final Report) March 1987.

[8]   R.O. Lokken and E.L. Courtright, "A Review of the Effects of Burn-up on the Thermal Conductivity of UO_2", Batelle Pacific Northwest Laboratory, BNWL-2270, Document prepared for the NRC.

[9]   "Properties of UO_2 and $(U,Gd)O_2$ Irradiated to High Burn-up", NFIR-II-RP-21-08.

[10]   C. Callu, D. Baron and J.M. Ruck, "EDF Fragment Relocation Model Based on the Displacement of Rigid Bodies", IAEA Conference on CANDU Fuel, Toronto, September 1997.

[11]   M. Oguma, "Cracking and Relocation Behavior of Nuclear Fuel Pellets During Rise to Power", *Nuclear Engineering and Design*, 76 (1983) pp. 33-45.

[12]   G. Capart, "Fission Gas Release from Oxide Fuels", BN7311-02, November 1973.

[13]   D. Baron, "FRAMATOME Experience on Fission Gas Release During Base and Transient Operating Conditions", ANS Williamsburg, April 1988.

[14]   A.H. Booth, "A Method of Calculating Fission Gas Diffusion from UO_2 Fuel and its Application to the X-2-F Loop Test", Canadian Report CRDC-721, September 1957.

[15] F. Gazarolli *et al.*, "Review of PWR Fuel Rod Waterside Corrosion Behaviour", EPRI NP-1472, August 1980.

[16] P. Billot, A. Giordano, "Comparison of Zircaloy Corrosion Models from the Evaluation of In-Reactor and Out-of-Pile Loop Performance", ASTM STP 1132, C.M. Euchen and A.M. Garde, Philadelphia, PA, 1991, pp. 539-565.

[17] I. Matsson, J.A. Turnbull, "The Integral Fuel Rod Behaviour Test IFA-597.3: Analysis of the Measurements", HWR-543, January 1998.

[18] S. Bourreau, B. Kapusta, P. Couffin, G.M. Decroix (CEA), J.C. Couty (EDF), E. Van Schel (FRAMATOME), "Temperature Measurements in High Burn-up UO$_2$ Fuel: EXTRAFORT Experiment", Communication to the OECD/NEA Seminar on Thermal Behaviour of Fuel, Cadarache, France (March 1998).

[19] T. Tverberg, M. Amaya, W. Wiesenack, "Analysis of the Thermal Behaviour of UO$_2$ and (U,Gd)O$_2$ Fuel in IFA-515", HWR-547, February 1998.

[20] J.A. Turnbull, P. Van Uffelen, S. Beguin, "The Influence of Fuel-to-Clad Gap and UO$_2$ Grain Size on Fission Gas Release in High Burn-up PWR Design Fuel Rods, IFA-519.9", HWR-548, January 1998.

[21] K. Lassmann, "The Radial Distribution of Plutonium in High Burn-up UO$_2$ Fuels", *JNM*, n°208 (1994), p. 223.

[22] A.T. Donaldson, "In-Reactor Creep Behaviour of Zircaloy Under Variable Loading Conditions in IFA-585", HWR-413, October 1994.

[23] L. Caillot, G. Delette, "Impact of Fuel Pellet Fragmentation on Pellet-Cladding Interaction in a PWR Fuel Rod: Results of the RECOR Experimental Programme", SMIRT 97 (C02/4), Lyon, August 1997.

[24] M. Vankeerberghen, "Recent Gas Flow Measurements in IFA-504 Rods", HWR-433, November 1995.

[25] S. Beguin, "The Lift-Off Experiment IFA-610.1 – Initial Results", HWR-544, January 1998.

[26] G. Delette, M. Charles, "Comportement des gaz de fission stables dans un combustible de REP", Compte rendu DTP/SECC n°95-003 CEA Internal Note (in French).

[27] D. Baron, S. Leclercq, J. Spino, S. Taheri, "Development of a Microindentation Technique to Determine the Fuel Mechanical Behaviour at High Burn-up", IAEA TCM on Advances in Pellet Technology for Improved Performance at High Burn-up, Tokyo, November 1996.

[28] P. Delobelle, P. Robinet, P. Geyer, P. Bouffioux and I. Le Pichon, "A Unified Model to Describe the Anisotropic Viscoplastic Behaviour of Zircaloy-4 Cladding Tubes", ASTM STP, Garmisch-Partenkirchen, September 1995.

[29] M. Charles, M. Bruet "Gap Conductance in a Fuel Rod, Modelling of the Furet and Contact Results", IAEA Specialists Meeting, Bowness-on-Windermere, April 1984.

Figure 1. Rod discretisation

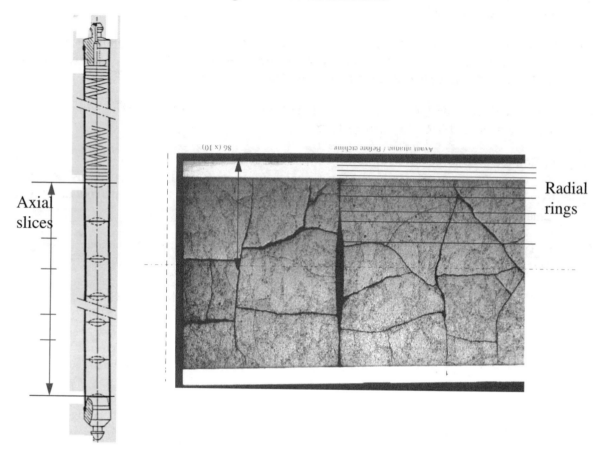

Axial slices

Radial rings

Figure 2. Overall flow diagram

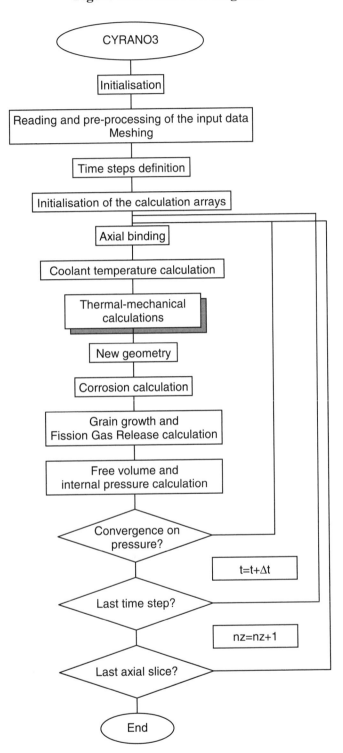

Figure 3. Principles of the fragmentation/relocation modelling

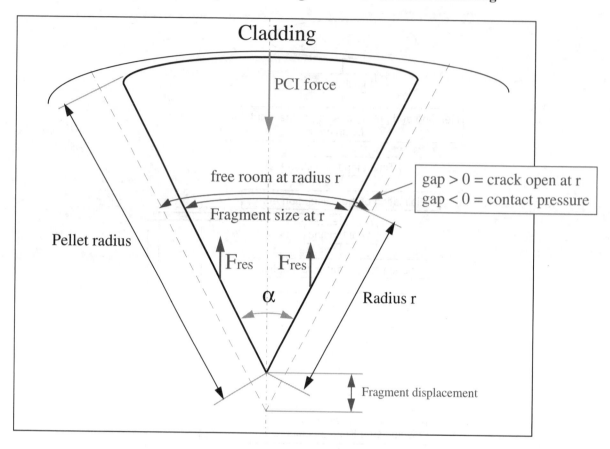

Figure 4. Wheatsheaf effect on a half pellet

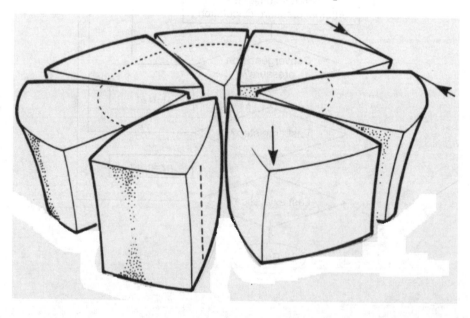

Figure 5. IFA-562, Rod 16 – comparison between measured temperature (C1) and calculated temperature with the Stora model (C2)

	FICHIER	VERSION	DATE	HEURE	LEGENDES	
C1:	exp56216.c3	-	00.00.00	00:00:00	-	-
C2:	stora.bin	2.5b	10. 2.98	9:42:55	IFA 562.2/.3/.	: TR 1 CR 1

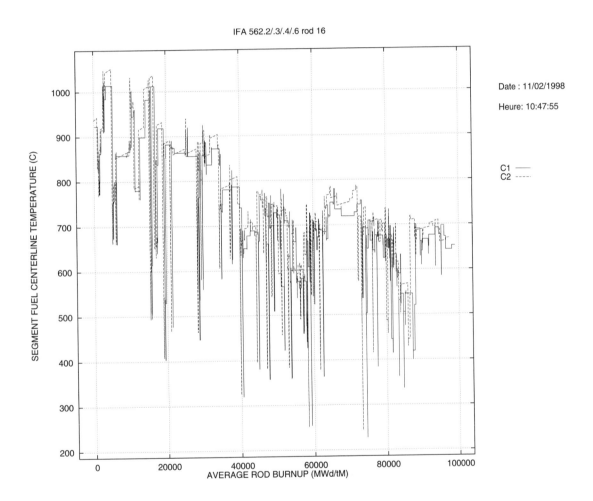

IFA 562.2/.3/.4/.6 rod 16

Date : 11/02/1998

Heure: 10:47:55

C1 ——
C2 -----

	FICHIER	VERSION	DATE	HEURE	LEGENDES
C1:	exp56216.c3	-	00.00.00	00:00:00	- -
C2:	bhu_3_lc.bin	2.5b	11. 2.98	9: 4:29	IFA 562.2/.3/. : TR 1 CR 1

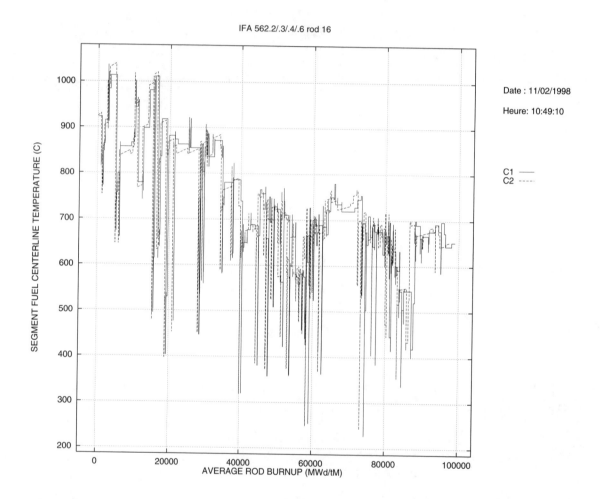

IFA 562.2/.3/.4/.6 rod 16

Date : 11/02/1998

Heure: 10:49:10

C1 ———
C2 ------

Figure 7. IFA-597.2 and .3, Rod 8 – comparison between measured temperature (C1) and calculated temperature with the Unified Baron-Hervé model without rim degradation (C2) and with rim degradation (C3)

	FICHIER	VERSION	DATE	HEURE	LEGENDES	
C1:	exp59723.c3	-	00.00.00	00:00:00	-	-
C2:	59723.bin	2.5b	10. 2.98	16:25:52	IFA 597.2	: TR 1 CR 1
C3:	59723_rim.bin	2.5b	11. 2.98	10:23:36	IFA 597.2	: TR 1 CR 1

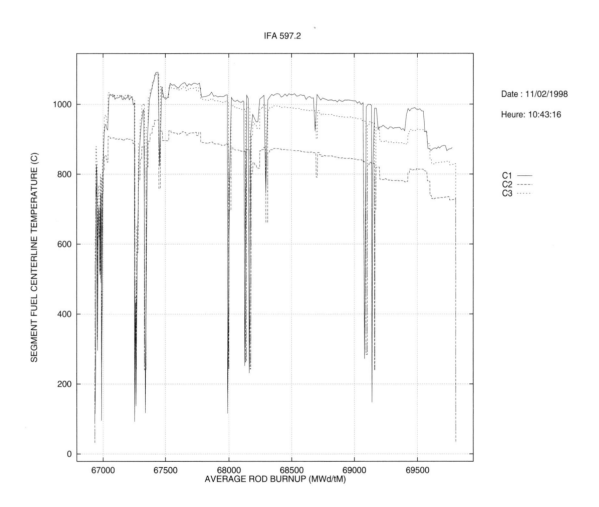

345

Figure 8. EXTRAFORT – Comparison of measured temperature (C1) and calculated temperature with the Stora model (C2)

	FICHIER	VERSION	DATE	HEURE	LEGENDES	
C1:	temp.txt	-	00.00.00	00:00:00	-	-
C2:	extraf3.bin	2.5b	12. 2.98	17:41:50	EXTRAFORT - C	: TR 1 CR 1

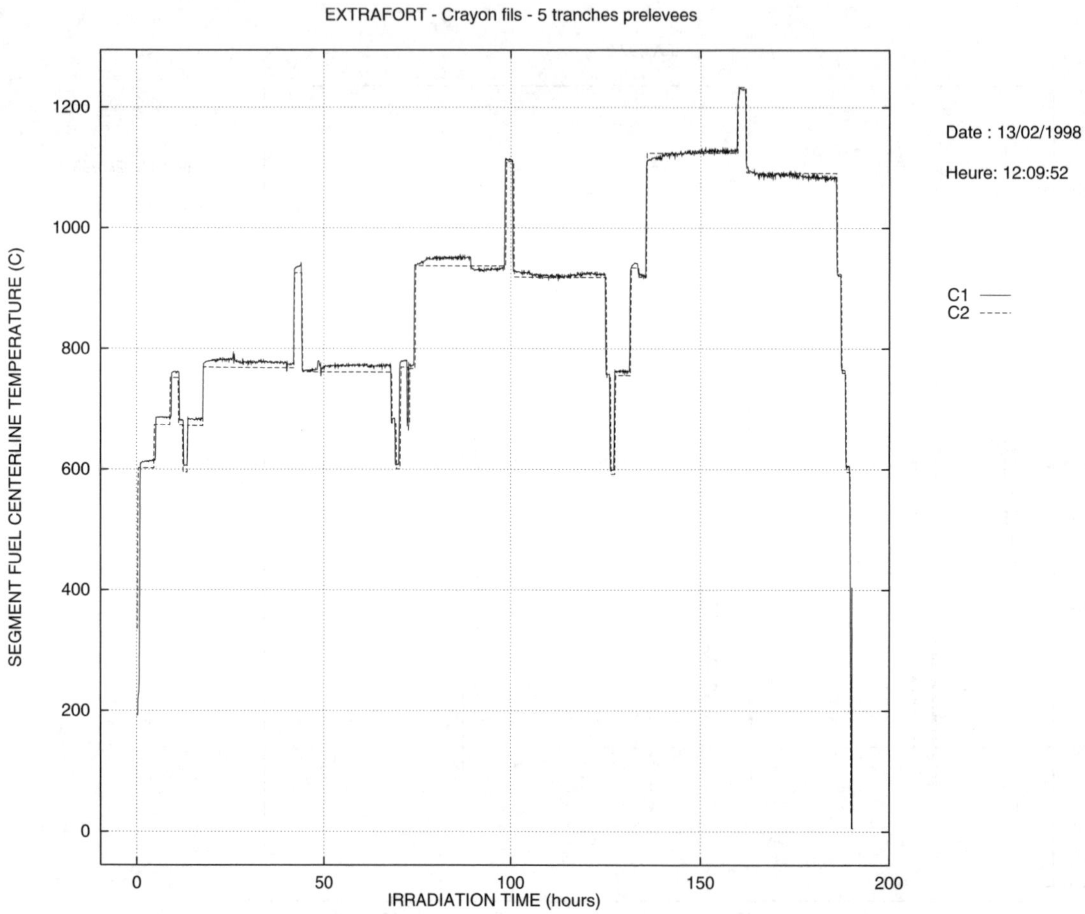

EXTRAFORT - Crayon fils - 5 tranches prelevees

Date : 13/02/1998

Heure: 12:09:52

C1 ———
C2 -------

:3POST-V4.1 (c) 1997 Electricite de France (EDF/DER/MTC)

Figure 9. IFA-515.10, UO$_2$ rod – comparison between measured temperature (C1) and calculated temperature with the Unified Baron-Hervé model (C2)

	FICHIER	VERSION	DATE	HEURE	LEGENDES	
C1:	exp5151.c3	-	00.00.00	00:00:00	-	-
C2:	5151.bin	2.5b	11. 2.98	8:59:11	IFA 515.10 rod	: TR 1 CR 1

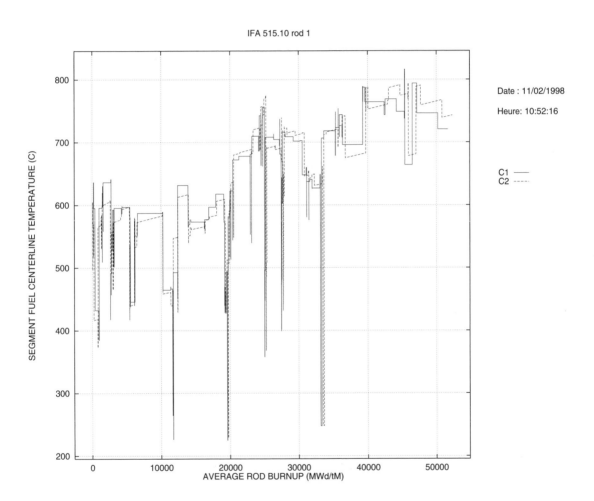

IFA 515.10 rod 1

Date : 11/02/1998

Heure: 10:52:16

C1 ——
C2 ------

Figure 10. IFA-515.10, gadolinia rod – comparison between measured temperature (C1) and calculated temperature with the Unified Baron-Hervé model (C2)

	FICHIER	VERSION	DATE	HEURE	LEGENDES
C1:	exp5152.c3	-	00.00.00	00:00:00	- -
C2:	5152.bin	2.5b	11. 2.98	8:53:25	IFA 515.10 rod : TR 1 CR 1

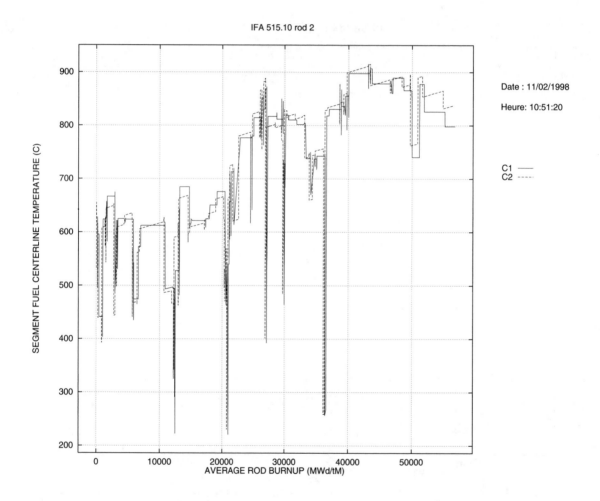

IFA 515.10 rod 2

Date : 11/02/1998

Heure: 10:51:20

C1 ——
C2 -----

348

Figure 11. IFA-519.9, Rod DH – comparison of measured FGR ratio (C1) and calculated FGR ratio with the CAPART model (C2) and the Baron-Maffeis model (C3)

	FICHIER	VERSION	DATE	HEURE	LEGENDES	
C1:	fgr519dh.c3	-	00.00.00	00:00:00	-	-
C2:	519dh_capart.bin	2.5b	12. 2.98	18: 4:28	IFA 519.9 rod	: CR 1
C3:	519dh2.bin	2.5b	13. 2.98	8:40:26	IFA 519.9 rod	: CR 1

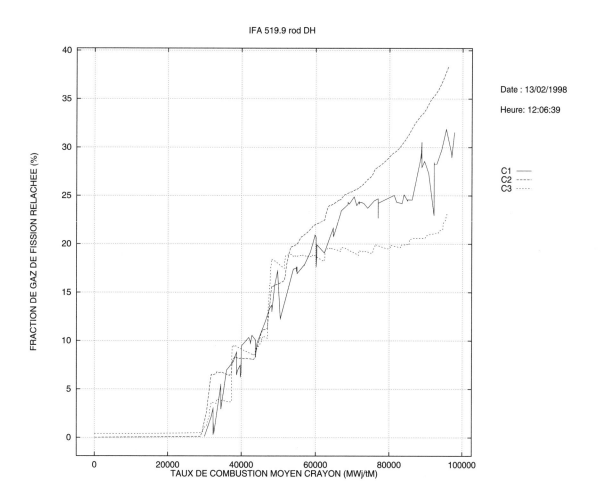

IFA 519.9 rod DH

Date : 13/02/1998

Heure: 12:06:39

C1 ——
C2 - - - -
C3 ·······

Figure 12. IFA-519.9, Rod DK – comparison of measured FGR ratio (C1) and calculated FGR ratio with the CAPART model (C2) and the Baron-Maffeis model (C3)

	FICHIER	VERSION	DATE	HEURE	LEGENDES		
C1:	fgr519dk.c3	-	00.00.00	00:00:00	-	-	
C2:	519dk_capart.bin	2.5b	12. 2.98	19: 9:27	IFA 519.9 rod	: CR	1
C3:	519dk75.bin	2.5b	13. 2.98	8:32:29	IFA 519.9 rod	: CR	1

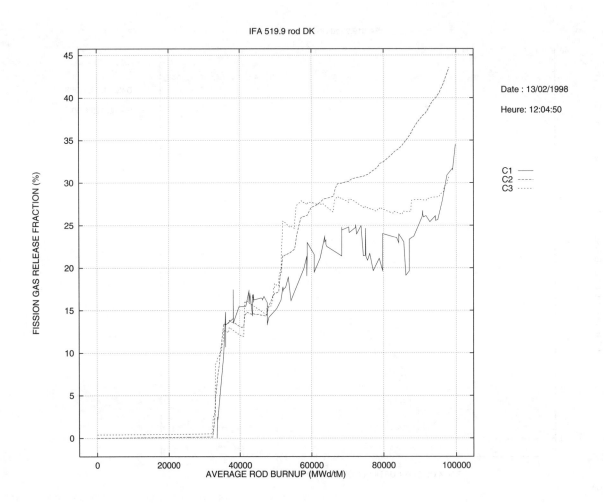

IFA 519.9 rod DK

Date : 13/02/1998

Heure: 12:04:50

C1 ——
C2 ------
C3 ······

Figure 13. IFA-562, Rod 13 – comparison between measured rod pressure (C1) and calculated rod pressure with the Baron-Maffeis model (C2)

	FICHIER	VERSION	DATE	HEURE	LEGENDES	
C1:	exp56213.c3	-	00.00.00	00:00:00	-	-
C2:	56213norm.c3	-	00.00.00	00:00:00	-	-

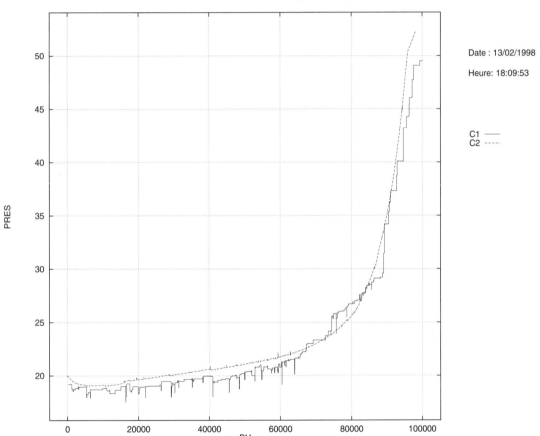

Pint normalise 2.4b rod 13 - Baron-Maffeis (facteur = 0.8)

Date : 13/02/1998

Heure: 18:09:53

C1 ——
C2 -----

Figure 14. IFA-585.1, Rod L – comparison of measured (C1) and calculated (C2) outer cladding diameter

	FICHIER	VERSION	DATE	HEURE	LEGENDES
C1:	diam585l.c3	-	00.00.00	00:00:00	- -
C2:	585l.bin	2.5b	13. 2.98	12:10:45	IFA-585.1 rod PW : TR 1 CR 1

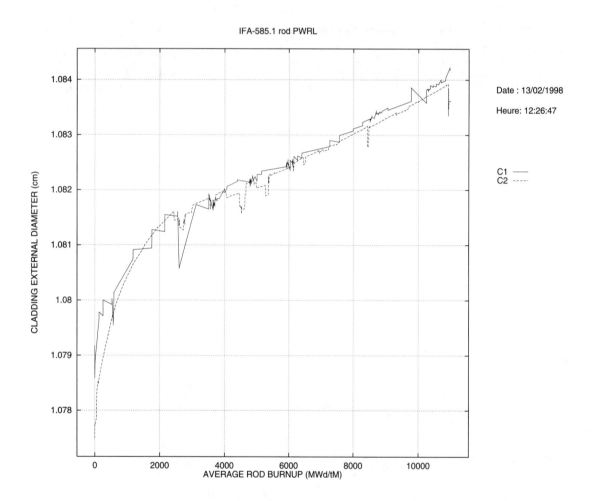

IFA-585.1 rod PWRL

Date : 13/02/1998

Heure: 12:26:47

C1 ——
C2 - - - -

C3POST-V4.1 (c) 1997 Electricite de France (EDF/DER/MTC)

Figure 15. IFA-585.1, Rod U – comparison of
measured (C1) and calculated (C2) outer cladding diameter

	FICHIER	VERSION	DATE	HEURE	LEGENDES
C1:	diam585u.c3	-	00.00.00	00:00:00	- -
C2:	585u.bin	2.5b	13. 2.98	12:11:16	IFA-585.1 rod PW : TR 1 CR 1

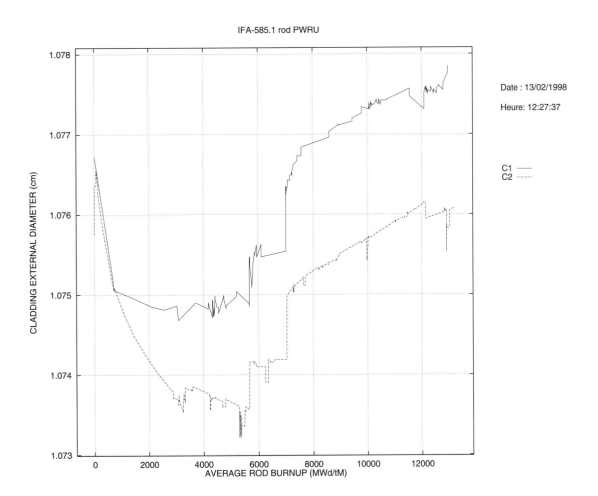

IFA-585.1 rod PWRU

Date : 13/02/1998

Heure: 12:27:37

C1 ——
C2 -----

Figure 16. The analytical RECOR experiment: power history

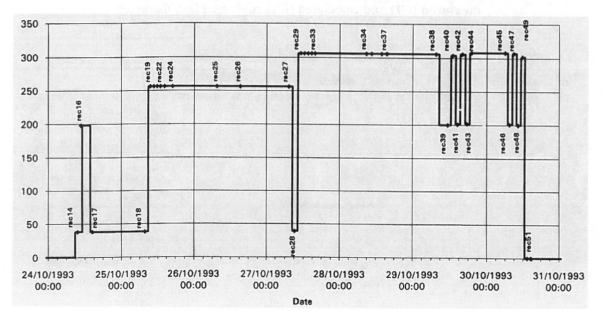

Figure 17. The analytical RECOR experiment: diameter evolution

Analytical experiment
Average cladding diameter evolution

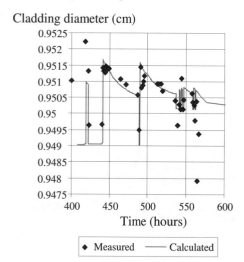

Analytical experiment
Average cladding diameter evolution at pellet-pellet location

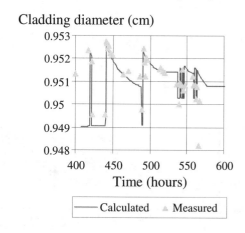

Figure 18. IFA-504, Rod 2 – comparison between measured hydraulic diameter and calculated pellet-to-clad diametral gap

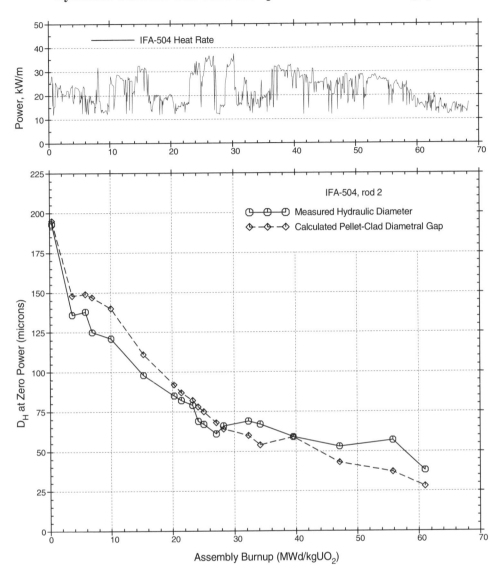

Figure 19. IFA-610 – rod pressure history

	FICHIER	VERSION	DATE	HEURE	LEGENDES
C1:	610lifttr.bin	2.5b	13. 2.98	12:51:50	IFA-610 Lift-off : CR 1

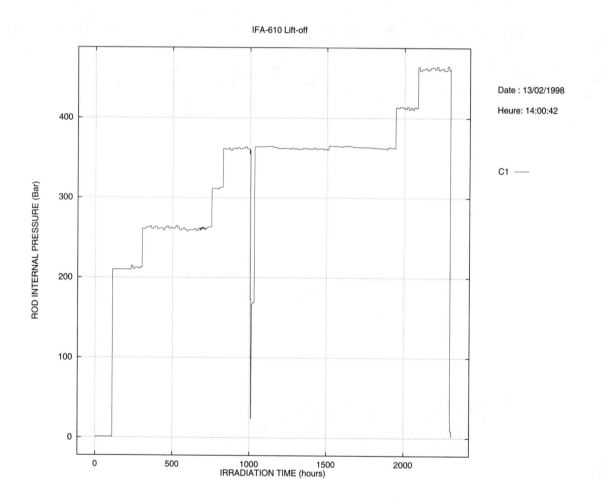

IFA-610 Lift-off

Date : 13/02/1998

Heure: 14:00:42

C1 ——

C3POST-V4.1 (c) 1997 Electricite de France (EDF/DER/MTC)

Figure 20. Comparison between measured (C1) and calculated fuel centre temperature with the Stora model, hollow pellets (C2) and solid pellets (C3)

	FICHIER	VERSION	DATE	HEURE	LEGENDES
C1:	exp6102.c3	-	00.00.00	00:00:00	- -
C2:	artr2.bin	2.5b	25. 2.98	10:30:34	IFA-610 Lift-off : TR 1 CR 1
C3:	arpl2.bin	2.5b	25. 2.98	10:35:22	IFA-610 Lift-off : TR 1 CR 1

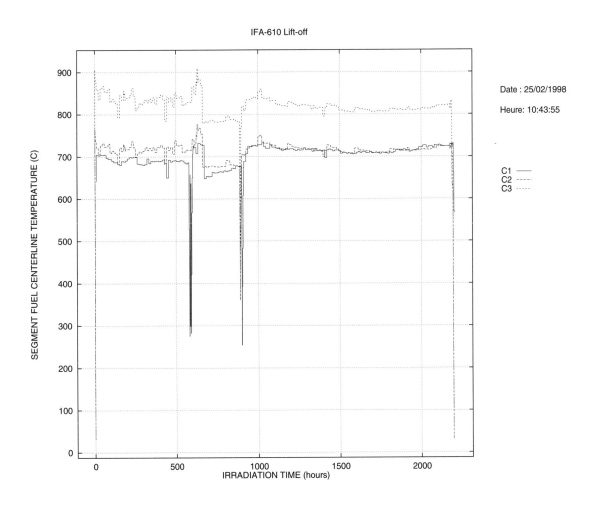

IFA-610 Lift-off

Date : 25/02/1998

Heure: 10:43:55

C1 ——
C2 - - -
C3 ·····

Figure 21. Illustration of the PCMI capabilities on part of the ramp tests database

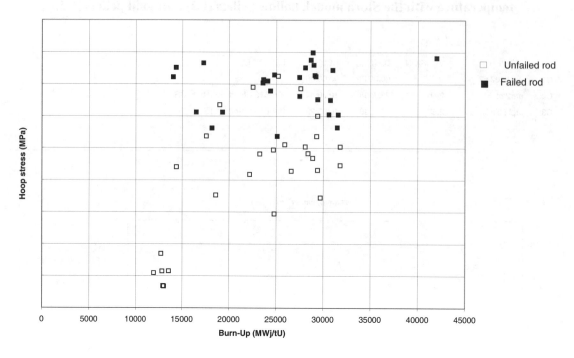

Figure 22. Comparison of calculated and measured oxide layers for low-tin Zy-4 claddings

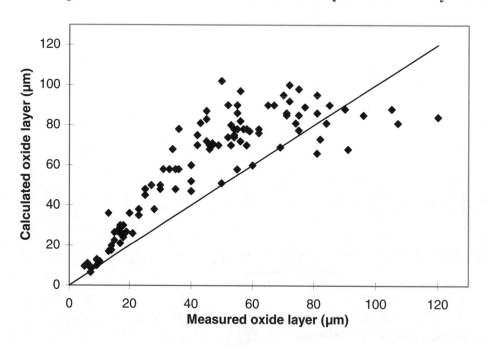

THERMAL ANALYSIS OF ULTRA-HIGH BURN-UP IRRADIATIONS EMPLOYING THE TRANSURANUS CODE

K. Lassmann, C.T. Walker, J. van de Laar
European Commission
Joint Research Centre
Institute for Transuranium Elements
P.O. Box 2340
D-76125 Karlsruhe

Abstract

A first tentative analysis of the Halden Ultra-High Burn-up experiment is presented employing the TRANSURANUS code. The TRANSURANUS high burn-up models are briefly explained with emphasis on the new TRANSURANUS-HWR burn-up model. At ultra-high burn-up, the trend of the temperature prediction is in disagreement with the experimental evidence and needs further clarification.

Introduction

In this paper an attempt is made to analyse measured fuel temperatures from the Ultra-High Burn-up Test performed in the OECD Halden Reactor Project [1,2]. In this test a considerable fuel temperature increase of more than 250°C from the beginning until the end of the irradiation at approximately 85000 MWd/tU was found. Two areas are of special interest.

a) The variation of the radial distributions of the fissile material with burn-up which determines the radial power density distribution (and hence the radial burn-up distribution) and the radial distribution of fission products such as Kr, Xe, Cs and Nd. Since the OECD Halden Reactor is a heavy water reactor, the light water reactor (LWR) burn-up equations of the TRANSURANUS code [3] were extended for heavy water reactor power stations (HWRs). Details of this work are given in Ref. [4].

b) The formation of the high burn-up structure (HBS) and its effect on the fuel centre temperature. According to a simple model [5], the formation of the HBS depends on the initial composition of fissile material, fuel geometry, section average burn-up and the burn-up threshold. The analysis in Ref. [4] showed that for a HWR the HBS starts at a section average burn-up of approximately 50000 MWd/tU (^{235}U enrichment = 5%) and 60000 MWd/tU (^{235}U enrichment = 10%) which is at least 10000 MWd/tU higher than in a LWR irradiation. Thus, for the analysis of the Ultra-High Burn-up Test the formation of the HBS must be considered.

The Halden Ultra-High Burn-up Test

Details of the test are given in Ref. [2]. The following design features make the test especially suited for the assessment of high burn-up models:

a) expansion thermometers measure the average fuel temperature without decalibration;

b) a small gap reduces the uncertainties of the gap conductance;

c) the temperatures stay well below the fission gas release threshold, i.e. the temperatures are not affected by Xe in the gap or presence of fission gas bubbles;

d) rapid burn-up accumulation is obtained by the use of small diameter pellets (5.92 mm) and a relatively high enrichment of 13%.

The linear rating started from ≈30-40 kW/m and decreased during the irradiation to ≈10 (15) kW/m. The four expansion thermometers gave very consistent results which will subsequently be compared with the predictions from the TRANSURANUS code. In order to focus on the degradation of the thermal conductivity of the fuel, a "normalisation" of the linear rating to 25 kW/m was performed by the experimentalists. This normalisation is based on the observation that the relation between temperature and linear rating is linear. However, the gradient varies with burn-up. The experimental results are shown in Figure 1. Clearly, a linear trend with burn-up is observed.

The TRANSURANUS high burn-up models

For the thermal analysis all the standard TRANSURANUS models were employed, for instance the standard relocation model and all standard high burn-up models. In the following the most relevant models are outlined.

The new TRANSURANUS-HWR burn-up model

The new TRANSURANUS-HWR burn-up model is an extension of the LWR burn-up model of the TRANSURANUS code [6]. As in several other models the radial distribution of plutonium is obtained from an empirical shape function. The basic equations are:

$$\frac{dN_{235}(r)}{dbu} = -\sigma_{a,235}N_{235}(r)A \tag{1}$$

$$\frac{dN_{238}(r)}{dbu} = -\sigma_{a,238}\overline{N}_{238}f(r)A$$

$$\frac{dN_{239}(r)}{dbu} = -\sigma_{a,239}N_{239}(r)A + \sigma_{c,238}\overline{N}_{238}f(r)A$$

$$\frac{dN_{j}(r)}{dbu} = -\sigma_{a,j}N_{j}(r)A + \sigma_{c,j-1}N_{j-1}(r)A$$

where $\sigma_{a,i}$ and $\sigma_{c,i}$ are the neutron absorption and capture cross-sections, respectively and the index j stands for the isotopes 240, 241 and 242. "A" is a conversion constant (for details see Ref. [6], Eq. 3) and $N_j(r)$ is the local concentration of the isotope j. The local concentration of ^{238}U is written as $\overline{N}_{238}f(r)$, where $f(r)$ is a normalised radial shape function which encapsulates the contribution of the resonance absorption to the total plutonium production. It is a function of the form:

$$f(r) = 1 + p_1 e^{-p_2(R-r)^{p_3}} \tag{2}$$

where R is the fuel radius and p_1, p_2, and p_3 are constants.

The equations outlined above are solved incrementally. For each average burn-up increment a new radial power density profile is calculated from which the radial burn-up profile is updated. Finally, the local concentrations of Kr, Xe, Cs and Nd are obtained by multiplying the local burn-up increment by the appropriate fission yields.

The first step in the development of a HWR burn-up model is the determination of the one group, spectrum-averaged cross-sections for neutron absorption and capture used in Eq. (1). For the development of the present simple model these were estimated from the neutron spectrum of the Halden HWR which differs significantly from that of a LWR; in the Halden Reactor the thermal flux is higher and the fast flux is lower than in a typical LWR.

The simple approach can also be used to estimate the relative contribution of the epithermal region to the total effective cross-section. For a LWR this number is approximately 0.44, whereas for a HWR a lower value of approximately 0.28 is obtained. This means that in a HWR less neutrons are captured in the epithermal region than in a LWR.

This quantitative information is used to modify the radial shape function, f(r), calculated with Eq. (2) for HWR conditions. The second term in this equation describes the relative contribution of neutron capture in the epithermal region and accordingly:

$$p_1^{HWR} = 0.64 p_1^{LWR} \qquad (3)$$

The modification of the cross-sections and the modification of the constant p_1 are the only differences compared with the LWR version and constitute the new TRANSURANUS-HWR burn-up model. This model is contained in the subroutine TUBRNP which can be run as a stand-alone program or as part of the TRANSURANUS code. When run as a stand-alone program, processes such as thermal fission gas release are not taken into account.

The new TRANSURANUS-HWR burn-up model was verified against 12 radial caesium profiles, 2 neodymium profiles, 7 plutonium profiles and 22 xenon profiles. With the exception of two Pu profiles, all the results for the radial distribution of Xe, Cs, Nd and Pu were obtained by electron probe microanalysis at the Institute for Transuranium elements carried out by one of the authors of this presentation(CTW).

In order to show the main difference between the widely used RADAR model (RAting Depression Analysis Routine) of Palmer et al. [7] and the new TRANSURANUS-HWR burn-up model, the average Pu concentration predicted by both models as a function of the section averaged burn-up for a ^{235}U enrichment of 3% is shown in Figure 2. Both the RADAR model and the new TRANSURANUS-HWR burn-up model give the local concentration of plutonium. The RADAR model, however, calculates only ^{239}Pu, whereas the TRANSURANUS model gives the total plutonium as the sum of ^{239}Pu, ^{240}Pu, ^{241}Pu and ^{242}Pu. Although the agreement in the prediction of the ^{239}Pu concentration is reasonable, it is evident that the total Pu cannot be predicted by the RADAR model at higher burn-up. Differences are also encountered in the predicted radial distribution of plutonium; the new TRANSURANUS-HWR burn-up model gives higher Pu concentrations near the fuel surface (Figure 3).

The verification has shown that the new TRANSURANUS-HWR burn-up model can predict reasonably well the radial Pu and burn-up profiles and hence the Xe, Cs and Nd profiles. In order to obtain an indication of whether the HBS also exists in fuel irradiated in HWRs, the local Xe concentration is plotted as a function of the local burn-up as calculated by the new model (Figure 4). Only Xe concentrations not affected by thermal fission gas release are considered. The result is very similar to that for LWR conditions (e.g. Ref. [5], Figure 4). A more or less linear behaviour up to the threshold burn-up is followed by a sharp decrease of Xe in the matrix which denotes the formation of the HBS.

Summary of the other TRANSURANUS high burn-up models

Degradation of the thermal conductivity with burn-up

For the degradation of the thermal conductivity of the fuel a simple model is used. The phononic term is written as:

$$\lambda_{phonon} = \frac{1}{a + a_x x + a_{bu} bu + bT} \qquad (4)$$

where a, a_x, a_{bu} and b are constants, bu is the burn-up and T is the temperature. The term $(a_x x)$ accounts for the effect of hypo-or hyperstoichiometry and $(a_{bu} bu)$ for the burn-up effect. The variable x is the absolute value of the deviation from stoichiometry, i.e. $|O/M - 2|$.

Formation of the HBS

The formation of the HBS is closely related to the local burn-up which depends in the present model on the initial composition of fissile material, fuel geometry and reactor type (LWR, HWR) [5]. Two burn-up thresholds can be identified:

a) A local threshold burn-up for the transition of the original grain structure to the formation of local HBS spots (approximately 60000 MWd/tU).

b) A local threshold for the fully developed HBS (approximately 75000 MWd/tU).

The thickness of the HBS is obtained from the radial location at which the threshold burn-up is first reached on the calculated burn-up profile.

Build-up porosity in the HBS

The properties of the HBS are different from the original UO_2 structure and need to be taken into account in the modelling of the fuel rod behaviour. Of specific interest is the porosity in the HBS which may reach up to 10-20%.

ITU data show that above the threshold for the formation of the HBS this porosity can be correlated linearly with the local burn-up (with an upper limit at a local burn-up of approximately 160000 MWd/t). Thus from the knowledge of the local burn-up the corresponding porosity value in the HBS can be calculated.

Xenon depletion of the matrix above the threshold burn-up

A simple model that describes the xenon depletion of the matrix (athermal release of Xe from the UO_2 grains) has been derived in Ref. [5]. This model has been incorporated into the TRANSURANUS code in incremental form. However, without a significant loss of accuracy the underlying differential equation can be integrated to obtain an analytical approximation which gives the results shown in Figure 4.

Results from the first tentative analysis of the Halden Ultra-High Burn-up Test

In order to be able to compare the predicted temperatures directly with the normalised temperature data, a constant linear rating of 25 kW/m was used. Several calculations were performed from which two results are presented: one result from a calculation in which the formation of a new pore structure in the HBS is considered, the other from a calculation in which the formation of a new pore structure in the HBS is neglected. The results obtained with the TRANSURANUS code are given in Figure 5.

It is obvious that in the calculation in which the formation of the new pore structure is taken into account, the trend of the temperature is in disagreement with the experimental data. Since it is reasonable to assume that the formation of new pores in the HBS has an influence on both the thermal and the mechanical behaviour of the fuel, the following questions need to answered:

1. Do the standard porosity correction factors for the thermal conductivity apply for the new pore structure with pore sizes of the order of 1-2 μm?

2. Is the thermal expansion of the fuel which is measured by the expansion thermometers affected at high burn-up by strong mechanical interaction between fuel and cladding?

REFERENCES

[1]   W. Wiesenack, "Review of Halden Reactor Project High Burn-up Fuel Data That Can be Used in Safety Analyses", *Nuclear Engineering and Design*, 172 (1997), pp. 83-92.

[2]   W. Wiesenack, "Assessment of UO_2 Conductivity Degradation Based on In-Pile Temperature Data", Proceedings of the 1997 International Topical Meeting on LWR Fuel Performance, Portland, Oregon (1997), pp. 507-511.

[3]   K. Lassmann, "TRANSURANUS: A Fuel Rod Analysis Code Ready for Use", *Journal of Nuclear Materials*, 188 (1992), pp. 295-302.

[4]   K. Lassmann, C.T. Walker, J. van de Laar, "Extension of the TRANSURANUS Burn-up Model to Heavy Water Reactor Conditions", *Journal of Nuclear Materials*, in print.

[5]   K. Lassmann, C.T. Walker, J. van de Laar, F. Lindström, "Modelling the High Burn-up UO_2 Structure in LWR Fuel", *Journal of Nuclear Materials*, 226 (1995), pp. 1-8.

[6]   K. Lassmann, C. O'Carroll, J. van de Laar, C.T. Walker, "The Radial Distribution of Plutonium in High Burn-up UO_2", *Journal of Nuclear Materials*, 208 (1994), pp. 223-231.

[7]   I.D. Palmer, K.W. Hesketh, P.A. Jackson, "A Model for Predicting the Radial Power Profile in a Fuel Pin", *Water Reactor Fuel Element Performance Computer Modelling*, Applied Science Publishers Ltd (1983), pp. 321-335.

Figure 1. Temperature data for the Halden Ultra-High Burn-up Test after normalisation to a linear rating of 25 kW/m

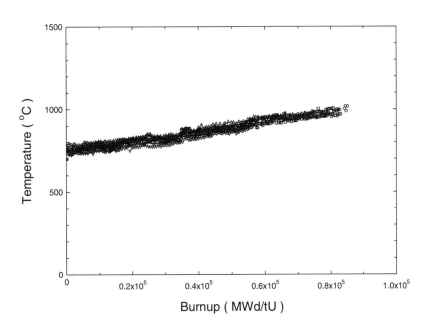

Figure 2. Average plutonium concentration for a ^{235}U enrichment of 3% as predicted by the RADAR model and the new TRANSURANUS-HWR burn-up model; the ^{239}Pu concentration predicted by the RADAR model is also the total plutonium in this case because the higher Pu isotopes are not considered by the RADAR model

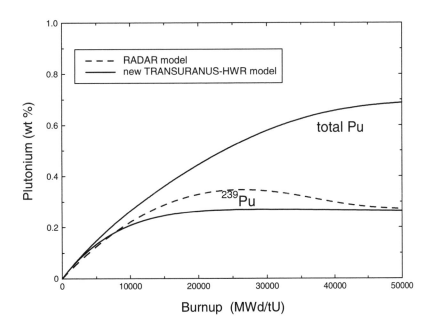

Figure 3. Comparison of the measured radial plutonium profile with that calculated by the RADAR model and the new TRANSURANUS-HWR burn-up model (Risø 2; fuel section M23-1-21$-3, 50000 MWd/tU, ^{235}U enrichment = 5.1%)

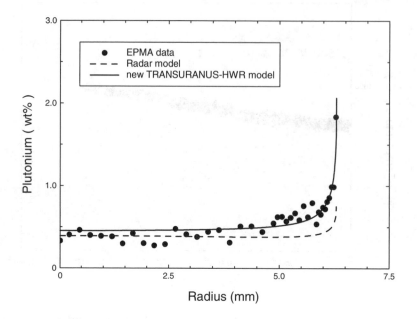

Figure 4. Comparison between predicted and measured local xenon concentrations as a function of the local burn-up. Only data points not affected by thermal fission gas release are considered. The threshold burn-up is set at 60000 MWd/tU.

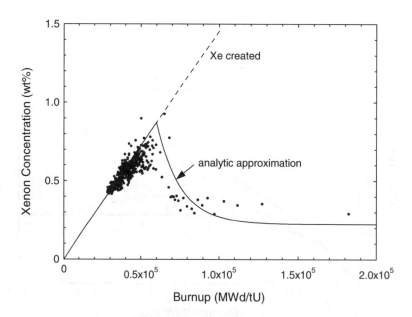

**Figure 5. Prediction of the fuel centre temperature as a
function of the section average burn-up for simplified conditions**

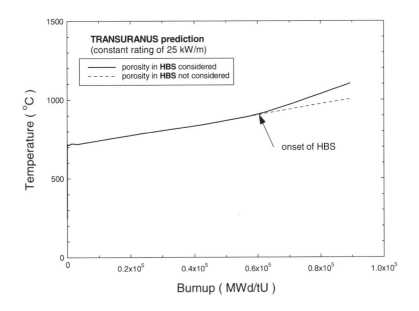

START-3 CODE GAP CONDUCTANCE MODELLING

U.Ê. Bibilashvily, À.V. Medvedev, S.Ì. Bogatyr, V.I. Kouznetsov, G.À. Khvostov
A.A. Bochvar All-Russia Research Institute of Inorganic Materials

Abstract

A model of integral fuel performance code START-3 for calculation of heat transfer across the fuel-cladding gap in nuclear fuel elements is presented. Comparisons are made between the prediction of gap conductance model and out-of-pile experiments, and the results are presented. The START-3 code has been used to predict central line temperature data available from the Halden reactor. The calculations have been made for two different fuel rod internal gas (helium, xenon).

Introduction

Any fuel performance code is a system of two interrelated main blocks: thermal and mechanical. Each consists of a large number of submodels. And the system is characterised by a plenty direct and feedback between these blocks. Therefore inexact modelling (account) of those or other processes (parameters) can result not only in quantitative, but also, in a number of cases, in qualitative mistakes. To the full, it is possible to attribute all above told, to account of heat coefficient from fuel to clad. The correct account of all factors influencing to process of a heat transfer, is a necessary condition of exact account of fuel temperatures fields, and consequently also of all other parameters influencing fuel rod serviceability The special urgency the modelling of process of a heat transfer gets with account of fuel rod, designed on achievement high burn-up (>45000 MWd/kgU). With such burn-up level begin to be displayed and to be accelerated such processes as structural change of a fuel composition (rim region), increase of fission gas release, and consequently by increase of the contents xenon and krypton in internal gas, degradation of fuel thermal properties (thermal conductivity, melting point) and approach of contact between fuel and clad. Intensive experimental and theoretical researches devoted to study of properties of materials and processes occurring in fuel rod with achievement high burn-up nowadays will be carried out. The present seminar is currently devoted to these problems.

The processes set forth above, certainly, in this or that measure influence the process of a heat transfer from fuel to clad and find the reflection with adaptation of fuel performance codes for accounts of fuel rod on high burn-up. The settlement code START-3 concerns to such integrated codes, one of which submodels and is the stated below model of a heat transfer GAP.

The model allows to determine heat transfer coefficient in various condition of fuel in relation to cladding: contact, gap. The contribution radiation component of coefficient of a heat transfer is also taken into account which, however for steady-state operation does not exceed 1%. Within the framework of one report to state all model representations accepted in the START-3 code, and concerning high burn-up it is obviously not possible. Therefore, in this case, the accent on consideration of questions of modelling of a gas component of heat transfer coefficient is made with high levels contents xenon and krypton in internal gas mixture of fuel rod, and also account of temperature jumps with fuel-clad contact.

In the report the comparisons between the prediction of gap conductance model and out-of-pile experiments, and the results are presented. The START-3 code has been used to predict central line temperature data available from the Halden reactor. The calculations have been made for two different fuel rod internal gas (helium, xenon). The agreement between model predictions and experimental data is satisfactory.

Model description

Heat transfer coefficient across fuel-clad can be expressed, in the assumption of absence convection, through the sum:

$$h = h_r + h_g + h_c \qquad (1)$$

where $\Rightarrow h_r$ – radiation heat transfer coefficient;

$\qquad \Rightarrow h_g$ – gas component;

$\qquad \Rightarrow h_c$ – contact component

The radiation heat transfer coefficient is given by:

$$h_r = K \frac{T_f^4 - T_c^4}{T_f - T_c}$$

(2)

where:

$$K = \sigma \frac{\varepsilon_f \cdot \varepsilon_c}{\varepsilon_f + \varepsilon_c - \varepsilon_f \cdot \varepsilon_c}$$

$\Rightarrow T_f, T_c$ – surfaces temperature of the fuel and the clad accordingly;
$\Rightarrow \varepsilon_f, \varepsilon_c$ – emissivities of the fuel and the clad accordingly;
$\Rightarrow \sigma$ – Stefan-Boltzmann constant.

The contribution radiation component of coefficient of a heat transfer is also taken into account which, however for steady-state operation does not exceed 1%. But under accident conditions this term may make a substantial contribution.

The gas component h_g is given by:

$$h_g = \frac{K}{\delta + \xi_f + \xi_c}$$

(3)

where $\Rightarrow \delta$ – the gap between fuel and clad, which is characterised by the sum of a nominal gap and composed causing influence of a roughness;
  $\Rightarrow \xi_f, \xi_c$ – temperature jumps distances on border of a gas mix both outside surface of fuel and internal clad surface;
  $\Rightarrow K$ – thermal conductivity of gas mixture.

Thermal conductivity of gas mixtures accepted in model is expressed in the classical Wasiljewa form [1]:

$$K = \sum_{j=1}^{n} \left[K_j \bigg/ \left(1 + \sum_{k=1}^{n} \left(\beta_{jk} \frac{x_k}{x_j} \right)_{j \neq k} \right) \right]$$

(4)

In Equation (4): $\Rightarrow K_j$ – thermal conductivity pure gas;
  $\Rightarrow \beta_{jk}$ – weighting factors;
  $\Rightarrow x_j$ – molar concentrations.

Nowadays is present much enough approximate, semi-empirical and empirical dependence for calculation of the weight factors [2-4]. With a choice of the equation for calculation of the weight factors we took into account the following moments: the best concurrence to experimental data in a wide range of temperatures and structures of gas mixes, especially for such compositions as He-Xe and He-Kr-Xe, minimum quantity of the initial data, which should be most reliable and accessible.

To all these requirements, from our point of view, most full satisfies dependence Mason [2], modified Ulibin [5] for a binary mix He-Xe:

$$\beta_{jk} = \frac{1.08}{2\sqrt{2}} \cdot \left(1 + \frac{M_j}{M_k}\right)^{-\frac{1}{2}} \left[1 + \left(\frac{K_j}{K_k}\right)^{\frac{1}{2}} \left(\frac{M_j}{M_k}\right)^{\frac{1}{4}}\right]^2 \tag{5}$$

where $\Rightarrow M$ – gas molecular weight;

$\qquad \Rightarrow K$ – thermal conductivity pure gas.

The temperature jump distances in model are calculated according to [7]:

$$\varsigma_f = \frac{5\pi^{\frac{1}{2}}}{8} \left(\sum_{i=1}^{n} XM_j \left\{ \frac{52}{25\pi} + \frac{1}{2} \left[1 - \alpha_j^{fc}\right] \right\} a_j^2 + \frac{1}{2} \frac{\left\{ \sum_{j=1}^{n} XM_j \left[2 - \alpha_j^{fc}\right] a_j \right\}^2}{\sum_{j=1}^{n} XM_j \alpha_j^{fc}} \right) L_j \tag{6}$$

In Equation (6):

$$XM_j = x_j \left(\frac{M}{M_j}\right)^{\frac{1}{2}}$$

$$a_j = \frac{K_j}{K} \left(\frac{M_j}{M}\right)^{\frac{1}{2}} \frac{1}{x_j + \sum_{j \neq i} \beta_{ji} x_i}$$

$\Rightarrow \alpha$ – accommodation coefficient (AC);

$\Rightarrow K$ – thermal conductivity;

$\Rightarrow L$ – mean free path.

The contact component of factor of a heat transfer in the submitted model is taken into account on an empirical dependence Lassmann-Pazdera, which empirical constants were picked up on the basis of processing 388 very different experiments:

$$\frac{h_c}{\lambda \cdot R} = b \cdot \left(\frac{P}{R^2 H}\right)^m \tag{7}$$

where:

$$\lambda = 2 \frac{\lambda_f \lambda_c}{\lambda_f + \lambda_c}$$

$$R = \sqrt{\frac{1}{2}\left(R_f^2 + R_c^2\right)}$$

$\Rightarrow H \qquad$ – Meyer hardness;

$\Rightarrow \lambda_f, \lambda_c$ – thermal conductivity of the fuel and the clad;

$\Rightarrow P \qquad$ – contact pressure;

$\Rightarrow b, m \quad$ – empirical constants;

$\Rightarrow R_f, R_c$ – roughness of the fuel and the clad.

Testing of the model

The testing of the model was carried out stage by stage. At the first stage the comparison of settlement dependence (5) with experimental data was carried out. In Figures 1 and 2 and in Table 1 the results of comparison settlement and experimental data for mixes He-Xe are submitted: and He-Kr-Xe with temperature 520 ^{i}Ñ [6]. For He-Xe mixture the maximal mistake makes 2.4%. As to ternary mixtures, only in one point from 20 mistakes has made 14.4%. Without the account of this point the maximal error has made 10.5%, minimal 1.02%, average 6.25%.

At the second stage of testing the check of reliability of account of temperature jumps was carried out (spent), the comparison calculations and experimental data was carried out by results of MPD experiments which have been carried out Chaddola and Loyalka [8]. With account of temperature jump distances, for helium, Ullman temperature dependence of AC coefficient was used [9]. The AC of argon was accepted constant and equal 0.8. The samples were considered smooth, therefore influence of a roughness on factor of a heat transfer was not taken into account, though, as such assumption fairly only for one series of samples (series ISM III) was marked in [8].

The received results are shown in Figures 3 and 4. As it is visible from Figure 3 experimental data rather strongly differ from calculated, that is caused by the above named reason, namely, influence of a roughness. Considering group of samples ISM III (see Figure 4) it is possible to note more satisfactory results. The maximal error has made 19% (one point). For other points the error does not exceed 10%, that allows to make about satisfactory concurrence of experiment to accounts, though it is necessary to note that fact, that the number of points (ISM III) is limited (for pure He-8, for a mix He-Ar-8).

As additional (implicit) check of a correctness of account of temperature jump distances it is possible to consider also tests of a contact component, which were carried out spent by results of out-of-pile experiments.

The comparison of accounts with generalised out-of-pile experiments is submitted in Figure 5. As the settlement dependence satisfactorily is visible from the diagrams describes results of experiments not only qualitatively, but also quantitatively.

In-reactor temperature measurements

The serviceability of model is confirmed also with calculations in-pile experiments, including FUMEX. By way of illustration on Figure 6 the results of account of fuel central temperature are given at the reactor start up. The experiment is carried out on Halden reactor [11] (description of the experiment see in Appendix A).

The accounts have shown, that strong dilution internal gas by the fission gas products can result in significant increase of temperature of fuel.

As it is visible from the diagram, with dilution close to 100% the increase of temperature can reach 300°C only at the expense of change of conditions of a heat transfer.

REFERENCES

[1]   A. Wassiljewa, *Physik. Z.*, 5, 737 (1904).

[2]   E.A. Mason, "Appropriate Formulae for Thermal Conductivity of Gas Mixtures", *Phys. Fluids.*, 2, 1958.

[3]   A.L. Lindsay and L.A. Bromley, *Ind. and Eng. Chem.*, 42:1 (1950) p. 1508.

[4]   R.S. Brokaw, *Journ. Chem. Phys.*, 42, 4, 1965.

[5]   S.A. Ulibin, *Teploenergetica N6*, 1962.

[6]   E. Thornton, *Proc. Phys. Soc.*, Vol. 76, pt 1, [1] 487.

[7]   S.K. Loyalka, *Nucl. Tech.*, Vol. 57, 1982.

[8]   V.K. Chandola, S.K. Loyalka, *Nucl. Tech.*, Vol. 56, 1982.

[9]   A. Ulman, R. Acharya and D.R. Olander, *J. Nucl. Mater.*, 51, 277 (1974).

[10]   K. Lassmann and Hohlefeld, *Nucl. Eng. Des.*, 103 (1987) 215.

[11]   IAEA letter Appendix 5, N 641-T1, 1995.

Table 1

MOLS FRACTION			MODEL	EXPERIMENT	DEVIATION
He	Kr	Xe			
0.219	0.086	0.695	80,53	82,09	-4,54651
0.486	0.057	0.457	175,9	178,5	-10,3015
0.709	0.032	0.259	313,8	317,2	-8,05797
0.863	0.016	0.121	476,4	479,4	-5,62992
0.248	0.119	0.633	89	90,59	-7,18238
0.480	0.082	0.438	174	176,5	-6,11702
0.742	0.041	0.217	343,7	347	-4,14365
0.865	0.021	0.114	480,1	483	-1,02459
0.215	0.217	0.568	80,72	82,01	-14,3946
0.507	0.136	0.357	189,6	191,9	-5,46798
0.706	0.081	0.213	316,3	319,1	-5,8702
0.859	0.039	0.102	475,3	477,9	-1,0559
0.227	0.394	0.379	88,07	89,03	-10,5226
0.479	0.266	0.255	182,3	183,9	-5,69231
0.729	0.138	0.133	343,2	345,3	-8,65079
0.856	0.073	0.071	477,9	479,7	-10,1685
0.245	0.593	0.162	101,7	102,2	-10,3509
0.519	0.378	0.103	211,6	212,4	-6,84211
0.743	0.202	0.055	365,4	366,4	-4,58334
0.865	0.106	0.029	497,7	498,5	-2,63672

Figure 1. Thermal conductivity of Helium-Xenon mixtures at 520°C

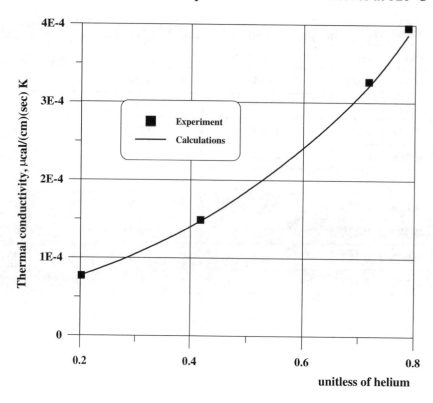

Figure 2. Thermal conductivity of He-Xe-Kr mixtures (model vs. experiment)

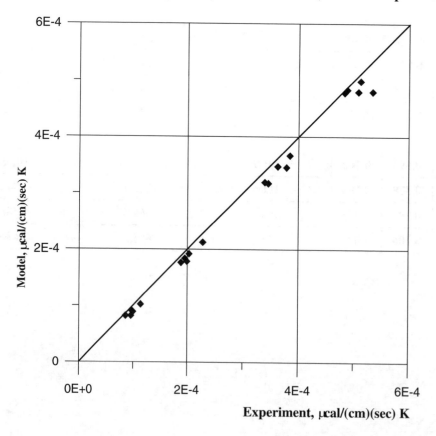

Figure 3. Gap conductance (model vs. experiment)

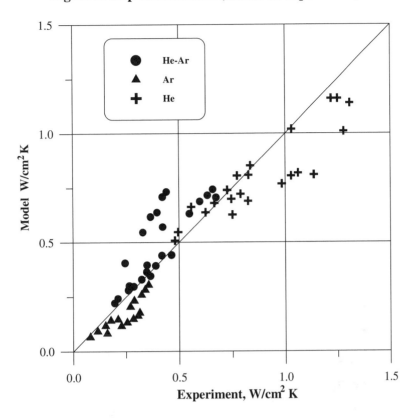

Figure 4. Gap conductance (model vs. experiment)

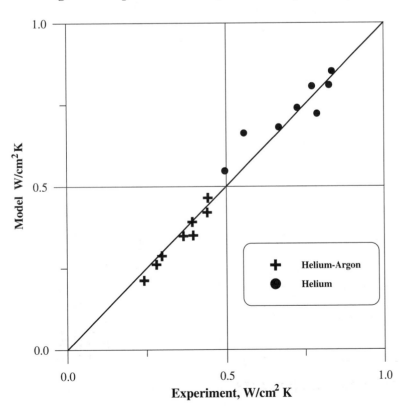

Figure 5. Comparison between experiments and the GAP model

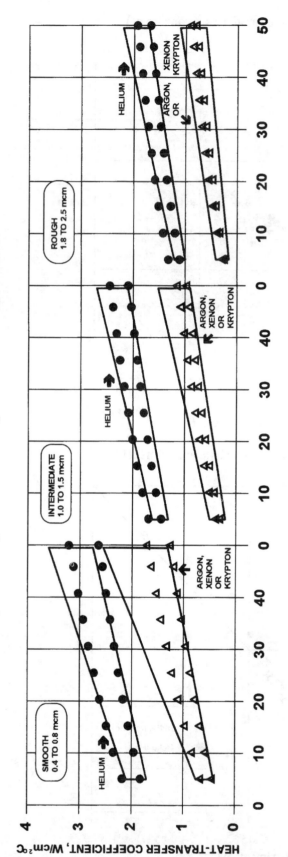

INTERFACIAL PRESSURE, MPa

☐ - experimental values, ● - calculated values (helium), Δ - calculated values (xenon).

Graphical summary of experimental values for heat-transfer coefficients between UO_2 and Zircaloy in various gases at atmospheric pressure. The values are for the three ranges of arithmetic-mean roughness heights, and for an interface temperature of 350 °C
J.A.L. ROBERTSON, A.M. ROSS, M.J.E. NOTLEY and J.R. MacEWAN "TEMPERATURE DISTRIBUTION IN UO_2 FUEL ELEMENTS
J. Nucl. Mat., Vol. 7 (1962), No. 3

Figure 6. Comparison between experiment and START-3 code prediction

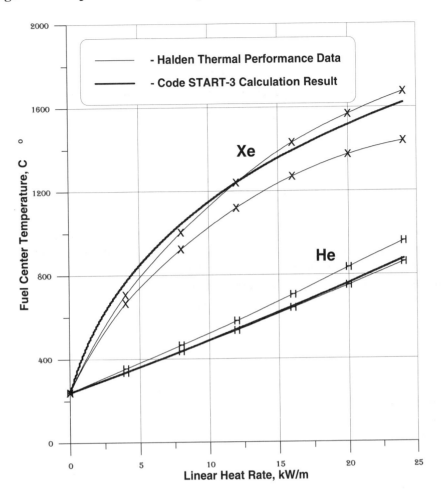

Appendix A

DATA ON FUEL CENTRELINE TEMPERATURES
ON START-UP WITH DIFFERENT FILL GAS COMPOSITIONS

The following data are from the start-up of a single assembly containing four short rods which could be flushed with different gases. Each rod contained thermocouples at the top and bottom of the fuel column. At the same axial level as the thermocouple tip were rings of three neutron detectors. In this way the local powers were recorded with high accuracy. Measurements of centre line temperature were recorded during several slow power ramps when the rods were filled with helium, argon or xenon at two atmospheres pressure. The data here are in the form of envelopes encompassing the readings of all eight thermocouples. In the main part, the spread represents the stochastic nature of pellet cracking and fragment relocation.

Fuel rod parameters

Clad material	Zr-2
Outside diameter (mm)	12.45
Inside diameter (mm)	10.79
Thickness (mm)	0.83
Fuel material	sintered UO_2 pellets
Outside diameter (mm)	10.59
Inner bore diameter for T/C insertion (mm)	1.8
Pellet length (mm)	$12.7 \pm I$
Dishing (mm^2)	$24 + 6$
Fuel to clad diametral gap - /im	200
Fuel density	95% TD, $10.35 \pm l$ g/cc
Enrichment	10 wt% ^{235}U
Fuel length (mm)	700
Coolant temperature (°C)	240
Coolant pressure - atmos.	34

The clad is cooled under boiling conditions.

LIST OF PARTICIPANTS

AUSTRIA

MENUT, Patrick
IAEA
NFC and WM
5 Wagramerstrasse
P.O. Box 100
A-1400 Vienne

Tel: +43 1 206022763
Fax: +43 1 2060722763
E-mail: p.j.m.menut@iaea.org

ONOUFRIEV, Vladimir
IAEA
NFC and WM
5 Wagramerstrasse
P.O. Box 100
A-1400 Vienne

Tel: +43 1 206022760/1
Fax: +43 1 20607
E-mail: v.onoufriev@iaea.org

BELGIUM

BAIRIOT, Hubert
FEX
Lijsterdreef 24
B-2400 Mol

Tel: +32 14 31 25 33 (private)
Fax: +32 14 32 09 52
E-mail:

LIPPENS, Marc
BELGONUCLÉAIRE
Avenue Oriane, 4
B-1200 Brussels

Tel: +32 2 774 0625
Fax: +32 2 774 0614
E-mail:

VAN UFFELEN, Paul
SCK/CEN
Fuel Research Unit
200 Boeretang
B-2400 Mol

Tel: +32 14 33 2207
Fax: +32 14 32 15 29
E-mail: pvuffele@sckcen.be

BULGARIA

ELENKOV, Dancho
Bulgarian Academy of Sciences
Institute for Nuclear Research and Nuclear
Energy
Tzarigradsko Schaussee 72
1784 Sofia

Tel: +359 2 74 31 306
Fax: +359 2 974 3121
E-mail: elenkov@inrne.acad.bg

CANADA

ARIMESCU, Viorel-Ioan
Atomic Energy of Canada, Ltd.
Chalk River National Laboratory
Chalk River
Ontario, KOJ 1JO

Tel: +1 613 584 8811 or 3384
Fax: +1 613 585 4200
E-mail: arimescui@aecl.ca

CHINA (REPUBLIC OF)

PENG, Chen
China Institute of Atomic Energy (CIAE)
P.O. Box 275 (64)
102413 Beijing

Tel:
Fax:
E-mail: wuzu10@mipsa.ciae.ac.cn

CZECH REPUBLIC

NOVOTNY, Michal
Nuclear Research Institute
25068 - Rez U Prahy

Tel:
Fax:
E-mail:

SVOBODA, Radek
Nuclear Research Institute
Dept. of Reactor Technology
Nuclear Power & Safety Division
25068 - Rez U Prahy

Tel: +420 2 6617 2449
Fax: +420 2 685 7960
E-mail: rad@nri.cz

VALACH, Mojmir
Nuclear Research Institute
Dept. of Reactor Technology
25068 - Rez U Prahy

Tel: +420 2 685 7958
Fax: +420 2 685 7960
E-mail: val@nri.cz

ZMITKO, Milan
Nuclear Research Institute
25068 - Rez U Prahy

Tel: +420 2 661 72453
Fax: +420 2 685 8191
E-mail: zmi@nri.cz

FINLAND

KELPPE, Seppo
VTT Energy
P.O. Box 1604
02044 VTT

Tel: +358 9 456 5026
Fax: +358 9 456 5000
E-mail: seppo.kelppe@vtt.fi

FRANCE

BALESTREIRI, Didier
CEA/CADARACHE
DRN/DEC/SDC/LCI
Bât. 315
F-13108 St. Paul-lez-durance Cedex

Tel: +33 (0)4 42 25 34 41
Fax: +33 (0)4 42 25 70 42
E-mail:

BARON, Daniel
EDF
Dépt. Mécanique et Technologie
des Composants
Les Renardières
Route de Sens, Ecuelles
F-77250 Moret sur Loing

Tel: +33 (0)1 60 73 61 05
Fax: +33 (0)1 60 73 61 49
E-mail: daniel.baron@der.edfgdf.fr

BEAUVY, Michel
CEA/CADARACHE
DRN/DEC/SPU/Epu
Bât. 737
F-13108 St. Paul-lez-Durance Cedex

Tel: +33 (0)4 42 25 48 48
Fax: +33 (0)4 42 25 48 48
E-mail:

BERNARD, Louis-Christian
FRAMATOME
10 rue Juliette Récamier
F-69456 Lyon Cedex 6

Tel: +33 (0)4 72 74 72 94
Fax: +33 (0)4 72 74 88 33
E-mail:

BESLU, Pierre
CEA/CADARACHE
DRN/DEC
Bât. 155
F-13108 St. Paul-lez-Durance Cedex

Tel: +33 (0)4 42 25 66 41
Fax: +33 (0)4 42 25 62 67
E-mail: beslu@decade.cea.fr

BLANPAIN, Patrick
FRAMATOME
10 rue Juliette Récamier
F-69456 Lyon Cedex

Tel: +33 (0)4 72 74 89 27
Fax: +33 (0)4 72 74 88 33
E-mail:

BONNAUD, Etienne
FRAMATOME
10 rue Juliette Récamier
F-69456 Lyon Cedex

Tel: +33 (0)4 72 74 85 54
Fax: +33 (0)4 72 74 88 33
E-mail:

BONNET, Christian
CEA/CADARACHE
DRN/DEC
Bât. 155
F-13108 St. Paul-lez-Durance Cedex

Tel: +33 (0)4 42 25 71 47
Fax: +33 (0)4 42 25 73 01
E-mail:

BOURREAU, Sandrine
CEA/SACLAY
DRN/DMT/SEMI
F-91191 Gif-sur-Yvette Cedex

Tel: +33 (0)1 69 08 48 81
Fax: +33 (0)1 69 08 26 81
E-mail: sandrine.bourreau@cea.fr

CAILLOT, Laurent
CEA/GRENOBLE
DRN/DTP/SECC
F-38041 Grenoble Cedex

Tel: +33 (0)4 76 88 32 43
Fax: +33 (0)4 76 88 51 51
E-mail: caillot@dtp.cea.fr

CAYET, Nicolas
EDF/SEPTEN
12-14 Avenue Dutriévoz
F-69828 Villeurbanne Cedex

Tel: +33 (0)4 72 82 73 34
Fax: +33 (0)4 72 82 77 13
E-mail: nicolas.cayet@edfgdf.fr

CHANTOIN, Pierre
CEA/CADARACHE
DRN/DEC/SDC/LEC
Bât. 315
F-13108 St. Paul-lez-Durance Cedex

Tel: +33 (0)4 42 25 76 86
Fax: +33 (0)4 42 25 48 78
E-mail: chantoin@come.cad.cea.fr

DELETTE, Gérard
CEA/GRENOBLE
DRN/DTP/SECC
F-38041 Grenoble Cedex

Tel: +33 (0)4 76 88 38 53
Fax: +33 (0)4 76 88 51 51
E-mail: delette@dtp.cea.fr

DESGRANGES, Lionel
CEA/CADARACHE
DRN/DEC/SDC/LEC
Bât. 316
F-13108 St. Paul-lez-Durance Cedex

Tel: +33 (0)4 42 25 31 59
Fax: +33 (0)4 42 25 48 78
E-mail: desgrang@come.cad.cea.fr

DURIEZ, Nicolas
CEA/CADARACHE
DRN/DEC/SPU/LPCA
Bât. 717
F-13108 St. Paul-lez-Durance Cedex

Tel: +33 (0)4 42 25 62 59
Fax: +33 (0)4 42 25 47 17
E-mail: veyrier@come.cad.cea.fr

GARCIA, Philippe
CEA/CADARACHE
DRN/DEC/SDC/LMC
Bât. 315
F-13108 St. Paul-lez-Durance Cedex

Tel: +33 (0)4 42 25 41 88
Fax: +33 (0)4 42 25 29 49
E-mail:

GUÉRIN, Yannick
CEA/CADARACHE
DRN/DEC/SDC
Bât. 315
F-13108 St. Paul-lez-Durance Cedex

Tel: +33 (0)4 42 25 79 14
Fax: +33 (0)4 42 25 47 47
E-mail: guerin@come.cad.cea.fr

LANSIART, Sylvie
CEA/SACLAY
DRN/DMT/SEMI
F-91191 Gif-sur-Yvette Cedex

Tel: +33 (0)1 69 08 23 80
Fax: +33 (0)1 69 08 26 81
E-mail: lansiart@dmt.cea.fr

LEMAIGNAN, Clément
CEA/GRENOBLE
DRN/DTP/SECC
85 x
F-38041 Grenoble Cedex

Tel: +33 (0)4 76 88 44 71
Fax: +33 (0)4 76 88 51 51
E-mail: lemaignan@cea.fr

LOZANO, Nathalie
CEA/CADARACHE
DRN/DEC/SDC/LEC
Bât. 315
F-13108 St. Paul-lez-Durance Cedex

Tel: +33 (0)4 42 25 31 59
Fax: +33 (0)4 42 25 48 78
E-mail: lozano@come.cad.cea.fr

MÉLIS, Jean-Claude
CEA/CADARACHE
IPSN/DRS
Bât. 250
F-13108 St. Paul-lez-Durance Cedex

Tel: +33 (0)4 42 25 37 22
Fax: +33 (0)4 42 25 29 71
E-mail: jean-claude.melis@ipsn.fr

NOIROT, Laurence
CEA/CADARACHE
DRN/DEC/SDC/LMC
Bât. 315
F-13108 St. Paul-lez-Durance Cedex

Tel: +33 (0)4 42 25 37 20
Fax: +33 (0)4 42 25 29 49
E-mail: lnoirot@come.cad.cea.fr

OBRY, Patrick
CEA/CADARACHE
DRN/DEC/SDC/LMC
Bât. 315
F-13108 St. Paul-lez-Durance Cedex

Tel: +33 (0)4 42 25 73 34
Fax: +33 (0)4 42 25 29 49
E-mail: obry@come.cad.cea.fr

PAGÈS, Jean Paul
CEA/CADARACHE
DRN/P
Bât. 707
F-13108 St. Paul-lez-Durance Cedex

Tel: +33 (0)4 42 25 72 28
Fax: +33 (0)4 42 25 48 47
E-mail:

PIRON, Jean-Paul
CEA/GRENOBLE
DRN/DTP/SECC
F-38041 Grenoble Cedex

Tel: +33 (0)4 76 88 94 77
Fax: +33 (0)4 76 88 51 51
E-mail: piron@dtp.cea.fr

PRADAL, Louis
TECHNICATOME
Centre de Cadarache
Bât. 403
BP N° 9
F-13115 St. Paul-lez-Durance

Tel: +33 (0)4 42 25 12 55
Fax:
E-mail:

QUINTARD, Michel
CNRS
LEPT-ENSAM
Esp. des Arts et Métiers
F-33405 Talence Cedex

Tel: +33 (0)5 56 84 54 03
Fax: +33 (0)5 56 84 54 01
E-mail: quintard@lept.ensam.u-bordeaux.fr

ROUAULT, Jacques
CEA/CADARACHE
DRN/DEC/SDC
Bât. 315
F-13108 St. Paul-lez-Durance Cedex

Tel: +33 (0)4 42 25 72 65
Fax: +33 (0)4 42 25 47 47
E-mail:

SARTORI, Enrico
OECD/NEA
Le Seine St. Germain
F-92130 Issy-les-Moulineaux

Tel: +33 (0)1 45 24 10 72
Fax: +33 (0)1 45 24 11 10
E-mail: sartori@nea.fr

STAICU, Michel Dragos
CEA/CADARACHE
DRN/DEC/SPU/Epu
Bât. 737
F-13108 St. Paul-lez-Durance Cedex

Tel: +33 (0)4 42 25 65 73
Fax: +33 (0)4 42 25 76 76
E-mail:

STRUZIK, Christine
CEA/CADARACHE
DRN/DEC/SDC/LMC
Bât. 315
F-13108 St. Paul-lez-Durance Cedex

Tel: +33 (0)4 42 25 23 35
Fax: +33 (0)4 42 25 29 49
E-mail:

TROTABAS, Maria
COGEMA
BC/CT
BP N° 4
F-78141 Velizy Cedex

Tel: +33 (0)1 39 26 38 31
Fax: +33 (0)1 39 26 27 53
E-mail:

VEYRIER, Nicolas
CEA/CADARACHE
DRN/DEC/SDC/LMC
Bât. 315
F-13108 St. Paul-lez-Durance Cedex

Tel: +33 (0)4 42 25 66 65
Fax: +33 (0)4 42 25 29 49
E-mail: veyrier@come.cad.cea.fr

GERMANY

LASSMANN, Klaus
Institute for Transuranium Elements
CEC Joint Research Centre
Postfach 2340
D-76125 Karlsruhe

Tel: +49 7247 951297
Fax: +49 7247 951379
E-mail: k.lassmann@itu.fzk.de

SONTHEIMER, Fritz
SIEMENS AG
KWU NBTM
Bunsenstrasse 43
D-91050 Erlangen

Tel: +49 9131 182856
Fax: +49 9131 184799
E-mail: fritz.sontheimer@erl19.seimens.de

HUNGARY

ZSOLDOS, Jeno
Hungarian Atomic Energy Authority
Nuclear Safety Directorate
Department for Licensing
Division for Licensing of Facilities
P.O. Box 676
Margit krt. 85
H-1539 Budapest 11

Tel: +36 75 318939
Fax: +36 75 311471
E-mail: h13478zso@ella.hu

JAPAN

NAKAMURA, Jinichi
JAERI
Fuel Reliability Laboratory
Departmenet of Reactor Safety Reserach
2-4 Shirakata-shirane
Tokai-mura, Naka-gun
Ibiraki-ken 319-11

Tel: +81 29 282 5295
Fax: +81 29 282 5323
E-mail: jinichi@popsvr.tokai.jaeri.go.jp

KOREA (REPUBLIC OF)

PARK, Kwangheon
Kyunghee University
Department of Nuclear Engineering
Kiheung
Yongin-gun
Kyungki-Do 449-701

Tel: +82 331 201 2536
Fax: +82 331 202 2410
E-mail: kpark@nms.kyunghee.ac.kr

NETHERLANDS

BAKKER, Klaas
E.C.N.
Netherlands Energy Reserach Foundation
P.O. Box 1
1755 ZG Petten

Tel: +31 224564386
Fax: +31 224563608
E-mail: k.bakker@ecn.nl

NORWAY

VITANZA, Carlo
Institutt for Energiteknikk
OECD Halden Reactor Project
OS Alle 13
P.O. Box 173
N-1751 Halden

Tel: +47 69 21 23 55
Fax: +47 69 21 22 01
E-mail: carlo.vitanza@hrp.no

WIESENACK, Wolfgang
Institutt for Energiteknikk
OECD Halden Reactor Project
OS Alle 13
P.O. Box 173
N-1751 Halden

Tel: +47 69 21 23 47 or 2200
Fax: +47 69 21 22 01
E-mail: wolfgang.wiesenack@hrp.no

RUSSIAN FEDERATION

KOUZNETSOV, Vladimir Ivanovich
Nuclear Fuel Department
St. Rogov 5
123060 Moscow

Tel: +7 095 190 8135
Fax: +7 095 196 6591
E-mail: bsm@bochvar.ru

NOVIKOV, Vladimir Vladimirovich
Institute of Inorganic Materials
Nuclear Fuel Department
St. Rogov 5
123060 Moscow

Tel: +7 095 190 8135
Fax: +7 095 196 6591
E-mail: novk@bochvar.ru

SWEDEN

MALÉN, Karl
Studsvik Nuclear AB
S-611822 Nyköping

Tel: +46 155 221670
Fax: +46 155 26 3156
E-mail: karl.malen@nuclear.studsvik.se

SWITZERLAND

OTT, Christophe
NOK
Nordostscheizerisch
Kraftwerke AG, KNW/std
Parkstrasse 23
CH-5401 Baden

Tel: +41 56 200 35 04
Fax: +41 56 200 35 94
E-mail: och@nok.ch

WALLIN, Hannu
Paul Scherrer Institute
PSI/OFLA/008
CH-5232 Villigen PSI

Tel: +41 56 310 4428
Fax:
E-mail: hannu.wallin@psi.ch

UNITED KINGDOM

GATES, Gary
British Nuclear Fuels plc
Fuel Engineering Department
Springfields Works
Salwick, Preston, Lancs PR4 OXJ

Tel: +44 1772 76 4078
Fax: +44 1772 76 3888
E-mail: gag2@bnfl.com

GOMME, Robin
AEA Technology
Windscale Laboratory
Seascale
Cumbria CA20 IPF

Tel: +44 19467 72330
Fax: +44 19467 72606
E-mail: robin.gomme@aeat.cu.uk

TURNBULL, Tony
Cherry-Lyn
The Green
Tockington
Bristol, BS 32 4NJ

Tel: +44 1454 620887
Fax: +44 1454 620887
E-mail:

UNITED STATES

BEYER, Carl E.
Pacific Northwest National Laboratory
Battelle Boulevard
P.O. Box 999
Richland, WA 99352

Tel: +1 509 372 4605
Fax: +1 509 372 4439
E-mail: ce_beyer@pnl.gov

BILLAUX, Michel
SIEMENS Power Corporation
Nuclear Division
2101 Horn Rapids Road
P.O. Box 130
Richland, WA 99352-0130

Tel: +1 509 375 8130
Fax: +1 509 375 8965
E-mail: michel_billaux@nfuel.com

YAGNIK, K. Suresh
EPRI Nuclear Power Division
EPRI Program Manager
3412 Hillview Avenue
P.O. Box 10412
Palo Alto, CA 94303

Tel: +1 650 855 2971
Fax: +1 650 855 1076
E-mail: syagnik@epri.com

VAN SWAM Leo F.P.
SIEMENS Power Corporation
Eng. and Manufacturing Facility
2101 Horn Rapids Road
P.O. Box 130
Richland, WA 99352-0130

Tel: +1 509 375 8170
Fax: +1 509 375 8402
E-mail: leo_vanswam@nfuel.com

OECD PUBLICATIONS, 2, rue André-Pascal, 75775 PARIS CEDEX 16

PRINTED IN FRANCE

(66 98 18 1 P) ISBN 92-64-16957-1 – No. 50427 1998